SCHAUM'S OUTLINE OF

PRINCIPLES AND PROBLEMS

of

PLANE GEOMETRY

with Coordinate Geometry

•

BY

BARNETT RICH, Ph.D.

Chairman, Department of Mathematics
Brooklyn Technical High School
New York City

•

SCHAUM'S OUTLINE SERIES

McGRAW-HILL BOOK COMPANY

New York, St. Louis, San Francisco, Toronto, Sydney

ISBN 07-052245-6

16 17 18 19 20 21 22 23 24 25 26 27 28 29 30 SH SH 7 5 4 3 2 1 0 6 9

Preface

The central purpose of this book is to provide maximum help for the student and maximum service for the teacher.

PROVIDING HELP FOR THE STUDENT:

This book has been designed to improve the learning of geometry far beyond that of the typical and traditional book in the subject. Students will find this text useful for these reasons:

(1) *Learning Each Rule, Formula and Principle*

Each rule, formula and principle is stated in simple language, is made to stand out in distinctive type, is kept together with those related to it, and is clearly illustrated by examples.

(2) *Learning Each Set of Solved Problems*

Each set of solved problems is used to clarify and apply the more important rules and principles. The character of each set is indicated by a title.

(3) *Learning Each Set of Supplementary Problems*

Each set of supplementary problems provides further application of rules and principles. A guide number for each set refers a student to the set of related solved problems. There are more than 2000 additional related supplementary problems. Answers for the supplementary problems have been placed in the back of the book.

(4) *Integrating the Learning of Plane Geometry*

In accordance with the syllabus of the State of New York, the book integrates plane geometry with arithmetic, algebra, numerical trigonometry, coordinate geometry and simple logic. To carry out this integration:

(a) A separate chapter is devoted to coordinate geometry.

(b) A separate chapter includes the complete proofs of all the required theorems together with the plan for each.

(c) A separate chapter explains fully 27 basic geometric constructions, including all the required constructions. Underlying geometric principles are provided for the constructions, as needed.

(d) Two separate chapters on methods of proof and improvement of reasoning present the simple and basic ideas of formal logic suitable for students at this stage.

(e) Throughout the book, algebra is emphasized as the major means of solving geometric problems through algebraic symbolism, algebraic equations and algebraic proof.

(5) *Learning Geometry Through Self-study*

The method of presentation in the book makes it ideal as a means of self-study. For the able student, this book will enable him to accomplish the work of the standard course of study in much less time. For the less able, the presentation of numerous illustrations and solutions provides the help needed to remedy weaknesses and overcome difficulties and in this way keep up with the class and at the same time gain a measure of confidence and security.

(6) *Extending Plane Geometry into Solid Geometry*

A separate chapter is devoted to the extension of two-dimensional plane geometry into three-dimensional solid geometry. It is especially important in this day and age that the student understand how the basic ideas of space are outgrowths of principles learned in plane geometry.

PROVIDING SERVICE FOR THE TEACHER:

Teachers of geometry will find this text useful for these reasons:

(1) *Teaching Each Chapter*

Each chapter has a central unifying theme. Each chapter is divided into two to ten major subdivisions which support its central theme. In turn, these chapter subdivisions are arranged in graded sequence for greater teaching effectiveness.

(2) *Teaching Each Chapter Subdivision*

Each of the chapter subdivisions contains the problems and materials needed for a complete lesson developing the related principles.

(3) *Making Teaching More Effective Through Solved Problems*

Through proper use of the solved problems, students gain greater understanding of the way in which principles are applied in varied situations. By solving problems, mathematics is learned as it should be learned — by doing mathematics. To ensure effective learning, solutions should be reproduced on paper. Students should seek the why as well as the how of each step. Once a student sees how a principle is applied to a solved problem, he is then ready to extend the principle to a related supplementary problem. Geometry is not learned through the reading of a textbook and the memorizing of a set of formulas. Until an adequate variety of suitable problems have been solved, a student will gain little more than a vague impression of plane geometry.

(4) *Making Teaching More Effective Through Problem Assignment*

The preparation of homework assignments and class assignments of problems is facilitated because the supplementary problems in this book are related to the sets of solved problems. Greatest attention should be given to the underlying principle and the major steps in the solution of the solved problems. After this, the student can reproduce the solved problems and then proceed to do those supplementary problems which are related to the solved ones.

OTHERS WHO WILL FIND THIS TEXT ADVANTAGEOUS:

This book can be used profitably by others beside students and teachers. In this group we include: (1) the parents of geometry students who wish to help their children through the use of the book's self-study materials, or who may wish to refresh their own memory of geometry in order to properly help their children; (2) the supervisor who wishes to provide enrichment materials in geometry, or who seeks to improve teaching effectiveness in geometry; (3) the person who seeks to review geometry or to learn it through independent self-study.

BARNETT RICH

Brooklyn Technical High School
April, 1963

CONTENTS

CONTENTS

CONTENTS

CONTENTS

Chapter 1

Lines, Angles, and Triangles

1. Historical Background of Geometry

The word *geometry* is derived from the Greek words *geos* meaning *earth* and *metron* meaning *measure*. The ancient Egyptians, Chinese, Babylonians, Romans and Greeks used geometry for surveying, navigation, astronomy and other practical occupations.

The Greeks sought to systematize the geometric facts that had been discovered by establishing logical reasons for them and relationships among them. The work of such men as Thales (600 B.C.), Pythagoras (540 B.C.), Plato (390 B.C.) and Aristotle (350 B.C.) in systematizing geometric facts and principles culminated in the geometry text *Elements*, written about 325 B.C. by Euclid. This most remarkable text has been in use for over 2000 years.

2. Undefined Terms of Geometry: Point, Line and Surface

A. Point, Line, and Surface are Undefined Terms

These terms begin the process of definition and underlie the definitions of all other geometric terms. However, meanings can be given to these undefined terms by means of descriptions. These descriptions which follow are not to be thought of as definitions.

B. Point

A point has position only. It has no length, width, or thickness.

A point is represented by a dot. Keep in mind however that the dot represents a point but is not a point, just as a dot on a map may represent a locality but is not the locality. A dot, unlike a point, has size.

A point is designated by a capital letter next to the dot, thus: $\cdot P$
$\cdot A$

C. Line

A line has length but has no width or thickness.

A line may be represented by the path of a piece of chalk on the blackboard or by a stretched rubber band.

A line is designated by the capital letters of any two of its points or by a small letter, thus: $\overline{A \quad B}$, $\frown_{C \quad D}$ or $\underline{\quad a \quad}$.

A line may be straight, curved, or a combination of these. To understand how lines differ, think of a line as being generated by a moving point.

A *straight line*, such as $\longrightarrow$, is generated by a point moving in the same direction.

A *curved line*, such as ⌒ , is generated by a point moving in a continuously changing direction.

A *broken line*, such as ∧ , is a combination of straight lines.

A straight line is unlimited in extent. It may be extended in either direction indefinitely.

A straight line is the shortest line between two points. Two straight lines intersect in a point.

D. Surface

A surface has length and width but no thickness. It may be represented by a blackboard, a side of a box or the outside of a sphere; these are representations of a surface but are not surfaces.

A *plane surface* or a *plane* is a surface such that a straight line connecting any two of its points lies entirely in it. A plane is a flat surface and may be represented by the surface of a flat mirror or the top of a desk.

Plane Geometry is the geometry that deals with plane figures that may be drawn on a plane surface. Unless otherwise stated, *figure* shall mean plane figure.

2.1 ILLUSTRATING UNDEFINED TERMS

Point, line, and surface are undefined terms. State which of these terms is illustrated by each of the following:

(*a*) A light ray   (*c*) A projection screen   (*e*) A stretched thread

(*b*) The top of a desk   (*d*) A ruler's edge   (*f*) The tip of a pin

   Ans. (*a*) line (*b*) surface (*c*) surface (*d*) line (*e*) line (*f*) point

3. Straight Line Segments

A straight line segment is the part of a straight line between two of its points. It is represented by the capital letters of these points or by a small letter. Thus AB or r represents the straight line segment between A and B.

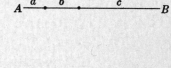

The expression *straight line segment* may be shortened to *line segment*, or *segment*, or even *line*, if the meaning is clear. Thus, line AB or AB means the straight line segment AB, unless otherwise stated.

Dividing a Line into Parts

If a line is divided into parts:

   1. The whole line equals the sum of its parts.

   2. The whole line is greater than any part.

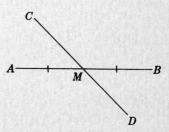

     Thus if AB is divided into three parts a, b and c, then $AB = a + b + c$. Also, AB is greater than a; this may be written $AB > a$.

If a line is divided into two equal parts:

   1. The point of division is the *midpoint of the line*.

   2. A line that crosses at the midpoint is said *to bisect the given line*.

     Thus if $AM = MB$, then M is the midpoint of AB, and CD bisects AB.

     Equal line segments may be shown by crossing them with the same number of strokes. Note that AM and MB are crossed by a single stroke.

3.1 NAMING LINE SEGMENTS and POINTS

(a) Name each line segment shown.

(b) Name the line segments that intersect at *A*.

(c) What other line segment can be drawn?

(d) Name the point of intersection of *CD* and *AD*.

(e) Name the point of intersection of *BC*, *AC* and *CD*.

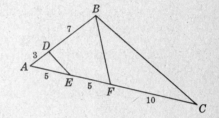

Solution:

(a) *AB*, *BC*, *CD*, *AC*, and *AD*. These segments may also be named by interchanging the letters; thus *BA*, *CB*, *DC*, *CA*, and *DA* are also correct.

(b) *AB*, *AC*, and *AD*   (c) *BD*   (d) *D*   (e) *C*

3.2 FINDING LENGTHS and POINTS of LINE SEGMENTS

(a) State the lengths of *AB*, *AC*, and *AF*.

(b) Name two midpoints.

(c) Name two line bisectors.

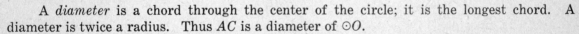

Solution:

(a) $AB = 3 + 7 = 10$, $AC = 5 + 5 + 10 = 20$, $AF = 5 + 5 = 10$.

(b) *E* is midpoint of *AF*, and *F* is midpoint of *AC*.

(c) *DE* is bisector of *AF*, and *BF* is bisector of *AC*.

4. Circles

A *circle* is a closed curve all of whose points are the same distance from the center. The symbol for circle is ⊙; for circles ⊙s. Thus ⊙*O* stands for the circle whose center is *O*.

The *circumference* is the distance around the circle. It contains 360°.

A *radius* is a line joining the center to a point on the circumference. From the definition of a circle, it follows that radii of a circle are equal. Thus *OA*, *OB* and *OC* are radii of ⊙*O* and *OA* = *OB* = *OC*.

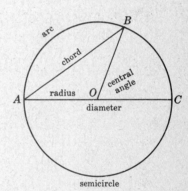

A *chord* is a line joining any two points on the circumference. Thus *AB* and *AC* are chords of ⊙*O*.

A *diameter* is a chord through the center of the circle; it is the longest chord. A diameter is twice a radius. Thus *AC* is a diameter of ⊙*O*.

An *arc* is a part of the circumference of a circle. The symbol for arc is ⌢. Thus $\overset{\frown}{AB}$ stands for arc *AB*. An arc of 1° is 1/360th of a circumference.

A *semicircle* is an arc equal to one-half of the circumference of a circle. A semicircle contains 180°. A diameter divides a circle into two semicircles. Thus diameter *AC* cuts ⊙*O* into two semicircles.

A *central angle* is an angle formed by two radii. Thus the angle between radii *OB* and *OC* is a central angle.

A *central angle of 1°* cuts off an arc of 1°. Thus if the central angle between *OE* and *OF* is 1°, then $\overset{\frown}{EF}$ is 1°.

Equal circles are circles having equal radii. Thus if *OE* = *O'G*, then circle *O* = circle *O'*.

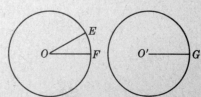

4.1   FINDING LINES and ARCS in a CIRCLE

In circle O: (a) find OC and AB, (b) find the number of degrees in $\widehat{AD}$, (c) find the number of degrees in $\widehat{BC}$.

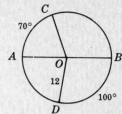

Solution:

(a)   Radius OC = radius OD = 12.   Diameter AB = 24.

(b)   Since semicircle ADB = 180°, $\widehat{AD}$ = 180° − 100° = 80°.

(c)   Since semicircle ACB = 180°, $\widehat{BC}$ = 180° − 70° = 110°.

5. *Angles*

An *angle* is the figure formed by two straight lines meeting at a point. The lines are the sides of the angle while the point is its vertex. The symbol for angle is ∠. The plural is ∡.

Thus AB and AC are the sides of the angle, shown in Fig. (a), and A is its vertex.

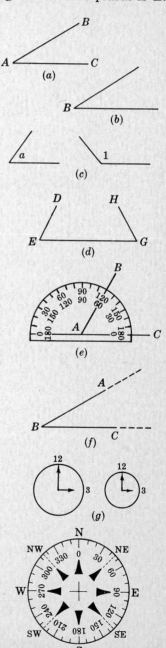

A.   Naming an Angle

An angle may be named in any of the following ways:

1. The vertex letter if there is only one angle having this vertex, as ∠B, in Fig. (b).

2. A small letter or a number placed between the sides of the angle and near the vertex, as ∠a or ∠1, in Fig. (c).

3. Three capital letters with the vertex letter between two others on the sides of the angle. Referring to Fig. (d), ∠E may be named ∠DEG or ∠GED. ∠G may be named ∠EGH or ∠HGE.

B.   Measuring the Size of an Angle

1. The size of an angle depends on the extent to which one side of the angle must be rotated or turned about the vertex until the turned side meets the other side.

    Thus the protractor in Fig. (e) shows that ∠A is 60°. If AC were rotated about the vertex A until it met AB, the amount of turn would be 60°.

    In using a protractor, be sure that the vertex of the angle is at the center and that one side is along the 0°-180° diameter.

2. The size of an angle does *not* depend on the length of the sides of the angle.

    Thus the size of ∠B, in Fig. (f), would not be changed if its sides AB and BC were made larger or smaller.

    No matter how large or small a clock is, the angle formed by its hands at 3 o'clock is 90°, as shown in Fig. (g).

3. The Navy Compass, shown in Fig. (h), is read clockwise from 0° to 360°, beginning with North. A rotation from N to E is a quarter turn or 90°, that from NE to E is an eighth turn or 45°.

C. Kinds of Angles:

1. **Acute Angle** — *An acute angle is an angle that is less than 90°.*

    Thus, $a°$ is less than 90°; this is symbolized by $a° < 90°$.

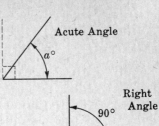

2. **Right Angle** — *A right angle is an angle that equals 90°.*

    Thus, rt. $\angle A = 90°$. The square corner denotes a right angle.

3. **Obtuse Angle** — *An obtuse angle is an angle that is more than 90° and less than 180°.*

    Thus, 90° is less than $b°$ and $b°$ is less than 180°; this is denoted by $90° < b° < 180°$.

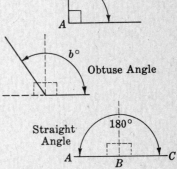

4. **Straight Angle** — *A straight angle is an angle that equals 180°.*

    Thus, st. $\angle B = 180°$. Note that the sides of a straight angle lie in the same straight line. However, do not confuse a straight angle with a straight line!

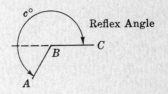

5. **Reflex Angle** — *A reflex angle is an angle that is more than 180° and less than 360°.*

    Thus, 180° is less than $c°$ and $c°$ is less than 360°; this is symbolized by $180° < c° < 360°$.

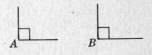

D. Additional Angle Facts:

1. *Equal angles are angles that have the same number of degrees.*

    Thus, rt. $\angle A =$ rt. $\angle B$ since each equals 90°.

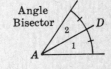

2. *A line that bisects an angle divides it into two equal parts.*

    Thus if AD bisects $\angle A$, then $\angle 1 = \angle 2$.

    Equal angles may be shown by crossing their arcs with the same number of strokes. Here the arcs of $\angle 1$ and $\angle 2$ are crossed by a single stroke.

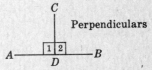

3. *Perpendiculars are lines that meet at right angles.*

    The symbol for perpendicular is ⊥; for perpendiculars, ⊥s. Thus if $CD \perp AB$, right angles 1 and 2 are formed.

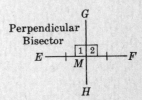

4. *A perpendicular bisector of a given line is both perpendicular to the line and bisects it.*

    Thus if GH is the ⊥ bisector of EF, then $\angle 1$ and $\angle 2$ are right angles and M is the midpoint of EF.

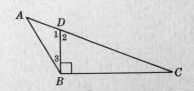

5.1 NAMING an ANGLE

Name the following angles in the diagram: (*a*) Two obtuse angles, (*b*) a right angle, (*c*) a straight angle, (*d*) an acute angle at D, (*e*) an acute angle at B.

Solution:

(*a*) $\angle ABC$, and $\angle ADB$ or $\angle 1$. The angles may also be named by reversing the order of the letters: $\angle CBA$ and $\angle BDA$.

(*b*) $\angle DBC$ (*c*) $\angle ADC$ (*d*) $\angle 2$ or $\angle BDC$ (*e*) $\angle 3$ or $\angle ABD$

5.2 ADDING and SUBTRACTING ANGLES

Find (a) $\angle AOC$, (b) $\angle BOE$, (c) obtuse $\angle AOE$.

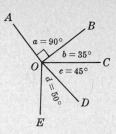

Solution:

(a) $\angle AOC = \angle a + \angle b = 90° + 35° = 125°$

(b) $\angle BOE = \angle b + \angle c + \angle d = 35° + 45° + 50° = 130°$

(c) $\angle AOE = 360° - (\angle a + \angle b + \angle c + \angle d) = 360° - 220° = 140°$

5.3 FINDING PARTS of ANGLES

Find (a) $\frac{2}{5}$ of a rt. $\angle$, (b) $\frac{2}{3}$ of a st. $\angle$, (c) $\frac{1}{2}$ of $31°$, (d) $\frac{1}{10}$ of $70°20'$.

Solution:

(a) $\frac{2}{5}(90°) = 36°$ (c) $\frac{1}{2}(31°) = 15\frac{1}{2}° = 15°30'$

(b) $\frac{2}{3}(180°) = 120°$ (d) $\frac{1}{10}(70°20') = \frac{1}{10}(70°) + \frac{1}{10}(20') = 7°2'$

5.4 FINDING ROTATIONS

In a half hour, what turn or rotation is made (a) by the minute hand, (b) by the hour hand?

What rotation is needed to turn: (c) from North to Southeast in a clockwise direction, (d) from Northwest to Southwest in a counterclockwise direction?

Solution:

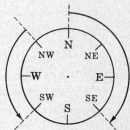

(a) In one hour, a minute hand completes a full circle of $360°$. Hence in a half hour it turns $180°$.

(b) In one hour, an hour hand turns $\frac{1}{12}$ of $360°$ or $30°$. Hence in a half hour it turns $15°$.

(c) Add turn of $90°$ from North to East and $45°$ from East to Southeast. $90° + 45° = 135°$.

(d) The turn from Northwest to Southwest is $\frac{1}{4}(360°) = 90°$.

5.5 FINDING ANGLES

Find the angle formed by the hands of the clock: (a) at 8 o'clock, (b) at 4:30 o'clock.

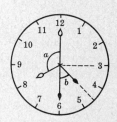

Solution:

(a) At 8 o'clock, $\angle a = \frac{1}{3}(360°) = 120°$.

(b) At 4:30 o'clock, $\angle b = \frac{1}{2}(90°) = 45°$.

5.6 APPLYING ANGLE FACTS

In the diagram shown: (a) Name two pairs of perpendicular lines, (b) find $\angle a$ if $\angle b = 42°$, (c) find $\angle AEB$ and $\angle CED$.

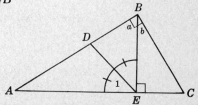

Solution:

(a) Since $\angle ABC$ is a right angle, $AB \perp BC$. Since $\angle BEC$ is a right angle, $BE \perp AC$.

(b) $\angle a = 90° - \angle b = 90° - 42° = 48°$.

(c) $\angle AEB = 180° - \angle BEC = 180° - 90° = 90°$. $\angle CED = 180° - \angle 1 = 180° - 45° = 135°$.

6. *Triangles*

A polygon is a closed plane figure bounded by straight line segments as sides. Thus Fig. (*a*) is a polygon. A polygon of 5 sides is a pentagon; it is named pentagon *ABCDE*, using its letters in order.

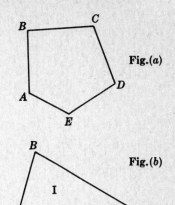

Fig.(*a*)

A triangle is a polygon having three sides. A vertex of a triangle is a point at which two of the sides meet. Vertices is the plural of vertex. The symbol for triangle is △: for triangles, ⧌.

A triangle may be named using its three letters in any order or using a Roman numeral placed inside of it. Thus the triangle shown in Fig. (*b*) is named △*ABC* or △I; its sides are *AB*, *AC* and *BC*; its vertices are *A*, *B* and *C*; its angles are ∠*A*, ∠*B* and ∠*C*.

Fig.(*b*)

A. *Classifying Triangles:*

Triangles are classified according to the equality of their sides or according to the kind of angles they have.

Triangles According to the Equality of their Sides

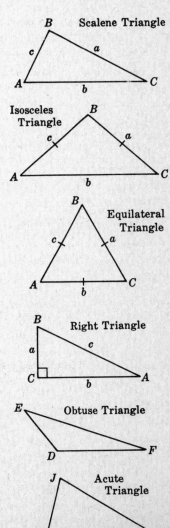

1. Scalene Triangle — *A scalene triangle is a triangle having no equal sides.*

    Thus in scalene triangle *ABC*, $a \neq b \neq c$. The small letter used for each side agrees with the capital letter of the angle opposite it. Also, $\neq$ means "is not equal to".

2. Isosceles Triangle — *An isosceles triangle is a triangle having at least two equal sides.*

    Thus in isosceles triangle *ABC*, $a = c$. These equal sides are called the *legs or arms* of the isosceles triangle; the remaining side is the *base b*. The angles on either side of the base are the *base angles*; the angle opposite the base is the *vertex angle*.

3. Equilateral Triangle — *An equilateral triangle is a triangle having three equal sides.*

    Thus in equilateral triangle *ABC*, $a = b = c$. Note that an equilateral triangle is also an isosceles triangle.

Triangles According to the Kind of Angles

1. Right Triangle — *A right triangle is a triangle having a right angle.*

    Thus in right triangle *ABC*, ∠*C* is the right angle. Side *c* opposite the right angle is the *hypotenuse*. The perpendicular sides, *a* and *b*, are the *legs or arms* of the right triangle.

2. Obtuse Triangle — *An obtuse triangle is a triangle having an obtuse angle.*

    Thus in obtuse triangle *DEF*, ∠*D* is the obtuse angle.

3. Acute Triangle — *An acute triangle is a triangle having three acute angles.*

    Thus in acute triangle *HJK*, ∠*H*, ∠*J* and ∠*K* are acute angles.

B.   *Special Lines in a Triangle*

1. **Angle Bisector of a Triangle** — *An angle bisec-tor of a triangle is a line that bisects an angle and extends to the opposite side.*

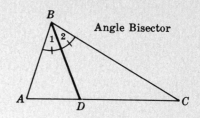

Angle Bisector

Thus *BD*, the angle bisector of ∠*B*, bisects ∠*B*, making ∠1 = ∠2.

2. **Median of a Triangle. Perpendicular Bisector of a Side** — *A median of a triangle is a line from a vertex to the midpoint of the opposite side.*

Thus *BM*, the median to *AC*, bisects *AC*, making *AM* = *MC*.

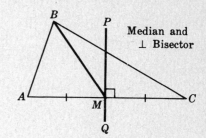

Median and ⊥ Bisector

A perpendicular bisector of a side of a triangle is a line that bisects and is perpen-dicular to a side.

Thus *PQ*, the perpendicular bisector of *AC*, bi-sects *AC* and is perpendicular to it.

3. **Altitude to a Side of a Triangle** — *An altitude of a triangle is a line from a vertex perpen-dicular to the opposite side.*

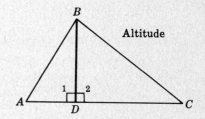

Altitude

Thus *BD*, the altitude to *AC*, is perpendicular to *AC* and forms right angles 1 and 2. Each angle bi-sector, median and altitude of a triangle extends from a vertex to the opposite side.

4. **Altitudes of Obtuse Triangle** — *In an obtuse triangle, the altitude drawn to either side of the obtuse angle falls outside the triangle.*

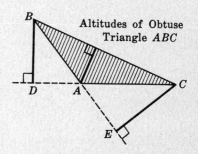

Altitudes of Obtuse Triangle *ABC*

Thus in obtuse triangle *ABC* (shaded), altitudes *BD* and *CE* fall outside the triangle. In each case, a side of the obtuse angle must be extended.

6.1   NAMING a TRIANGLE and its PARTS

In Fig. 1, name (*a*) an obtuse triangle, (*b*) two right triangles and the hypotenuse and legs of each. (*c*) In Fig. 2, name two isosceles triangles; also name the legs, base and vertex angle of each.

Solution:

(*a*) Since ∠*ADB* is an obtuse angle, △*ADB* or △II is obtuse.

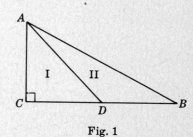

Fig. 1

(*b*) Since ∠*C* is a right angle, △I and △*ABC* are right tri-angles. In △I, *AD* is the hypotenuse and *AC* and *CD* are the legs. In △*ABC*, *AB* is the hypotenuse and *AC* and *BC* are the legs.

(*c*) Since *AD* = *AE*, △*ADE* is an isosceles triangle. In △*ADE*, *AD* and *AE* are the legs, *DE* is the base and ∠*A* is the vertex angle.

Since *AB* = *AC*, △*ABC* is an isosceles triangle. In △*ABC*, *AB* and *AC* are the legs, *BC* is the base and ∠*A* is the vertex angle.

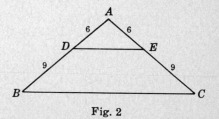

Fig. 2

6.2 SPECIAL LINES in a TRIANGLE

Name the equal lines and angles in the adjacent diagram shown: (*a*) if *AE* is the altitude to *BC*, (*b*) if *CG* bisects ∠*ACB*, (*c*) if *KL* is the perpendicular bisector of *AD*, (*d*) if *DF* is the median to *AC*.

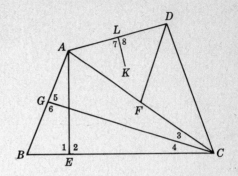

Solution:

(*a*) Since $AE \perp BC$, $\angle 1 = \angle 2$.

(*b*) Since *CG* bisects ∠*ACB*, $\angle 3 = \angle 4$.

(*c*) Since $LK \perp$ bisector of *AD*, $AL = LD$ and $\angle 7 = \angle 8$.

(*d*) Since *DF* is median to *AC*, $AF = FC$.

7. *Pairs of Angles*

A. *Kinds of Pairs of Angles:*

1. Adjacent Angles — *Adjacent angles are two angles which have the same vertex and a common side between them.*

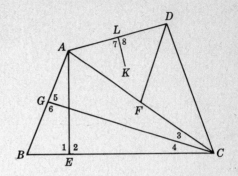

Adjacent Angles

Thus, the entire angle of $c°$ has been cut into two adjacent angles of $a°$ and $b°$. These adjacent angles have same vertex *A*, and a common side *AD* between them. Here, $a° + b° = c°$.

2. Vertical Angles — *Vertical angles are two non-adjacent angles formed by two intersecting lines.*

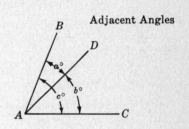

Vertical Angles

Thus, ∠1 and ∠3 are vertical angles formed by intersecting lines *AB* and *CD*. Also, ∠2 and ∠4 are another pair of vertical angles formed by the same lines.

3. Complementary Angles — *Complementary angles are two angles whose sum equals* **90°.**

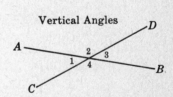

Complementary Angles

Fig. 1   Fig. 2

Thus, in Fig. 1 the angles of $a°$ and $b°$ are adjacent complementary angles. However, in Fig. 2 the complementary angles are non-adjacent. In each case, $a° + b° = 90°$. Either of two complementary angles is said to be the complement of the other.

4. Supplementary Angles — *Supplementary angles are two angles whose sum equals* **180°.**

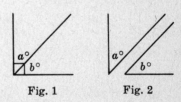

Supplementary Angles

Fig. 3   Fig. 4

Thus, in Fig. 3 the angles of $a°$ and $b°$ are adjacent supplementary angles. However, in Fig. 4 the supplementary angles are non-adjacent. In each case, $a° + b° = 180°$. Either of two supplementary angles is said to be the supplement of the other.

B.   Principles of Pairs of Angles:

Pr. 1:   *If an angle of c° is cut into two adjacent angles of a° and b°, then a° + b° = c°.*

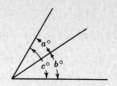

Thus if $a° = 25°$ and $b° = 35°$, then $c° = 25° + 35° = 60°$.

Pr. 2:   *Vertical angles are equal.*

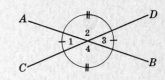

Thus if *AB* and *CD* are straight lines, then $\angle 1 = \angle 3$ and $\angle 2 = \angle 4$. Hence, if $\angle 1 = 40°$, $\angle 3 = 40°$; in such a case, $\angle 2 = \angle 4 = 140°$.

Pr. 3:   *If two complementary angles contain a° and b°, then a° + b° = 90°.*

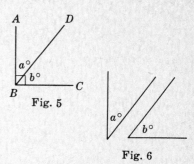

Thus if angles of a° and b° are complementary and $a° = 40°$, then $b° = 50°$ (Fig. 5 or Fig. 6)

Fig. 5

Fig. 6

Pr. 4:   *Adjacent angles are complementary if their exterior sides are perpendicular to each other.*

Thus in Fig. 5, a° and b° are complementary since their exterior sides *AB* and *BC* are perpendicular to each other.

Pr. 5:   *If two supplementary angles contain a° and b°, then a° + b° = 180°.*

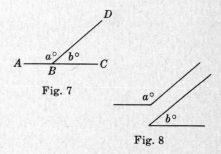

Thus if angles of a° and b° are supplementary and $a° = 140°$, then $b° = 40°$ (Fig. 7 or Fig. 8).

Fig. 7

Fig. 8

Pr. 6:   *Adjacent angles are supplementary if their exterior sides lie in the same straight line.*

Thus in Fig. 7, a° and b° are supplementary angles since their exterior sides *AB* and *BC* lie in the same straight line *AC*.

Pr. 7:   *If supplementary angles are equal, each of them is a right angle. (Equal supplementary angles are right angles.)*

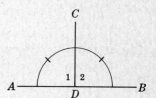

Thus if $\angle 1$ and $\angle 2$ are both equal and supplementary, then each of them is a right angle.

7.1   NAMING PAIRS of ANGLES

(*a*) In Fig. 1, name two pairs of supplementary angles.

(*b*) In Fig. 2, name two pairs of complementary angles.

(*c*) In Fig. 3, name two pairs of vertical angles.

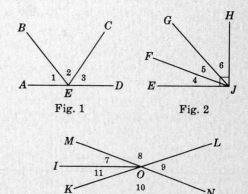

Fig. 1   Fig. 2

Fig. 3

Solution:

(*a*) Since their sum is 180°, the supplementary angles are
(*1*) $\angle 1$ and $\angle BED$, (*2*) $\angle 3$ and $\angle AEC$.

(*b*) Since their sum is 90°, the complementary angles are
(*1*) $\angle 4$ and $\angle FJH$, (*2*) $\angle 6$ and $\angle EJG$.

(*c*) Since *KL* and *MN* are intersecting lines, the vertical angles are (*1*) $\angle 8$ and $\angle 10$, (*2*) $\angle 9$ and $\angle MOK$.

7.2 FINDING PAIRS of ANGLES

Find the angles in each:

(a) The angles are supplementary and the larger is twice the smaller.

(b) The angles are complementary and the larger is 20° more than the smaller.

(c) The angles are adjacent and form an angle of 120°. The larger is 20° less than three times the smaller.

(d) The angles are vertical and complementary.

Solutions:

In each, x is a number only. This number indicates the number of degrees contained in the angle. Hence, if $x = 60$, the angle equals 60°.

(a)
Let x = smaller angle, $2x$ = larger angle.
Pr. 5: $x + 2x = 180$, $3x = 180$, $x = 60$.   *Ans.* 60° and 120°
$2x = 120$.

(b)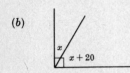
Let x = smaller angle, $x + 20$ = larger angle.
Pr. 3: $x + (x + 20) = 90$, $2x + 20 = 90$, $x = 35$.   *Ans.* 35° and 55°
$x + 20 = 55$.

(c)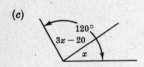
Let x = smaller angle and $3x - 20$ = larger angle.
Pr. 1: $x + (3x - 20) = 120$, $4x - 20 = 120$, $x = 35$.   *Ans.* 35° and 85°
$3x - 20 = 85$.

(d)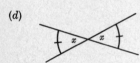
Let x = each of the equal vertical angles.   **(Pr. 2)**
Pr. 3: $x + x = 90$, $2x = 90$, $x = 45$.   *Ans.* 45° each

7.3 FINDING a PAIR of ANGLES USING TWO UNKNOWNS

If two angles are represented by a and b, obtain two equations for each problem; then find the angles.

	Equations	Angles
(a) The angles are adjacent forming an angle of 88°. One is 36° more than the other.	*Ans.* $a + b = 88$ $a = b + 36$	62°, 26°
(b) The angles are complementary. One is twice as large as the other.	*Ans.* $a + b = 90$ $a = 2b$	60°, 30°
(c) The angles are supplementary. One is 60° less than twice the other.	*Ans.* $a + b = 180$ $a = 2b - 60$	100°, 80°
(d) The angles are two angles of a triangle whose third angle is 40°. The difference of the angles is 24°.	*Ans.* $a + b = 140$ $a - b = 24$	82°, 58°

Supplementary Problems

The numbers in parentheses at the end of each first line of a supplementary problem refer to the set of solved problems which contains problems of the same kind. Refer to these for help.

1. Point, line and surface are undefined terms. Which of these is illustrated by (a) the tip of a sharpened pencil, (b) the shaving edge of a blade, (c) a sheet of paper, (d) a side of a box, (e) the crease of a folded paper, (f) the junction of two roads on a map? (2.1)

2. (a) Name the line segments that intersect at E.
   (b) Name the line segments that intersect at D.
   (c) What other line segments can be drawn?
   (d) Name the point of intersection of AC and BD. (3.1)

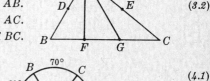

3. (a) Find the length of AB if AD is 8 and D is the midpoint of AB.
   (b) Find the length of AE if AC is 21 and E is the midpoint of AC.
   (c) Name two line bisectors if F and G are the trisection points of BC. (3.2)

4. (a) Find OB if diameter AD = 36.
   (b) Find $\overset{\frown}{AE}$ if E is the midpoint of semicircle AED.
   Find the number of degrees in
   (c) $\overset{\frown}{CD}$, (d) $\overset{\frown}{AC}$, (e) $\overset{\frown}{AEC}$ (4.1)

5. Name the following angles in the diagram: (5.1)
   (a) An acute angle at B
   (b) An acute angle at E
   (c) A right angle
   (d) Three obtuse angles
   (e) A straight angle

6. (a) Find ∠ADC if ∠c = 45° and ∠d = 85°, (5.2)
   (b) Find ∠AEB if ∠e = 60°,
   (c) Find ∠EBD if ∠a = 15°,
   (d) Find ∠ABC if ∠b = 42°.

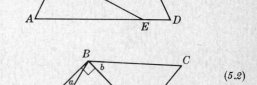

7. Find (a) $\frac{5}{6}$ of a rt. ∠, (b) $\frac{2}{9}$ of a st. ∠, (c) $\frac{1}{3}$ of 31°, (d) $\frac{1}{5}$ of 45°55′. (5.3)

8. What turn or rotation is made
   (a) by an hour hand in 3 hours, (b) by the minute hand in $\frac{1}{3}$ of an hour? (5.4)
   What rotation is needed to turn from
   (c) West to Northeast in a clockwise direction,
   (d) East to South in a counterclockwise direction,
   (e) Southwest to Northeast in either direction?

9. Find the angle formed by the hand of the clock (5.5)
   (a) at 3 o'clock, (b) at 10 o'clock, (c) at 5:30 o'clock, (d) at 11:30 o'clock.

10. In the diagram shown: (5.6)

   (a) Name two pairs of perpendicular lines.

   (b) Find $\angle BCD$ if $\angle 4$ is $39°$.

   If $\angle 1 = 78°$, find (c) $\angle BAD$, (d) $\angle 2$, (e) $\angle CAE$.

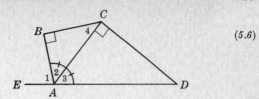

11. (a) In Fig. 1, name three right triangles and the hypotenuse and legs of each.   (6.1)

   In Fig. 2, name

   (b) two obtuse triangles and

   (c) two isosceles triangles. Also, name the legs, the base and the vertex angle of each.

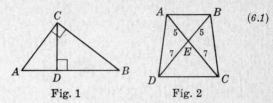

Fig. 1   Fig. 2

12. Name the equal lines and angles   (6.2)

   (a) if PR is $\perp$ bisector of AB

   (b) if BF bisects $\angle ABC$

   (c) if CG is an altitude to AD

   (d) if EM is a median to AD.

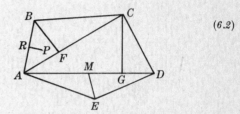

13. State the relationship between each pair of angles: (7.1)

   (a) $\angle 1$ and $\angle 4$ (d) $\angle 4$ and $\angle 5$

   (b) $\angle 3$ and $\angle 4$ (e) $\angle 1$ and $\angle 3$

   (c) $\angle 1$ and $\angle 2$ (f) $\angle AOD$ and $\angle 5$

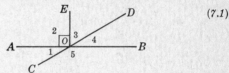

14. Find the angles in each: (7.2)

   (a) The angles are complementary and the smaller is $40°$ less than the larger.

   (b) The angles are complementary and the larger is four times the smaller.

   (c) The angles are supplementary and the smaller is one-half of the larger.

   (d) The angles are supplementary and the larger is $58°$ more than the smaller.

   (e) The angles are supplementary and the larger is $20°$ less than three times the smaller.

   (f) The angles are adjacent and form an angle of $140°$. The smaller is $28°$ less than the larger.

   (g) The angles are vertical and supplementary.

15. If two angles are represented by a and b, obtain two equations for each problem; then find the angles:

   (a) The angles are adjacent and form an angle of $75°$. Their difference is $21°$.   (7.3)

   (b) The angles are complementary. One is $10°$ less than three times the other.

   (c) The angles are supplementary. One is $20°$ more than four times the other.

Chapter 2

Methods of Proof

1. Proof by Deductive Reasoning

A. Deductive Reasoning is Proof

Deductive reasoning enables us to obtain true or acceptably true conclusions provided the statements from which they are deduced or derived are true or accepted as true. It consists of three steps as follows:

1. Making a *general statement* referring to a whole set or class of things, such as the class of dogs:

    All dogs are quadrupeds (have four feet).

2. Making a *particular statement* about one or some of the members of the set or class referred to in the general statement:

    All greyhounds are dogs.

3. Making a *deduction* that follows logically when the general statement is applied to the particular statement:

    All greyhounds are quadrupeds.

Deductive reasoning is called *syllogistic reasoning* since the three types of statements constitute a syllogism. In a syllogism the general statement is the major premise, the particular statement is the minor premise, and the deduction is the conclusion. Thus in the above syllogism:

1. The major premise is: *All dogs are quadrupeds.*
2. The minor premise is: *All greyhounds are dogs.*
3. The conclusion is: *All greyhounds are quadrupeds.*

Using a circle, as in the adjoining diagram, to represent each set or class will help us understand the relationships involved in deductive or syllogistic reasoning.

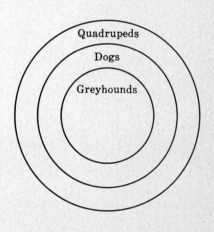

1. Since the major premise or general statement states that all dogs are quadrupeds, the circle representing dogs must be inside that for quadrupeds.

2. Since the minor premise or particular statement states that all greyhounds are dogs, the circle representing greyhounds must be inside that for dogs.

3. The conclusion is obvious. Since the circle of greyhounds must be inside the circle of quadrupeds, the only possible conclusion is that greyhounds are quadrupeds.

B.   *Observation, Measurement, and Experimentation are not Proof*

    1. *Observation cannot serve as proof.* Appearances may be misleading. Eyesight, as in the case of a color blind person, may be defective. Thus in each of the figures, *AB* does not seem to equal *CD* although it actually does.

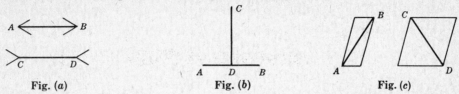

          Fig. (*a*)   Fig. (*b*)   Fig. (*c*)

    2. *Measurement cannot serve as proof.* It applies only to the limited number of cases involved. The conclusion it provides is not exact but approximate, depending on the precision of the instrument and the care of the observer. In measurement, allowance should be made for possible error equal to half the smallest unit of measurement used. Thus if an angle is measured to the nearest degree, an allowance of half a degree of error should be made.

    3. *Experiment cannot serve as proof.* Its conclusions are only probable ones. The degree of probability depends on the particular situations or instances examined in the process of experimentation. Thus it is probable that the dice are loaded if ten successive 7's are rolled with the pair, and the probability is much greater if twenty successive 7's are rolled; however, neither probability is a certainty.

1.1   USING CIRCLES to DETERMINE GROUP RELATIONSHIPS

    In (*a*) to (*e*) each letter, such as *A*, *B* and *R*, represents a set or group. Complete each statement. Show how circles may be used to represent each set or group.

(*a*) If *A* is *B* and *B* is *C*, then (?).   (*d*) If *C* is *D* and *E* is *C*, then (?).

(*b*) If *A* is *B* and *B* is *E* and *E* is *R*, then (?).   (*e*) If squares (*S*) are rectangles (*R*) and rec-

(*c*) If *X* is *Y* and (?), then *X* is *M*.   tangles are parallelograms (*P*), then (?).

Solution:

    (*a*) *A* is *C*   (*b*) *A* is *R*   (*c*) *Y* is *M*   (*d*) *E* is *D*   (*e*) Squares are parallelograms

1.2   COMPLETING a SYLLOGISM

    Write the statement needed to complete each syllogism.

Major Premise (General Statement)	Minor Premise (Particular Statement)	Conclusion (Deduced Statement)
(*a*) A cat is a domestic animal.	Fluffy is a cat.	(?)
(*b*) All men must die.	(?)	Jan must die.
(*c*) Vertical angles are equal.	$\angle c$ and $\angle d$ are vertical angles.	(?)
(*d*) (?)	A square is a rectangle.	A square has equal diagonals.
(*e*) An obtuse triangle has only one obtuse angle.	(?)	$\triangle ABC$ has only one obtuse angle.

Solution:

(*a*) Fluffy is a domestic animal, (*b*) Jan is a man, (*c*) $\angle c = \angle d$, (*d*) A rectangle has equal diagonals,

(*e*) $\triangle ABC$ is an obtuse triangle.

2. *Assumptions: Axioms and Postulates*

The entire structure of proof in geometry must rest upon or begin with some unproved general statements, called assumptions. These are statements which we must willingly assume or accept as true in order to be able to deduce other statements.

Assumptions are either axioms or postulates:

1. An axiom is an assumption applicable to mathematics in general. Thus, *a quantity may be substituted for its equal in an expression or equation* may be used in algebra as well as geometry.

2. A postulate is an assumption applicable to a particular branch of mathematics, such as geometry. Thus, *two straight lines can intersect in one and only one point* applies specifically to geometric figures.

The following list of axioms and postulates should be thoroughly learned.

AXIOMS

Ax. 1 *Things equal to the same or equal things are equal to each other.*

Thus the value of a dime is equal to the value of two nickels, since each value is 10¢.

Ax. 2 *A quantity may be substituted for its equal in any expression or equation.* (Substitution Axiom)

Thus if $x = 5$ and $y = x + 3$, then by substituting 5 for x, $y = 5 + 3 = 8$.

Ax. 3 *The whole equals the sum of its parts.*

Thus the total value of a dime, a nickel and a penny is 16¢.

Ax. 4 *Any quantity equals itself.* (Identity)

Thus $x = x$, $\angle A = \angle A$, $AB = AB$, etc.

Ax. 5 *If equals are added to equals, the sums are equal.* (Addition Axiom)

$$
\begin{array}{lll}
(1) & 7 \text{ dimes} = 70¢ & (2) \quad x + y = 12 \\
\text{Add:} & \underline{2 \text{ dimes} = 20¢} & \text{Add:} \quad \underline{x - y = 8} \\
& 9 \text{ dimes} = 90¢ & \phantom{\text{Add:}} \quad 2x = 20
\end{array}
$$

Ax. 6 *If equals are subtracted from equals, the differences are equal.* (Subtraction Axiom)

$$
\begin{array}{lll}
(1) & 7 \text{ dimes} = 70¢ & (2) \quad x + y = 12 \\
\text{Subtract:} & \underline{2 \text{ dimes} = 20¢} & \text{Subtract:} \quad \underline{x - y = 8} \\
& 5 \text{ dimes} = 50¢ & \phantom{\text{Subtract:}} \quad 2y = 4
\end{array}
$$

Ax. 7 *If equals are multiplied by equals, the products are equal.* (Multiplication Axiom)

Thus if the price of one book is $2, the price of three books is $6.

Special multiplication axiom: Doubles of equals are equal.

Ax. 8 *If equals are divided by equals, the quotients are equal.* (Division Axiom)

Thus if the price of one pound of butter is 80¢ then, at the same rate, the price of a quarter of a pound is 20¢.

Special division axiom: Halves of equals are equal.

Ax. 9 *Like powers of equals are equal.*

Thus if $x = 5$, then $x^2 = 5^2$ or $x^2 = 25$.

Ax. 10 *Like roots of equals are equal.*

Thus if $y^3 = 27$, then $y = \sqrt[3]{27} = 3$.

POSTULATES

Post. 1 *One and only one straight line can be drawn between any two points.*

Thus AB is the only straight line that can be drawn between A and B.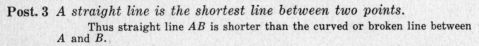

Post. 2 *Two straight lines can intersect in one and only one point.*

Thus only P is the point of intersection of AB and CD.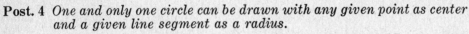

Post. 3 *A straight line is the shortest line between two points.*

Thus straight line AB is shorter than the curved or broken line between A and B.

Post. 4 *One and only one circle can be drawn with any given point as center and a given line segment as a radius.*

Thus only circle A can be drawn with A as center and AB as a radius.

Post. 5 *Any geometric figure can be moved without change in size or shape.*

Thus $\triangle$I can be moved to a new position without a change in its size or shape.

Post. 6 *A straight line segment has one and only one midpoint.*

Thus only M is the midpoint of AB.

Post. 7 *An angle has one and only one bisector.*

Thus only AD is the bisector of $\angle A$.

Post. 8 *Through any point on a line, one and only one perpendicular can be drawn to the line.*

Thus only $PC \perp AB$ at point P on AB.

Post. 9 *Through any point outside a line, one and only one perpendicular can be drawn to the given line.*

Thus only PC can be drawn $\perp AB$ from point P outside AB.

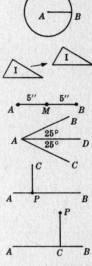

2.1 APPLYING AXIOM 1: Things equal to the same or equal things are equal to each other.

In each, what conclusion follows when Axiom 1 is applied to the given data?

(a) Given: $a = 10$, $b = 10$, $c = 10$

(b) Given: $a = 25$, $a = c$

(c) Given: $a = b$, $c = b$

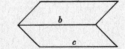

(d) Given: $\angle 1 = 40°$, $\angle 2 = 40°$, $\angle 3 = 40°$

(e) Given: $\angle 1 = \angle 2$, $\angle 3 = \angle 1$

(f) Given: $\angle 3 = \angle 1$, $\angle 2 = \angle 3$

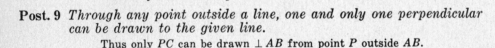

Solution:

(a) Since a, b and c each equals 10, then $a = b = c$.

(b) Since c and 25 each equals a, then $c = 25$.

(c) Since a and c each equals b, then $a = c$.

(d) Since $\angle 1$, $\angle 2$ and $\angle 3$ each equals 40°, then $\angle 1 = \angle 2 = \angle 3$

(e) Since $\angle 2$ and $\angle 3$ each equals $\angle 1$, then $\angle 2 = \angle 3$.

(f) Since $\angle 1$ and $\angle 2$ each equals $\angle 3$, then $\angle 1 = \angle 2$.

2.2 APPLYING AXIOM 2: A quantity may be substituted for its equal in any expression or equation.

In each, what conclusion follows when Axiom 2 is applied to the given data?

(a) Evaluate $2a + 2b$ when $a = 4$ and $b = 8$.

(b) Find x if $3x + 4y = 35$ and $y = 5$.

(c) Given: $\angle 1 + \angle B + 2 = 180°$

$\angle 1 = \angle A$, $\angle 2 = \angle C$

Solution:

(a) Substitute 4 for a and 8 for b:

$2a + 2b$

$2(4) + 2(8)$

$8 + 16 = 24$ *Ans.*

(b) Substitute 5 for y:

$3x + 4y = 35$

$3x + 4(5) = 35$

$3x + 20 = 35$

$3x = 15$, $x = 5$ *Ans.*

(c) Substitute $\angle A$ for $\angle 1$ and $\angle C$ for $\angle 2$:

$\angle 1 + \angle B + \angle 2 = 180°$

$\angle A + \angle B + \angle C = 180°$ *Ans.*

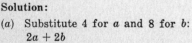

2.3   APPLYING AXIOM 3: The whole equals the sum of its parts.

In each, state the conclusions that follow when Axiom 3 is applied to the given data.

(a)

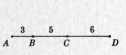

(b)

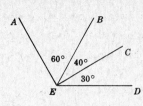

(c)

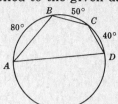

Solution:

(a) $AC = 3 + 5 = 8$
     $BD = 5 + 6 = 11$
     $AD = 3 + 5 + 6 = 14$

(b) $\angle AEC = 60° + 40° = 100°$
     $\angle BED = 40° + 30° = 70°$
     $\angle AED = 60° + 40° + 30° = 130°$

(c) $\overset{\frown}{AC} = 80° + 50° = 130°$
     $\overset{\frown}{BD} = 50° + 40° = 90°$
     $\overset{\frown}{AD} = 80° + 50° + 40° = 170°$

2.4   APPLYING AXIOMS 4, 5, and 6: Identity, addition and subtraction axioms

In each, state a conclusion that follows when Axioms 4, 5 and 6 are applied to the given data:

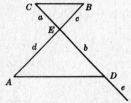

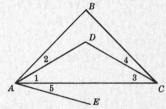

(a) Given: $a = e$

(b) Given: $a = c$, $b = d$

(c) Given: $\angle BAC = \angle DAE$

(d) Given: $\angle BAC = \angle BCA$,
     $\angle 1 = \angle 3$

Solution:

(a) 1. $a = e$ 1. Given
     2. $b = b$ 2. Iden.
     3. $a + b = b + e$ 3. Add. Ax.
     4. $CD = EF$ 4. Subst.

(c) 1. $\angle BAC = \angle DAE$ 1. Given
     2. $\angle 1 = \angle 1$ 2. Iden.
     3. $\angle BAC - \angle 1 = \angle DAE - \angle 1$ 3. Subt. Ax.
     4. $\angle 2 = \angle 5$ 4. Subst.

(b) 1. $a = c$ 1. Given
     2. $b = d$ 2. Given
     3. $a + b = c + d$ 3. Add. Ax.
     4. $CD = AB$ 4. Subst.

(d) 1. $\angle BAC = \angle BCA$ 1. Given
     2. $\angle 1 = \angle 3$ 2. Given
     3. $\angle BAC - \angle 1 = \angle BCA - \angle 3$ 3. Subt. Ax.
     4. $\angle 2 = \angle 4$ 4. Subst.

2.5   APPLYING AXIOMS 7 and 8: Multiplication and division axioms

In each, state the conclusions that follow when the multiplication and division axioms are applied to the given data:

(a)

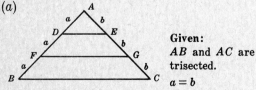

Given:
AB and AC are
trisected.
$a = b$

(b)

Given: $\angle A = \angle C$
$\angle 1 = \frac{1}{2}\angle A$
$\angle 2 = \frac{1}{2}\angle C$

Solution:

(a) If $a = b$, then $2a = 2b$ since doubles of equals
     are equal.   Hence, $AF = DB = AG = EC$.

     Also, $3a = 3b$, using Multiplication Axiom.
Hence, $AB = AC$.

(b) If $\angle A = \angle C$, then $\frac{1}{2}\angle A = \frac{1}{2}\angle C$ since halves of
     equals are equal.   Hence, $\angle 1 = \angle 2$.

2.6 APPLYING AXIOMS to STATEMENTS

Complete each sentence and state the axiom that applies.

(a) If Harry and Alice are the same age today, then in ten years (?).

(b) Since 32°F and 0°C are the temperatures for the freezing point of water, then (?).

(c) If Henry and John are the same weight now and each reduces 20 lb., then (?).

(d) If two stocks of equal value triple in value, then (?).

(e) If two ribbons of equal size are cut into five equal parts, then (?).

(f) If Joan and Agnes are the same height as Anne, then (?).

(g) If two air conditioners of the same price are each discounted 10%, then (?).

Solution:

(a)	they will be the same age.	Add. Ax.	(e) their parts will be the same size.	Div. Ax.
(b)	$32°F = 0°C$	Ax. 1	(f) Joan and Agnes are of the same height.	Ax. 1
(c)	they will be the same weight.	Subt. Ax.	(g) they will have the same price.	Subt. Ax.
(d)	they will have the same value.	Mult. Ax.		

2.7 APPLYING POSTULATES

State the postulate needed to correct each of the following diagrams and its accompanying statement:

(a) 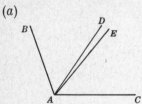 (b) 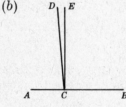 (c) 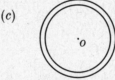 (d)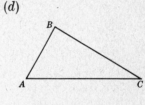

AD and AE bisect $\angle A$ CD and $CE \perp AB$ Both circles have O as a center and the same radius. $AB + BC = AC$

Solution:

(a) Post. 7   (b) Post. 8   (c) Post. 4   (d) Post. 3 (AC is less than the sum of AB and BC.)

3. Basic Angle Theorems

A *theorem* is a statement to be proved. Each of the following basic theorems requires the use of definitions, axioms or postulates for its proof.

Note. The term "principle" (abbreviated Pr.) includes important geometric statements such as theorems, axioms, postulates, and definitions.

Pr. 1: *All right angles are equal.*

Thus in Fig. 1, $\angle A = \angle B$.

Fig. 1

Pr. 2: *All straight angles are equal.*

Thus in Fig. 2, $\angle C = \angle D$.

Fig. 2

Pr. 3: *Complements of the same or of equal angles are equal.*

This is a combination of the following two principles:

(a) Complements of the same angle are equal.

Thus in Fig. 3, $\angle a = \angle b$. Each is the complement of $\angle x$.

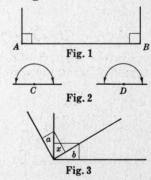

Fig. 3

(b) Complements of equal angles are equal.

Thus in Fig. 4, $\angle c = \angle d$. Their complements are the equal angles, $\angle x$ and $\angle y$.

Fig. 4

Pr. 4: *Supplements of the same or of equal angles are equal.*

This is a combination of the following two principles:

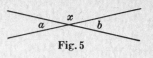

Fig. 5

(a) Supplements of the same angle are equal.

Thus in Fig. 5, $\angle a = \angle b$. Each is the supplement of $\angle x$.

(b) Supplements of equal angles are equal.

Thus in Fig. 6, $\angle c = \angle d$. Their supplements are the equal angles, $\angle x$ and $\angle y$.

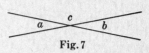

Fig. 6

Pr. 5: *Vertical angles are equal.*

Thus in Fig. 7, $\angle a = \angle b$. This follows from Pr. 4, since $\angle a$ and $\angle b$ are supplements of the same angle, $\angle c$.

Fig. 7

3.1 APPLYING BASIC THEOREMS: Principles 1 to 5.

State the basic angle theorem needed to prove $\angle a = \angle b$ in each case.

(a)
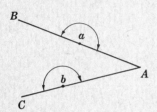

(b)
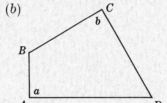

(c)
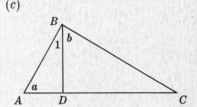

Given:
AB and AC are straight lines.

Solution:

(a) Since AB and AC are st. lines, $\angle a$ and $\angle b$ are st. $\angle$. Hence, $\angle a = \angle b$.

Ans. All straight angles are equal.

Given: $BA \perp AD$,
$BC \perp CD$

(b) Since $BA \perp AD$ and $BC \perp CD$, $\angle a$ and $\angle b$ are rt. $\angle$. Hence, $\angle a = \angle b$.

Ans. All right angles are equal.

Given: $AB \perp BC$
$\angle a$ comp. $\angle 1$

(c) Since $AB \perp BC$, $\angle B$ is a rt. $\angle$, making $\angle b$ comp. $\angle 1$. Since $\angle a$ comp. $\angle 1$, $\angle a = \angle b$.

Ans. Complements of the same angle are equal.

4. *Determining Hypothesis and Conclusion*

A. *Statement Forms: (1) Subject-Predicate Form or (2) If-Then Form*

The statements "A heated metal expands" and "If a metal is heated, then it expands" are two forms of the same idea. The following table shows how each form may be divided into its two important parts, the *hypothesis* which tells *what is given* and the *conclusion* which tells *what is to be proved*. Note that in the if-then form, the word *then* may be omitted.

FORMS	HYPOTHESIS (what is given)	CONCLUSION (what is to be proved)
1. **Subject-Predicate Form** A heated metal expands.	Hypothesis is Subject A heated metal	Conclusion is Predicate expands
2. **If-Then Form** If a metal is heated, then it expands.	Hypothesis is If-Clause If a metal is heated	Conclusion is Then-Clause then it expands

B.   Converse of a Statement

The converse of a statement is formed by interchanging the hypothesis and conclusion. Hence to form the converse of an if-then statement, interchange the if-clause and then-clause. In the case of the subject-predicate form, interchange subject and predicate.

Thus the converse of *Triangles are polygons* is *Polygons are triangles*.   Also, the converse of *If a metal is heated, then it expands* is *If a metal expands, then it is heated*.   Note in each of these cases that the statement is true but that its converse need not necessarily be true.

Pr. 1: *The converse of a true statement is not necessarily true.*

> Thus the statement *Triangles are polygons* is true.   Its converse need not be true.

Pr. 2: *The converse of a definition is always true.*

> Thus the converse of the definition *A triangle is a polygon of three sides* is *A polygon of three sides is a triangle*.   Both the definition and its converse are true.

4.1   DETERMINING the HYPOTHESIS and CONCLUSION of a STATEMENT in SUBJECT-PREDICATE FORM

Determine the hypothesis and conclusion of each statement.

STATEMENTS	ANSWERS Hypothesis (Subject)	Conclusion (Predicate)
(a) Perpendiculars form right angles.	Perpendiculars	form right angles
(b) Complements of the same angle are equal.	Complements of the same angle	are equal
(c) An equilateral triangle is equiangular.	An equilateral triangle	is equiangular
(d) A right triangle has only one right angle.	A right triangle	has only one right angle
(e) A triangle is not a quadrilateral.	A triangle	is not a quadrilateral
(f) An alien has no right to vote.	An alien	has no right to vote

4.2   DETERMINING the HYPOTHESIS and CONCLUSION of a STATEMENT in IF-THEN FORM

Determine the hypothesis and conclusion of each statement.

STATEMENTS	ANSWERS Hypothesis (if-clause)	Conclusion (then-clause)
(a) If a line bisects an angle, then it divides the angle into two equal parts.	If a line bisects an angle	then it divides the angle into two equal parts
(b) A triangle has an obtuse angle if it is an obtuse triangle.	If it is an obtuse triangle	(then) a triangle has an obtuse angle
(c) If a student is sick, he should not go to school.	If a student is sick	(then) he should not go to school
(d) A student, if he wishes to pass, must study regularly.	If he wishes to pass	(then) a student must study regularly

4.3 FORMING CONVERSES and DETERMINING THEIR TRUTH

State whether the given statement is true. Then form its converse and state whether this is necessarily true.

(a) A quadrilateral is a polygon.   (b) An obtuse angle is greater than a right angle.

(c) Florida is a state of the United States.   (d) If you are my pupil, then I am your teacher.

(e) An equilateral triangle is a triangle that has all equal sides.

Solution:

(a) Statement is true. Its converse, *A polygon is a quadrilateral,* is not necessarily true; it might be a triangle.

(b) Statement is true. Its converse, *An angle greater than a right angle is an obtuse angle,* is not necessarily true; it might be a straight angle.

(c) Statement is true. Its converse *A state of the United States is Florida,* is not necessarily true; it might be any one of the other 49 states.

(d) Statement is true. Its converse, *If I am your teacher, then you are my pupil,* is also true.

(e) The statement, a definition, is true. Its converse, *A triangle that has all equal sides is an equilateral triangle,* is also true.

5. *Proving a Theorem*

In the proof of theorems, use accepted symbols and abbreviations.

Procedure:	Prove: All right angles are equal.
1. Divide the theorem into its hypothesis (what is given) and its conclusion (what is to be proved): (See note 1.)	(Hypothesis) (Conclusion) All right angles are equal
2. On one side, make a marked diagram: (See note 2.) 3. On the other side, state what is given and what is to be proved: (See note 3.)	Given: $\angle A$ and $\angle B$ are rt. $\angle$. To Prove: $\angle A = \angle B$.
4. Present a plan: (See note 4.)	Plan: Since each angle equals 90°, the angles are equal, using Ax. 1: Things equal to the same thing are equal to each other.
5. On left, provide statements in successively numbered steps: (See note 5.) 6. On right, provide a reason for each statement: (See note 6)	**Statements** ⎪ **Reasons** 1. $\angle A$ and $\angle B$ are rt. $\angle$. ⎪ 1. Given 2. $\angle A$ and $\angle B$ each $= 90°$. ⎪ 2. A rt. $\angle = 90°$. 3. $\angle A = \angle B$. ⎪ 3. Things $=$ to same thing $=$ each other.

Note 1. Underline the hypothesis by using a single line, and the conclusion by using a double line.

Note 2. Markings on the diagram should include helpful symbols such as square corners for right angles, cross marks for equalities, questions marks for parts to be proved equal, etc.

Note 3. The *Given* and *To Prove* must refer to the figures and letters of the diagram.

Note 4. The plan is very advisable but not an essential part of the proof. It should state the major methods of proof to be used.

Note 5. The last statement is the one to be proved. Statements must refer to the figures and letters of the diagram.

Note 6. A reason must be given for each statement. Acceptable reasons in the proof of a theorem are:

      (1) Given facts   (3) Axioms   (5) Assumed theorems

      (2) Definitions   (4) Postulates   (6) Theorems previously proved

A. Syllogistic Proof of a Theorem

In a syllogistic proof of a theorem such as *All right angles are equal*, statements and reasons are classified as major premises, minor premises or conclusions, as follows:

	First Syllogism	Second Syllogism
Major Premise (General statement)	A right angle equals 90°.	Things equal to the same thing are equal to each other.
Minor Premise (Particular statement)	∠A and ∠B are right angles.	∠A and ∠B each equals 90°.
Conclusion (Deduced statement)	∠A and ∠B each equals 90°.	∠A = ∠B.

The second syllogism above follows the first syllogism. Note that the conclusion of the first syllogism becomes the minor premise of the second. By using the procedure previously shown for proving a theorem, such repetitions of conclusion and minor premise are avoided.

5.1 PROVING a THEOREM

Procedure:	**Prove:** Supplements of equal angles are equal.
1. Divide the theorem into its hypothesis and conclusion:	(Hypothesis)　　　　　　(Conclusion) <u>Supplements of equal angles</u> <u>are equal.</u>
2. Make a marked diagram on one side: 3. State what is given and what is to be proved on the other side:	**Given:** ∠a sup. ∠1, ∠b sup. ∠2 ∠1 = ∠2 **To Prove:** ∠a = ∠b
4. Present a plan:	**Plan:** Using the subtraction axiom, the equal angles may be subtracted from the equal pairs of supplementary angles. The equal remainders are the required angles.

5. Provide statements on the left: 6. Provide reasons for each statement to the right of the statement:	**Statements**	**Reasons**
	1. ∠a sup. ∠1, ∠b sup. ∠2	1. Given
	2. ∠a + ∠1 = 180°, ∠b + ∠2 = 180°.	2. Sup. ∠ are ∠ whose sum = 180°.
	3. ∠a + ∠1 = ∠b + ∠2	3. Things = same thing are = each other.
	4. ∠1 = ∠2	4. Given
	5. ∠a = ∠b	5. If =s − =s, diff. are =.

Supplementary Problems

1. In (a) to (e), each letter such as C, D or R represents a set or group. Complete each statement.　　　　　　　　　　　　　　　　　　　　　(1.1)

   (a) If A is B and B is H, then (?).

   (b) If C is D and P is C, then (?).

   (c) If (?) and B is R, then B is S.

   (d) If E is F, F is G and G is K, then (?).

   (e) If G is H, H is R and (?), then A is R.

   (f) If triangles are polygons and polygons are geometric figures, then (?).

   (g) If a rectangle is a parallelogram and a parallelogram is a quadrilateral, then (?).

2. Indicate the statement needed to complete each syllogism. (1.2)

Major Premise (General Statement)	Minor Premise (Particular Statement)	Conclusion (Deduced Statement)
(a) Even numbers are divisible by 2.	Numbers ending in an even number are even numbers.	(?)
(b) Metals are heavier than gases.	(?)	Gold and silver are heavier than gases.
(c) (?)	Those who get 65 or over will pass the examination.	Those who get 65 or over will pass the course.
(d) Straight angles are equal.	(?)	$\angle a$ and $\angle d$ are equal.
(e) (?)	Triangles are polygons.	Triangles have as many angles as sides.

3. In each, state the conclusion which follows when Axiom 1 is applied to the given data. (2.1)

(a) $a = 7$, $c = 7$, $f = 7$
(b) $b = 15$, $b = g$
(c) $f = h$, $h = a$
(d) $a = c$, $c = f$, $f = h$
(e) $b = d$, $d = g$, $g = e$

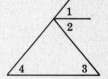

(f) $\angle 1 = 50°$, $\angle 3 = 50°$, $\angle 4 = 50°$
(g) $\angle 1 = 55°$, $\angle 4 = \angle 1$
(h) $\angle 1 = \angle 2$, $\angle 2 = \angle 3$, $\angle 3 = \angle 4$
(i) $\angle 2 = \angle 3$, $\angle 3 = \angle 1$, $\angle 1 = \angle 4$
(j) $\angle 1 = \angle 4$, $\angle 2 = \angle 4$, $\angle 3 = \angle 4$

4. In each, state the conclusion which follows when the substitution axiom, Axiom 2, is applied to the given data. (2.2)

(a) Evaluate $a^2 + 3a$ when $a = 10$.
(b) Evaluate $x^2 - 4y$ when $x = 4$ and $y = 3$.
(c) Does $b^2 - 8 = 17$ when $b = 5$?
(d) Find x if $x + y = 20$ and $y = x + 3$.
(e) Find y if $x + y = 20$ and $y = 3x$.

(f) Find x if $5x - 2y = 24$ and $y = 3$.
(g) Find x if $x^2 + 3y = 45$ and $y = 3$.
(h) Find $\angle b$ if $\angle a + \angle b = 90°$ and $\angle a = 27°$.
(i) Find $\angle y$ if $\angle x + \angle y + \angle z = 180°$, $\angle x = \angle y$ and $\angle z = 80°$.

5. In each, state the conclusion that follows when Axiom 3 is applied to the given data. (2.3)

(a)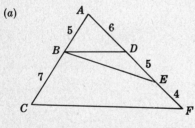

(b)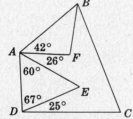

(c)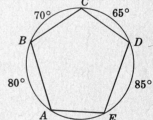

6. In each, state a conclusion involving two new equals that follows when Axiom 4, 5 and 6 (Identity, Addition and Subtraction Axioms) are applied to the given data. (2.4)

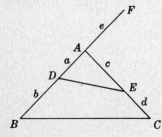

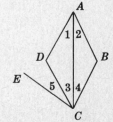

(a) Given: $b = e$
(b) Given: $b = c$, $a = d$
(c) Given: $\overparen{AB} = \overparen{CF}$
(d) Given: $\overparen{AB} = \overparen{EF}$, $\overparen{BC} = \overparen{DE}$
(e) Given: $\angle 4 = \angle 5$
(f) Given: $\angle 1 = \angle 3$, $\angle 2 = \angle 4$

7. In the adjoining figure, AD and BC are trisected.   (2.5)

   (a) If $AD = BC$, why does $AE = BF$?
   (b) If $EG = FH$, why does $AG = BH$?
   (c) If $GD = HC$, why does $AD = BC$?
   (d) If $ED = FC$, why does $EG = FH$?

8. In the adjacent figure, $\angle BCD$ and $\angle ADC$ are trisected.   (2.5)

   (a) If $\angle BCD = \angle ADC$, why does $\angle FCD = \angle FDC$?
   (b) If $\angle 1 = \angle 2$, why does $\angle BCD = \angle ADC$?
   (c) If $\angle 1 = \angle 2$, why does $\angle ADF = \angle BCF$?
   (d) If $\angle EDC = \angle ECD$, why does $\angle 1 = \angle 2$?

9. Complete each and state the axiom that applies.   (2.6)

   (a) If Bill and Dick earn the same amount of money each hour and their rate of pay is increased by the same amount, then (?).
   (b) In the past year, those stocks have tripled in value. If they had the same value last year, then (?).
   (c) A week ago, there were two classes that had the same register. If the same number of pupils were dropped in each, then (?).
   (d) Since 100°C and 212°F are the temperatures for the boiling point of water, then (?).
   (e) If two boards have the same length and each is cut into four equal parts, then (?).
   (f) Since he has $2,000 in Bank A, $3,000 in Bank B and $5,000 in Bank C, then (?).
   (g) If three quarters and four nickels are compared with three quarters and two dimes, (?).

10. Answer each of the following by stating the basic angle theorem needed.   (3.1)

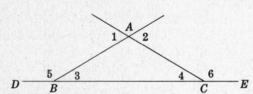

   (a) Why does $\angle 1 = \angle 2$?
   (b) Why does $\angle DBC = \angle ECB$?
   (c) If $\angle 3 = \angle 4$, why does $\angle 5 = \angle 6$?
   (d) If $AF \perp DE$ and $GC \perp DE$, why does $\angle 7 = \angle 8$?
   (e) If $AF \perp DE$, $GC \perp DE$ and $\angle 11 = \angle 12$, why does $\angle 9 = \angle 10$?

11. Determine the hypothesis and conclusion of each statement.   (4.1 and 4.2)

   (a) Stars twinkle.
   (b) Jet planes are the speediest.
   (c) Water boils at 212° Farhrenheit.
   (d) If it is the American flag, its colors are red, white and blue.
   (e) You cannot learn geometry if you fail to do homework in the subject.
   (f) A batter goes to first base if the umpire calls a fourth ball.
   (g) If A is B's brother and C is B's son, then A is C's uncle.
   (h) An angle bisector divides the angle into two equal parts.
   (i) A line is trisected if it is divided into three equal parts.
   (j) A pentagon has five sides and five angles.
   (k) Some rectangles are squares.
   (l) Angles do not become larger if their sides are made longer.
   (m) Angles, if they are equal and supplementary, are right angles.
   (n) The figure cannot be a polygon if one of its sides is not a straight line.

12. State the converse of each of the following true statements. State whether the converse is necessarily true. (a) Half a right angle is an acute angle. (b) An obtuse triangle is a triangle having one obtuse angle. (c) If the umpire called a third strike, then the batter is out. (d) If I am taller than you, then you are shorter than I. (e) If I am heavier than you, then our weights are unequal.   (4.3)

13. Prove each of the following. (a) Straight angles are equal. (b) Complements of equal angles are equal. (c) Vertical angles are equal.   (5.1)

Chapter 3

Congruent Triangles

1. Congruent Triangles

(*a*) **Congruent figures** are figures which have the same size and the same shape; they are the exact duplicates of each other. Such figures can be made to coincide so that their corresponding parts will fit together. The aim of mass production is to manufacture exact duplicates or congruent figures.

Thus two circles having the same radius are congruent circles.

(*b*) **Congruent triangles** are triangles which have the same size and the same shape.

If two triangles are congruent, their corresponding sides and angles are equal. The symbol for "is congruent to" is $\cong$. This symbol is a combination of $=$ for the same size and $\sim$ for the same shape.

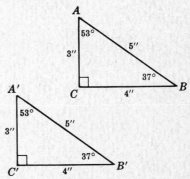

Thus congruent triangles ABC and $A'B'C'$, in the adjacent figure, have equal corresponding sides ($AB = A'B'$, $BC = B'C'$ and $AC = A'C'$) and equal corresponding angles ($\angle A = \angle A'$, $\angle B = \angle B'$ and $\angle C = \angle C'$).

Read $\triangle ABC \cong \triangle A'B'C'$ as "Triangle ABC is congruent to triangle A-prime, B-prime, C-prime."

Note in the congruent triangles how corresponding equal parts may be located. Corresponding sides lie opposite equal angles, and corresponding angles lie opposite equal sides.

Basic Principles of Congruent Triangles

Pr. 1: *If two triangles are congruent, then their corresponding parts are equal.* (Corresponding parts of congruent triangles are equal.)

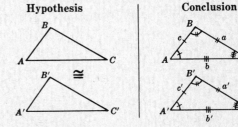

Thus if $\triangle ABC \cong \triangle A'B'C'$, then $\angle A = \angle A'$, $\angle B = \angle B'$, $\angle C = \angle C'$, $a = a'$, $b = b'$ and $c = c'$.

Methods of Proving that Triangles are Congruent

Pr. 2: (s.a.s. = s.a.s) *If two sides and the included angle of one triangle equal the corresponding parts of the other, then the triangles are congruent.*

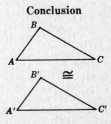
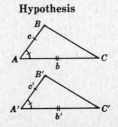

Thus if $b = b'$, $c = c'$ and $\angle A = \angle A'$, then $\triangle ABC \cong \triangle A'B'C'$.

Pr. 3: (a.s.a. = a.s.a.) *If two angles and the included side of one triangle equal the corresponding parts of the other, then the triangles are congruent.*

Thus if ∠A = ∠A′, ∠C = ∠C′ and b = b′, then △ABC ≅ △A′B′C′.

Hypothesis | Conclusion

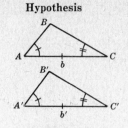
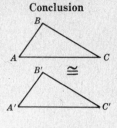

Pr. 4: (s.s.s. = s.s.s.) *If three sides of one triangle equal three sides of another, then the triangles are congruent.*

Thus if a = a′, b = b′ and c = c′, then △ABC ≅ △A′B′C′.

Hypothesis | Conclusion

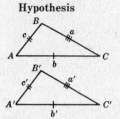
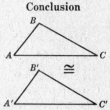

1.1   SELECTING CONGRUENT TRIANGLES

From each of the following, select congruent triangles and state the congruency principle.

(a)

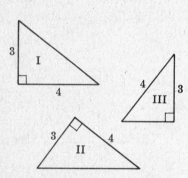

(b)

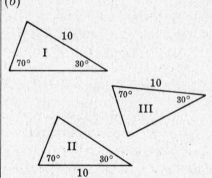

(c)

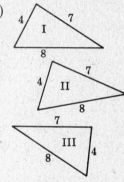

Solution:

(a) △I ≅ △II   s.a.s. = s.a.s.

In △III, the right angle is not between 3 and 4.

(b) △II ≅ △III   a.s.a. = a.s.a.

In △I, side 10 is not between 70° and 30°.

(c) △I ≅ △II ≅ △III
s.s.s. = s.s.s.

1.2   DETERMINING the REASON for CONGRUENCY of TRIANGLES

In each, △I can be proved congruent to △II. Make a diagram showing the equal parts of both triangles and state the congruency principle.

(a)

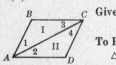

Given: ∠1 = ∠4
        ∠2 = ∠3

To Prove:
        △I ≅ △II

(b)

Given: BE = EC
        AE = ED

To Prove:
        △I ≅ △II

(c)

Given: Isos. △ABC
        Isos. △ADC
AC is common base.

To Prove:
        △I ≅ △II

Solution:

(a)

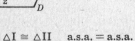

AC is a common side of both △.

(b)

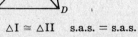

∠1 and ∠2 are vertical angles.

(c)

BD is a common side of both △.

△I ≅ △II   a.s.a. = a.s.a. | △I ≅ △II   s.a.s. = s.a.s. | △I ≅ △II   s.s.s. = s.s.s.

1.3 FINDING ADDITIONAL PARTS NEEDED to PROVE CONGRUENT TRIANGLES

In each, state the additional parts needed to prove △I ≅ △II by the congruency principle indicated below the diagram.

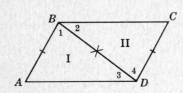

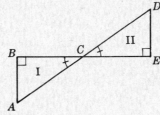

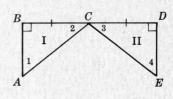

(a) By s.s.s. = s.s.s.

(b) By s.a.s. = s.a.s.

(c) By a.s.a. = a.s.a.

(d) By a.s.a. = a.s.a.

(e) By s.a.s. = s.a.s.

Solution:

(a) If $AD = BC$, then △I ≅ △II by s.s.s. = s.s.s.

(b) If $\angle 1 = \angle 4$, then △I ≅ △II by s.a.s. = s.a.s.

(c) If $BC = CE$, then △I ≅ △II by a.s.a. = a.s.a.

(d) If $\angle 2 = \angle 3$, then △I ≅ △II by a.s.a. = a.s.a.

(e) If $AB = DE$, then △I ≅ △II by s.a.s. = s.a.s.

1.4 SELECTING CORRESPONDING PARTS of CONGRUENT TRIANGLES

In each, the equal parts needed to prove △I ≅ △II are marked. State the remaining parts that are equal.

(a)

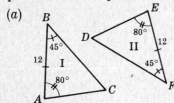

(b)

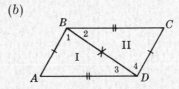

(c)

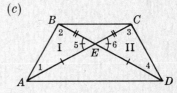

Solution: Equal corresponding sides lie opposite equal angles.

Equal corresponding angles lie opposite equal sides.

(a) Opposite 45°; $AC = DE$.

Opposite 80°; $BC = DF$.

Opposite 12; $\angle C = \angle D$.

(b) Opp. AB and CD; $\angle 3 = \angle 2$.

Opp. BC and AD; $\angle 1 = \angle 4$.

Opp. common side BD;

$\angle A = \angle C$.

(c) Opp. AE and ED; $\angle 2 = \angle 3$.

Opp. BE and EC; $\angle 1 = \angle 4$.

Opp. $\angle 5$ and $\angle 6$; $AB = CD$.

1.5 APPLYING ALGEBRA to EQUAL CORRESPONDING PARTS of CONGRUENT TRIANGLES

In each, find x and y.

(a)

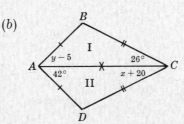

(b)

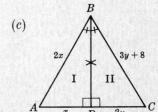

(c)

Solution:

(a) Since △I ≅ △II, corresponding angles are equal.

Hence, $2x = 24$ or $x = 12$, and $3y = 60$ or $y = 20$.

(b) Since △I ≅ △II, corresponding angles are equal.

Hence, $x + 20 = 26$ or $x = 6$, and $y - 5 = 42$ or $y = 47$.

(c) Since △I ≅ △II, corresponding sides are equal. Then

(1) $2x = 3y + 8$ and (2) $x = 2y$

Substituting $2y$ for x in (1), we obtain $2(2y) = 3y + 8$ or $y = 8$. Then $x = 2y = 16$.

1.6 PROVING a CONGRUENCY PROBLEM

Given: $BF \perp DE$
         $BF \perp AC$
         $\angle 3 = \angle 4$

To Prove: $AF = FC$

Plan: Prove $\triangle I \cong \triangle II$

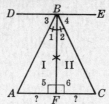

PROOF: Statements	Reasons
1. $BF \perp AC$	1. Given
2. $\angle 5 = \angle 6$	2. $\perp$s form rt. $\angle$. Rt. $\angle$ are =.
3. $BF = BF$	3. Identity
4. $BF \perp DE$	4. Given
5. $\angle 1$ is the complement of $\angle 3$. $\angle 2$ is the complement of $\angle 4$.	5. Adjacent angles are complementary if exterior sides are $\perp$ to each other.
6. $\angle 3 = \angle 4$	6. Given
7. $\angle 1 = \angle 2$	7. Complements of = $\angle$ are =.
8. $\triangle I \cong \triangle II$	8. a.s.a. = a.s.a.
9. $AF = FC$	9. Corresponding parts of congruent $\triangle$ are =.

1.7 PROVING a CONGRUENCY PROBLEM STATED in WORDS

Prove: If the opposite sides of a quadrilateral are equal and a diagonal is drawn, equal angles are formed between the diagonal and the sides.

Given: Quadrilateral $ABCD$
         $AB = CD$, $BC = AD$
         AC is a diagonal.

To Prove: $\angle 1 = \angle 4$, $\angle 2 = \angle 3$

Plan: Prove $\triangle I \cong \triangle II$

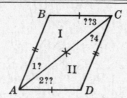

PROOF: Statements	Reasons
1. $AB = CD$, $BC = AD$	1. Given
2. $AC = AC$	2. Identity
3. $\triangle I \cong \triangle II$	3. s.s.s. = s.s.s.
4. $\angle 1 = \angle 4$, $\angle 2 = \angle 3$	4. Corresponding parts of congruent $\triangle$ are =.

2. *Isosceles and Equilateral Triangles*

Principles of Isosceles and Equilateral Triangles

Pr. 1: *If two sides of a triangle are equal, the angles opposite these sides are equal.* (Base angles of an isosceles triangle are equal.)

Thus in $\triangle ABC$, if $AB = BC$, then $\angle A = \angle C$. (A proof of Principle 1 is given in Chapter 16.)

Hypothesis	Conclusion

Pr. 2: *If two angles of a triangle are equal, the sides opposite these angles are equal.*

    Thus in $\triangle ABC$, if $\angle A = \angle C$, then $AB = BC$.

    (Pr. 2 is the converse of Pr. 1. A proof of Principle 2 is given in Chapter 16.)

Hypothesis   Conclusion

Pr. 3: *An equilateral triangle is equiangular.*

    Thus in $\triangle ABC$, if $AB = BC = CA$, then $\angle A = \angle B = \angle C$.

    (Pr. 3 is a corollary of Pr. 1. A *corollary* of a theorem is another theorem whose proof follows readily from the theorem.)

Pr. 4: *An equiangular triangle is equilateral.*

    Thus in $\triangle ABC$, if $\angle A = \angle B = \angle C$, then $AB = BC = CA$.

    (Pr. 4 is the converse of Pr. 3 and a corollary of Pr. 2.)

2.1 APPLYING PRINCIPLES 1 and 3:

In a triangle, equal angles are opposite equal sides.

In each, state the equal angles that are opposite equal sides of a triangle.

(a)   (b)   (c)   (d)

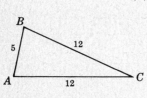
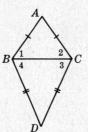
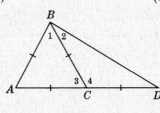
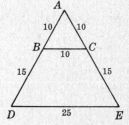

Solution:

(a) Since $AC = BC$,   (b) Since $AB = AC$,   (c) Since $AB = AC = BC$,   (d) Since $AB = BC = AC$,
        $\angle A = \angle B$. $\angle 1 = \angle 2$. $\angle A = \angle 1 = \angle 3$. $\angle A = \angle ACB = \angle ABC$.

                  Since $BD = CD$,   Since $BC = CD$,   Since $AE = AD = DE$,
                       $\angle 3 = \angle 4$. $\angle 2 = \angle D$. $\angle A = \angle D = \angle E$.

2.2 APPLYING PRINCIPLES 2 and 4:

In a triangle, equal sides are opposite equal angles.

In each, state the equal sides that are opposite equal angles of a triangle.

(a)   (b)   (c)   (d)

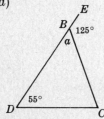
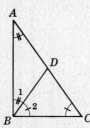
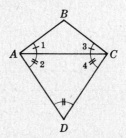
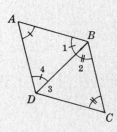

Solution:

(a) Since $\angle a = 55°$,   (b) Since $\angle A = \angle 1$,   (c) Since $\angle 1 = \angle 3$,   (d) Since $\angle A = \angle 1 = \angle 4$,
        $\angle a = \angle D$. $AD = BD$. $AB = BC$. $AB = BD = AD$.

   Hence, $BC = CD$.   Since $\angle 2 = \angle C$,   Since $\angle 2 = \angle 4 = \angle D$,   Since $\angle 2 = \angle 3$,
                     $BD = CD$. $CD = AD = AC$. $BD = CD$.

2.3 APPLYING ISOSCELES TRIANGLE PRINCIPLES

In each, △I can be proved congruent to △II. Make a diagram showing the equal parts of both triangles and state the congruency principle.

(a)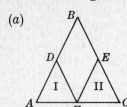

Given: $AB = BC$
          $AD = EC$
F is midpoint of AC.
To Prove: $\triangle I \cong \triangle II$

(b)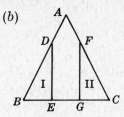

Given: $AB = AC$
          BC is trisected at E
and G.
          $DE \perp BC$
          $FG \perp BC$
To Prove: $\triangle I \cong \triangle II$

Solution:

(a)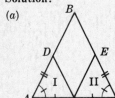

Since $AB = BC$,
          $\angle A = \angle C$.
$\triangle I \cong \triangle II$
s.a.s. = s.a.s.

(b)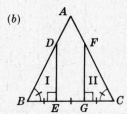

Since $AB = AC$,
          $\angle B = \angle C$.
$\triangle I \cong \triangle II$
a.s.a. = a.s.a.

2.4 PROVING an ISOSCELES TRIANGLE PROBLEM

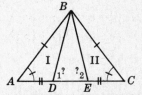

Given: $AB = BC$
          AC is trisected at D and E.
To Prove: $\angle 1 = \angle 2$
Plan: Prove $\triangle I \cong \triangle II$ to obtain $BD = BE$.

PROOF: Statements	Reasons
1. AC is trisected at D and E.	1. Given
2. $AD = EC$	2. To trisect is to divide into three equal parts.
3. $AB = BC$	3. Given
4. $\angle A = \angle C$	4. In a △, ∡ opposite = sides are =.
5. $\triangle I \cong \triangle II$	5. s.a.s. = s.a.s.
6. $BD = BE$	6. Corresponding parts of congruent △ are =.
7. $\angle 1 = \angle 2$	7. Same as 4.

2.5 PROVING an ISOSCELES TRIANGLE PROBLEM STATED in WORDS

Prove: The bisector of the vertex angle of an isosceles triangle is a median to the base.

Solution:

<u>The bisector of the vertex angle of an isosceles triangle is a median to the base.</u>

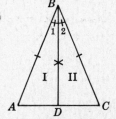

Given: Isosceles $\triangle ABC$ $(AB = BC)$
          BD bisects $\angle B$.
To Prove: BD is a median to AC.
Plan: Prove $\triangle I \cong \triangle II$ to obtain $AD = DC$.

PROOF: Statements	Reasons
1. $AB = BC$	1. Given
2. BD bisects $\angle B$.	2. Given
3. $\angle 1 = \angle 2$	3. To bisect is to divide into two equal parts.
4. $BD = BD$	4. Identity
5. $\triangle I \cong \triangle II$	5. s.a.s. = s.a.s.
6. $AD = DC$	6. Corresponding parts of congruent △ are =.
7. BD is a median to AC.	7. A line from a vertex of a △ bisecting opposite side is a median.

Supplementary Problems

1. From each, select congruent triangles and state the congruency principle.   *(1.1)*

(a)

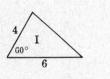   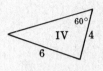

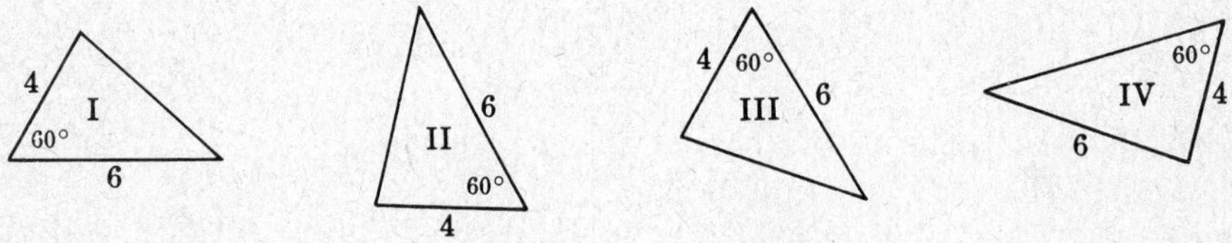

(b)

(c)

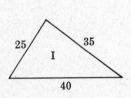 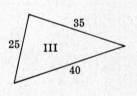

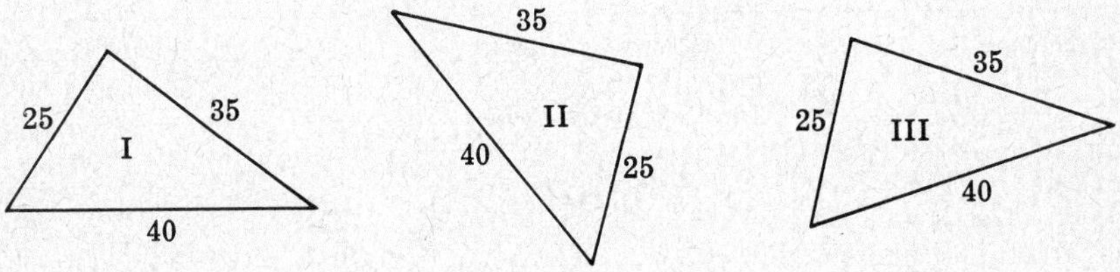

2. In each, △I can be proved congruent to △II. State the congruency principle.   *(1.2)*

(a)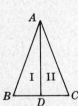

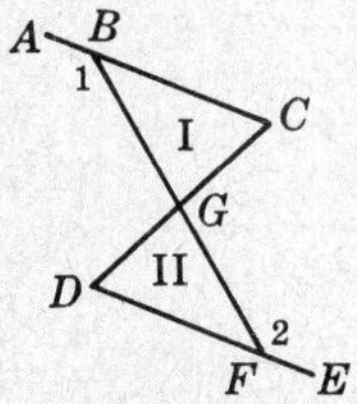

Given: ∠1 = ∠2
  G is midpoint of *BF*.
To Prove: △I ≅ △II

(b)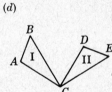

Given: *AB* ⊥ *BE*
  EF ⊥ *BE*
  BC = *DE*
  AB = *EF*
To Prove: △I ≅ △II

(c)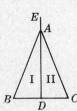

Given: *AB* = *AC*
  AD is median to *BC*.
To Prove: △I ≅ △II

(d)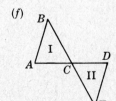

Given: *BC* ⊥ *CE*
  AC ⊥ *CD*
  AC = *CD*
  BC = *CE*
To Prove: △I ≅ △II

(e)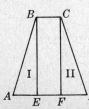

Given: ∠*EAB* = ∠*EAC*
  AD ⊥ *BC*
To Prove: △I ≅ △II

(f)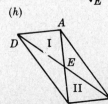

Given: *AD* and *BE* bisect
  each other.
To Prove: △I ≅ △II

(g)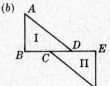

Given: *BE* ⊥ *AD*
  CF ⊥ *AD*
  BE = *CF*
  AD is trisected.
To Prove: △I ≅ △II

(h)

Given: *AD* ⊥ *AC*
  BC ⊥ *AC*
  BD bisects *AC*.
To Prove: △I ≅ △II

3. In each, state the additional parts needed to prove △I ≅ △II by the congruency principle indicated below the diagram. *(1.3)*

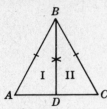

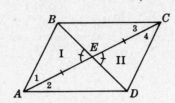

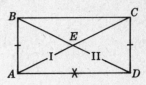

△I is △ABD, △II is △ACD.
Triangles overlap each other.

(a) s.s.s. = s.s.s.

(b) s.a.s. = s.a.s.

(c) a.s.a. = a.s.a.

(d) s.a.s. = s.a.s.

(e) s.s.s. = s.s.s.

(f) s.a.s. = s.a.s.

4. In each, the equal parts needed to prove △I ≅ △II are marked. State the remaining parts that are equal. *(1.4)*

(a)

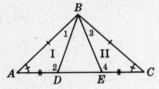

(b)

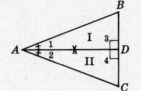

(c)

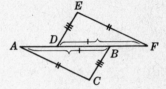

5. In each, find x and y. *(1.5)*

(a)

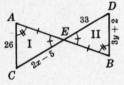

(b)

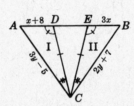

(c)

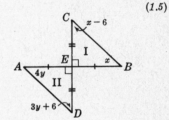

6. Prove each of the following. *(1.6)*

(a) and (b)

(a) **Given:** $BD \perp AC$
　　　　 D is midpoint of AC.
　　To Prove: $AB = BC$

(b) **Given:** BD is altitude to AC.
　　　　 BD bisects $\angle B$.
　　To Prove: $\angle A = \angle C$

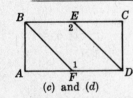

(c) and (d)

(c) **Given:** $\angle 1 = \angle 2$, $BF = DE$
　　　　 BF bisects $\angle B$.
　　　　 DE bisects $\angle D$.
　　　　 $\angle B$ and $\angle D$ are rt. $\angle$.
　　To Prove: $AB = CD$

(d) **Given:** $BC = AD$
　　　　 E is midpoint of BC.
　　　　 F is midpoint of AD.
　　　　 $AB = CD$, $BF = DE$
　　To Prove: $\angle A = \angle C$

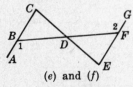

(e) and (f)

(e) **Given:** $\angle 1 = \angle 2$
　　　　 CE bisects BF.
　　To Prove: $\angle C = \angle E$

(f) **Given:** BF and CE bisect each other.
　　To Prove: $BC = EF$

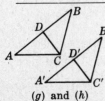

(g) and (h)

(g) **Given:** $CD = C'D'$, $AD = A'D'$
　　　　 CD is altitude to AB.
　　　　 $C'D'$ is altitude to $A'B'$.
　　To Prove: $\angle A = \angle A'$

(h) **Given:** CD bisects $\angle C$.
　　　　 $C'D'$ bisects $\angle C'$.
　　　　 $\angle C = \angle C'$, $\angle B = \angle B'$,
　　　　 $BC = B'C'$.
　　To Prove: $CD = C'D'$

7. Prove each of the following.   (1.7)

   (*a*) If a line bisects an angle of a triangle and is perpendicular to the opposite side, then it bisects that side.

   (*b*) If the diagonals of a quadrilateral bisect each other, then its opposite sides are equal.

   (*c*) If the base and a leg of one isosceles triangle equal the base and a leg of another isosceles triangle, then their vertex angles are equal.

   (*d*) Lines drawn from a point on the perpendicular bisector of a given line to the ends of the given line are equal.

   (*e*) If the legs of one right triangle are equal respectively to the legs of another, their hypotenuses are equal.

8. In each, state the equal angles that are opposite equal sides of a triangle.   (2.1)

(*a*)   (*b*)   (*c*)

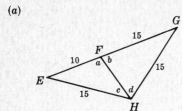   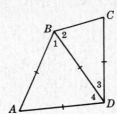   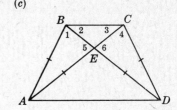

9. In each, state the equal sides that are opposite equal angles of a triangle.   (2.2)

(*a*)   (*b*)   (*c*)

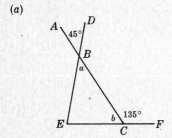   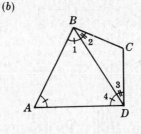   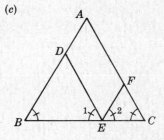

10. In each, two triangles are to be proved congruent. Make a diagram showing the equal parts of both triangles and state the reason for congruency.   (2.3)

(*a*)   (*b*)   (*c*)

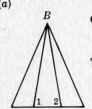

Given:
  $AD = CE$
  $\angle 1 = \angle 2$

To Prove:
  $\triangle ABD \cong \triangle CBE$

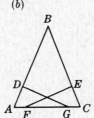

Given:
  $AB = BC$
  $DG \perp AB$
  $EF \perp BC$
  $BD = BE$

To Prove:
  $\triangle AGD \cong \triangle CFE$

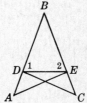

Given:
  $\angle 1 = \angle 2$
  $AD = EC$

To Prove:
  $\triangle ABE \cong \triangle BCD$

11. In each, $\triangle I$, $\triangle II$ and $\triangle III$ can be proved congruent. Make a diagram showing the equal parts and state the reason for congruency.   (2.3)

(*a*)   (*b*)

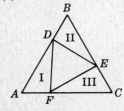

Given:
  $\triangle ABC$ is equilateral.
  $AF = BD = CE$

To Prove:
  $\triangle I \cong \triangle II \cong \triangle III$

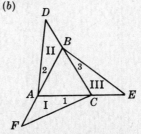

Given:
  $\triangle ABC$ is equilateral.
  AF, BD and CE are extensions of the sides of $\triangle ABC$.
  $\angle 1 = \angle 2 = \angle 3$

To Prove:
  $\triangle I \cong \triangle II \cong \triangle III$

12. Prove each of the following.   (2.4)

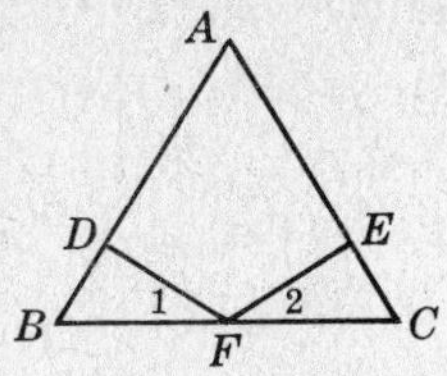

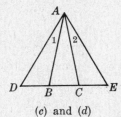

(a) **Given:** $AB = AC$
         F is midpoint of BC.
         $\angle 1 = \angle 2$
    To Prove: $FD = FE$

(b) **Given:** $AB = AC$
         $AD = AE$
         $FD \perp AB, FE \perp AC$
    To Prove: $BF = FC$

(a) and (b)

(c) **Given:** $AB = AC$
         $\angle A$ is trisected.
    To Prove: $AD = AE$

(d) **Given:** $AB = AC$
         $DB = BC$
         $CE = BC$
    To Prove: $AD = AE$

(c) and (d)

13. Prove each of the following.   (2.5)

(a) The median to the base of an isosceles triangle bisects the vertex angle.

(b) If the bisector of an angle of a triangle is also an altitude to the opposite side, then the other two
sides of the triangle are equal.

(c) If a median to a side of a triangle is also an altitude to that side, then the triangle is isosceles.

(d) In an isosceles triangle, the medians to the legs are equal.

(e) In an isosceles triangle, the bisectors of the base angles are equal.

Chapter 4

Parallel Lines, Distances, and Angle Sums

1. *Parallel Lines*

Parallel lines are straight lines which lie in the same plane and do not intersect however far they are extended. The symbol for parallel is ∥; thus $AB \parallel CD$ is read, "AB is parallel to CD." In diagrams, arrows are used to indicate that lines are parallel.

A *transversal* of two or more lines is a line that cuts across these lines. Thus EF is a transversal of AB and CD.

Interior angles formed by two lines cut by a transversal are the angles between the two lines, while *exterior angles* are those on the outside. Thus, of the eight angles formed by AB and CD cut by EF: the interior angles are $\angle 1$, $\angle 2$, $\angle 3$ and $\angle 4$; the exterior angles are $\angle 5$, $\angle 6$, $\angle 7$ and $\angle 8$.

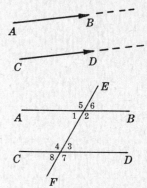

A. *Pairs of Angles Formed by Two Lines Cut by a Transversal*

1. *Corresponding angles* of two lines cut by a transversal are angles on the same side of the transversal and on the same side of the lines. Thus $\angle 1$ and $\angle 2$ are corresponding angles of AB and CD cut by transversal EF. Note that the two angles are to the right of the transversal and below the lines.

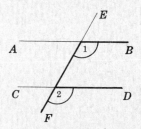

In the case of two parallel lines cut by a transversal, the sides of two corresponding angles form a capital F in varying positions as shown below.

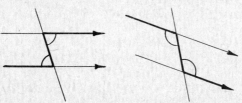 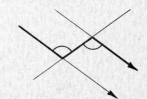

2. *Alternate interior angles* of two lines cut by a transversal are non-adjacent angles between the two lines and on opposite sides of the transversal. Thus $\angle 1$ and $\angle 2$ are alternate interior angles of AB and CD cut by EF.

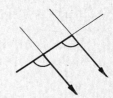

In the case of parallel lines cut by a transversal, the sides of two alternate interior angles form a capital Z or N in varying positions as shown below.

 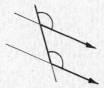 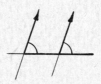 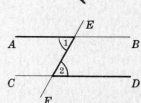

3. *Interior angles on the same side of the transversal* can be readily located by noting the capital U formed by their sides when the lines cut by the transversal are parallel.

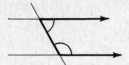 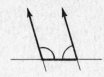 

B. Principles of Parallel Lines

Pr. 1: Parallel Line Postulate — *Through a given point not on a given line, one and only one line can be drawn parallel to a given line.*

 Thus, either l_1 or l_2 but not both may be parallel to l_3.

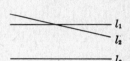

C. Proving that Lines are Parallel

Pr. 2: *Two lines are parallel if a pair of corresponding angles are equal.*

 Thus, $l_1 \parallel l_2$ if $\angle a = \angle b$.

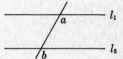

Pr. 3: *Two lines are parallel if a pair of alternate interior angles are equal.*

 Thus, $l_1 \parallel l_2$ if $\angle c = \angle d$.

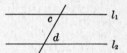

Pr. 4: *Two lines are parallel if a pair of interior angles on the same side of a transversal are supplementary.*

 Thus, $l_1 \parallel l_2$ if $\angle e$ and $\angle f$ are supplementary.

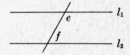

Pr. 5: *Lines are parallel if they are perpendicular to the same line.* (Perpendiculars to the same line are parallel.)

 Thus, $l_1 \parallel l_2$ if l_1 and l_2 are each perpendicular to l_3.

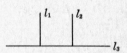

Pr. 6: *Lines are parallel if they are parallel to the same line.* (Parallels to the same line are parallel.)

 Thus, $l_1 \parallel l_2$ if l_1 and l_2 are each parallel to l_3.

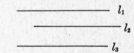

D. Properties of Parallel Lines

Pr. 7: *If two lines are parallel, each pair of corresponding angles are equal.* (Corresponding angles of parallel lines are equal.)

 Thus, if $l_1 \parallel l_2$, then $\angle a = \angle b$.

Pr. 8: *If two lines are parallel, each pair of alternate interior angles are equal.* (Alternate interior angles of parallel lines are equal.)

 Thus, if $l_1 \parallel l_2$, then $\angle c = \angle d$.

Pr. 9: *If two lines are parallel, each pair of interior angles on the same side of the transversal are supplementary.*

 Thus, if $l_1 \parallel l_2$, $\angle e$ and $\angle f$ are supplementary.

Pr. 10: *If lines are parallel, a line perpendicular to one of them is perpendicular to the others also.*

Thus, if $l_1 \parallel l_2$ and $l_3 \perp l_1$, then $l_3 \perp l_2$.

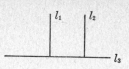

Pr. 11: *If lines are parallel, a line parallel to one of them is parallel to the others also.*

Thus, if $l_1 \parallel l_2$ and $l_3 \parallel l_1$, then $l_3 \parallel l_2$.

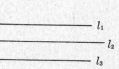

Pr. 12: *If the sides of two angles are respectively parallel to each other, the angles are either equal or supplementary.*

Thus, if $l_1 \parallel l_3$ and $l_2 \parallel l_4$, then $\angle a = \angle b$ and $\angle a$ and $\angle c$ are supplementary.

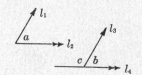

1.1 NUMERICAL APPLICATIONS of PARALLEL LINES

In each find x and y.

(a)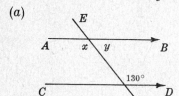

(b)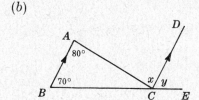

(c)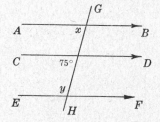

(d)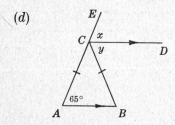

(e)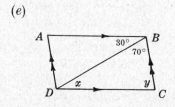

(f)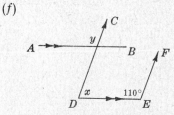

Solution:

(a) $x = 130$ (Pr. 8)   (b) $x = 80$ (Pr. 8)   (c) $x = 75$ (Pr. 7)
     $y = 180 - 130 = 50$ (Pr. 9) $y = 70$ (Pr. 7) $y = 180 - 75 = 105$ (Pr. 9)

(d) $x = 65$ (Pr. 7)   (e) $x = 30$ (Pr. 8)   (f) $x = 180 - 110$ (Pr. 9)
     Since $\angle B = \angle A$, $\angle B = 65°$. $y = 180 - (30 + 70)$ (Pr. 9) $x = 70$
     Hence, $y = 65$ (Pr. 8) $y = 80$ $y = 110$ (Pr. 12)

1.2 APPLYING PARALLEL LINE PRINCIPLES and their CONVERSES

In each, the first statement is given. State the parallel line principle needed as the reason for each of the remaining statements.

(a) 1. $\angle 1 = \angle 2$ 1. Given
     2. $AB \parallel CD$ 2. (?)
     3. $\angle 3 = \angle 4$ 3. (?)

(c) 1. $\angle 5$ sup. $\angle 4$ 1. Given
     2. $EF \parallel GH$ 2. (?)
     3. $\angle 3 = \angle 6$ 3. (?)

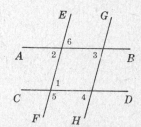

(b) 1. $\angle 2 = \angle 3$ 1. Given
     2. $EF \parallel GH$ 2. (?)
     3. $\angle 4$ sup. $\angle 5$ 3. (?)

(d) 1. $EF \perp AB$, $GH \perp AB$,   1. Given
        $EF \perp CD$
     2. $EF \parallel GH$ 2. (?)
     3. $CD \perp GH$ 3. (?)

Solution:

(a) 2. Pr. 3   (b) 2. Pr. 2   (c) 2. Pr. 4   (d) 2. Pr. 5
     3. Pr. 7   3. Pr. 9   3. Pr. 8   3. Pr. 10

1.3 ALGEBRAIC APPLICATIONS of PARALLEL LINES

In each, find x and y. Provide the reason for each equation obtained from the diagram.

(a)

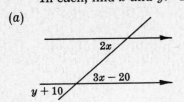

(b)

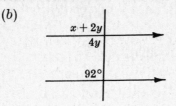

(c)

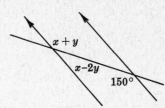

Solution:

(a) $3x - 20 = 2x$ (Pr. 8)
     $x = 20$
     $y + 10 = 2x$ (Pr. 7)
     $y + 10 = 40$ (Subst.)
     $y = 30$

(b) $4y = 180 - 92 = 88$ (Pr. 9)
     $y = 22$
     $x + 2y = 92$ (Pr. 7)
     $x + 44 = 92$ (Subst.)
     $x = 48$

(c) (1) $x + y = 150$ (Pr. 8)
   (2) $x - 2y = 30$ (Pr. 9)
     $3y = 120$ (Subt. Ax.)
     $y = 40$
   $x + 40 = 150$ [Subst. in (1)]
     $x = 110$

1.4 PROVING a PARALLEL LINE PROBLEM

Given: $AB = AC$
       $AE \parallel BC$

To Prove: AE bisects $\angle DAC$

Plan: Show that $\angle 1$ and $\angle 2$ equal the equal angles B and C.

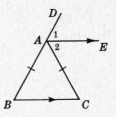

PROOF: Statements	Reasons
1. $AE \parallel BC$	1. Given
2. $\angle 1 = \angle B$	2. Corresponding $\angle$ of $\parallel$ lines are $=$.
3. $\angle 2 = \angle C$	3. Alternate interior $\angle$ of $\parallel$ lines are $=$.
4. $AB = AC$	4. Given
5. $\angle B = \angle C$	5. In a $\triangle$, $\angle$ opposite $=$ sides are $=$.
6. $\angle 1 = \angle 2$	6. Things $=$ to $=$ things are $=$ to each other.
7. AE bisects $\angle DAC$.	7. To divide into two equal parts is to bisect.

1.5 PROVING a PARALLEL LINE PROBLEM STATED in WORDS

Prove: If the diagonals of a quadrilateral bisect each other, the opposite sides are parallel.

Given: Quad. $ABCD$
       AC and BD bisect each other.

To Prove: $AB \parallel CD$
        $AD \parallel BC$

Plan: Prove $\angle 1 = \angle 4$ by showing $\triangle I \cong \triangle II$.
      Prove $\angle 2 = \angle 3$ by showing $\triangle III \cong \triangle IV$

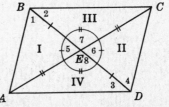

PROOF: Statements	Reasons
1. AC and BD bisect each other.	1. Given
2. $BE = ED$, $AE = EC$	2. To bisect is to divide into two equal parts.
3. $\angle 5 = \angle 6$, $\angle 7 = \angle 8$	3. Vertical $\angle$ are $=$.
4. $\triangle I \cong \triangle II$, $\triangle III \cong \triangle IV$	4. s.a.s. $=$ s.a.s.
5. $\angle 1 = \angle 4$, $\angle 2 = \angle 3$	5. Corresponding parts of congruent $\triangle$ are $=$.
6. $AB \parallel CD$, $BC \parallel AD$	6. Lines cut by a transversal are $\parallel$ if alternate interior $\angle$ are $=$.

2. Distances

Each of the following situations involves the distance between two geometric figures. *In each of these cases, the distance is a straight line segment which is the shortest line between the figures.*

A. Distances Between Two Geometric Figures

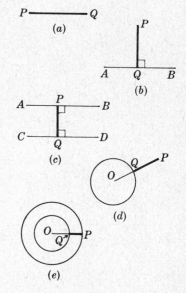

1. The distance **between two points,** such as P and Q, is the line segment PQ, as shown in Fig. (a).

2. The distance **between a point and a line,** such as P and AB in Fig. (b), is the line segment PQ, the perpendicular from the point to the line.

3. The distance **between two parallels,** such as AB and CD, is the line segment PQ, a perpendicular between the two parallels, as shown in Fig. (c).

4. The distance **between a point and a circle,** such as P and circle O in Fig. (d), is PQ, the line segment of OP between the point and the circle.

5. The distance **between two concentric circles,** such as the two circles whose center is O, is PQ, the line segment of the large radius between the two circles, as shown in Fig. (e).

B. Distance Principles

Pr. 1: *If a point is on the perpendicular bisector of a line segment, then it is equidistant from the ends of the line segment.*

   Thus if P is on CD, the $\perp$ bisector of AB, then $PA = PB$.

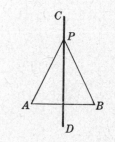

Pr. 2: *If a point is equidistant from the ends of a line segment, then it is on the perpendicular bisector of the line segment.* (Pr. 2 is the converse of Pr. 1.)

   Thus if $PA = PB$, then P is on CD, the $\perp$ bisector of AB.

Pr. 3: *If a point is on the bisector of an angle, then it is equidistant from the sides of the angle.*

   Thus if P is on AB, the bisector of $\angle A$, then $PQ = PR$ where PQ and PR are the distances of P from the sides of the angle.

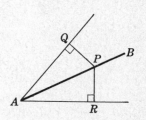

Pr. 4: *If a point is equidistant from the sides of an angle, then it is on the bisector of the angle.* (Pr. 4 is the converse of Pr. 3.)

   Thus if $PQ = PR$ where PQ and PR are the distances of P from the sides of $\angle A$, then P is on AB, the bisector of $\angle A$.

Pr. 5: *Two points each equidistant from the ends of a line segment determine the perpendicular bisector of the line segment.* (The line joining the vertices of two isosceles triangles having a common base is the perpendicular bisector of the base.)

   Thus if $PA = PB$ and $QA = QB$, then P and Q determine CD, the $\perp$ bisector of AB.

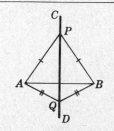

Pr. 6: *The perpendicular bisectors of the sides of a triangle meet in a point which is equidistant from the vertices of the triangle.*

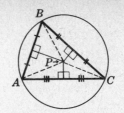

Thus if P is the intersection of the $\perp$ bisectors of the sides of $\triangle ABC$, then $PA = PB = PC$. P is the center of the circumscribed circle and is the *circumcenter* of $\triangle ABC$.

Pr. 7: *The bisectors of the angles of a triangle meet in a point which is equidistant from the sides of the triangle.*

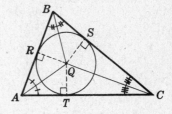

Thus if Q is the intersection of the bisectors of the angles of $\triangle ABC$, then $QR = QS = QT$ where these are the distances from Q to the sides of $\triangle ABC$. Q is the center of the inscribed circle and is the *incenter* of $\triangle ABC$.

2.1   FINDING DISTANCES

In each of the following, find the distance and indicate the kind of distance involved.

Find the distance
(a) from P to A
(b) from P to CD

(c) from A to BC
(d) from AB to CD

Find the distance
(e) from P to inner circle O
(f) from P to outer circle O
(g) between the concentric circles

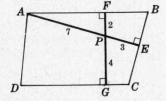

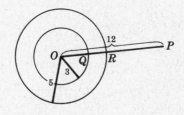

Solution:

(a) $PA = 7$, dist. between two points.

(b) $PG = 4$, dist. from a point to a line.

(c) $AE = 10$, dist. from a point to a line.

(d) $FG = 6$, dist. between two parallel lines.

(e) $PQ = 12 - 3 = 9$, dist. from a point to a circle.

(f) $PR = 12 - 5 = 7$, dist. from a point to a circle.

(g) $QR = 5 - 3 = 2$, dist. between two concentric circles.

2.2   LOCATING a POINT SATISFYING GIVEN CONDITIONS

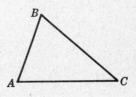

(a) Locate P, a point on BC and equidistant from A and C.
(b) Locate Q, a point on AB and equidistant from BC and AC.
(c) Locate R, the center of the circumscribed circle of $\triangle ABC$.
(d) Locate S, the center of the inscribed circle of $\triangle ABC$.

Solution:

(a)

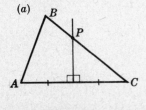

(Using Pr. 1)

(b)

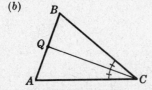

(Using Pr. 3)

(c)

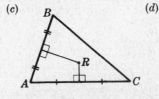

(Using Pr. 6)

(d)

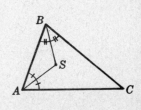

(Using Pr. 7)

2.3 APPLYING PRINCIPLES 2 and 4

In each, describe P, Q and R as equidistant points and locate them on a bisector.

(a)

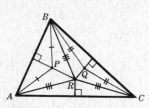

(b)

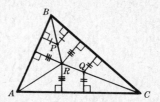

Solution:

(a) Since P is equidistant from A and B, it is on ⊥ bisector of AB. Since Q is equidistant from B and C, it is on ⊥ bisector of BC. Since R is equidistant from A, B and C, it is on ⊥ bisectors of AB, BC and AC.

(b) Since P is equidistant from AB and BC, it is on bisector of $\angle B$. Since Q is equidistant from AC and BC, it is on bisector of $\angle C$. Since R is equidistant from AB, BC and AC, it is on the bisectors of $\angle A$, $\angle B$ and $\angle C$.

2.4 APPLYING PRINCIPLES 1, 3, 6, and 7

In each, describe P, Q and R as equidistant points. Also, describe R as the center of a circle.

(a)

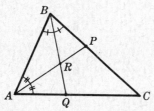

(b)

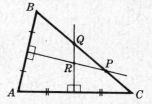

Solution:

(a) Since P is on bisector of $\angle A$, it is equidistant from AB and AC. Since Q is on bisector of $\angle B$, it is equidistant from AB and BC.

Since R is on bisectors of $\angle A$ and $\angle B$, it is equidistant from AB, BC and AC. R is the incenter of $\triangle ABC$, i.e. the center of its inscribed circle.

(b) Since P is on ⊥ bisector of AB, it is equidistant from A and B. Since Q is on ⊥ bisector of AC, it is equidistant from A and C.

Since R is on ⊥ bisectors of AB and AC, it is equidistant from A, B and C. R is the circumcenter of $\triangle ABC$, i.e. the center of its circumscribed circle.

2.5 APPLYING PRINCIPLE 5

In each, find two points equidistant from the ends of a line segment and find the perpendicular bisector determined by the two points.

(a)

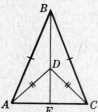

(b)

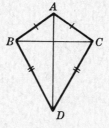

(c)

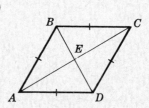

Solution:

(a) B and D are equidistant from A and C. Hence BE is ⊥ bisector of AC.

(b) A and D are equidistant from B and C. Hence AD is ⊥ bisector of BC.

(c) B and D are equidistant from A and C. Hence BD is ⊥ bisector of AC.

A and C are equidistant from B and D. Hence AC is ⊥ bisector of BD.

3. Sum of the Angles of a Triangle

The angles of any triangle may be cut off, as in Fig. 1, and then fitted together as shown in Fig. 2. The three angles will form a straight angle.

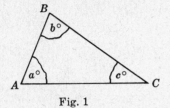

Fig. 1

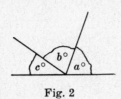

Fig. 2

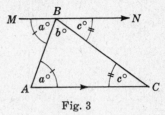

Fig. 3

We can prove that the sum of the angles of a triangle equals 180° by drawing a line through one vertex of the triangle parallel to the side opposite the vertex. In Fig. 3, MN is drawn through B, parallel to AC. Note that the straight angle at B equals the sum of the angles of $\triangle ABC$; that is, $a° + b° + c° = 180°$. Each pair of equal angles is a pair of alternate interior angles of parallel lines.

A. Interior and Exterior Angles of a Polygon

An exterior angle of a polygon is formed whenever one of its sides is extended through a vertex. If each of the sides of a polygon is extended, as shown in Fig. 4, there will be an exterior angle formed at each vertex. Each of these exterior angles is the supplement of its adjacent interior angle.

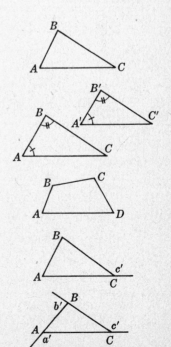

Fig. 4

Thus in the case of pentagon $ABCDE$, there will be 5 exterior angles, one at each vertex. Note that each exterior angle is the supplement of an adjacent interior angle. For example, $\angle a + \angle a' = 180°$.

B. Angle Sum Principles

Pr. 1: *The sum of the angles of a triangle equals a straight angle or 180°.*

Thus in $\triangle ABC$, $\angle A + \angle B + \angle C = 180°$.

Pr. 2: *If two angles of one triangle equal respectively two angles of another triangle, the remaining angles are equal.*

Thus in $\triangle ABC$ and $\triangle A'B'C'$, if $\angle A = \angle A'$ and $\angle B = \angle B'$, then $\angle C = \angle C'$.

Pr. 3: *The sum of the angles of a quadrilateral equals 360°.*

Thus in quadrilateral $ABCD$, $\angle A + \angle B + \angle C + \angle D = 360°$.

Pr. 4: *Each exterior angle of a triangle equals the sum of its two non-adjacent interior angles.*

Thus in $\triangle ABC$, $\angle c' = \angle A + \angle B$.

Pr. 5: *The sum of the exterior angles of a triangle equals 360°.*

Thus in $\triangle ABC$, $\angle a' + \angle b' + \angle c' = 360°$.

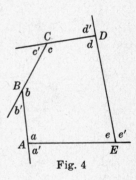

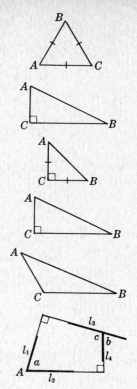

Pr. 6:   *Each angle of an equilateral triangle equals 60°.*

   Thus if △*ABC* is equilateral, then ∠*A* = 60°, ∠*B* = 60° and ∠*C* = 60°.

Pr. 7:   *The acute angles of a right triangle are complementary.*

   Thus in rt. △*ABC*, if ∠*C* = 90°, then ∠*A* + ∠*B* = 90°.

Pr. 8:   *Each acute angle of an isosceles right triangle equals 45°.*

   Thus in isos. rt. △*ABC*, if ∠*C* = 90°, then ∠*A* = 45° and ∠*B* = 45°.

Pr. 9:   *A triangle can have no more than one right angle.*

   Thus in rt. △*ABC*, if ∠*C* = 90°, then ∠*A* and ∠*B* cannot be rt. ∡.

Pr. 10:   *A triangle can have no more than one obtuse angle.*

   Thus in obtuse △*ABC*, if ∠*C* is obtuse, then ∠*A* and ∠*B* cannot be obtuse angles.

Pr. 11:   *Two angles are equal or supplementary if their sides are respectively perpendicular to each other.*

   Thus if $l_1 \perp l_3$ and $l_2 \perp l_4$, then ∠*a* = ∠*b* and ∠*a* and ∠*c* are supplementary.

3.1   NUMERICAL APPLICATIONS of ANGLE SUM PRINCIPLES

In each, find *x* and *y*.

(a)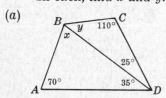

(b)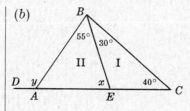

(c)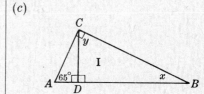

(d)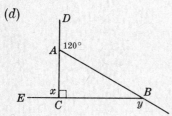

(e)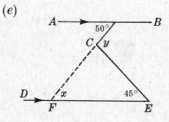

(f)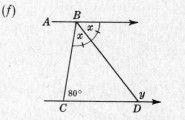

Solution:

(a) $x + 35 + 70 = 180$ **(Pr. 1)**
          $x = 75$
      $y + 110 + 25 = 180$ **(Pr. 1)**
          $y = 45$

Check: The sum of the angles of quad. *ABCD* should equal 360°.
   $70 + 120 + 110 + 60 \overset{?}{=} 360$
          $360 = 360$

(d)   Since $DC \perp EB$, $x = 90$.
   $x + y + 120 = 360$ **(Pr. 5)**
   $90 + y + 120 = 360$
          $y = 150$

(b)   *x* is ext. ∠ of △I.
   $x = 30 + 40$ **(Pr. 4)**
   $x = 70$
   y is an ext. ∠ of △*ABC*.
   $y = ∠B + 40$ **(Pr. 4)**
   $y = 85 + 40 = 125$

(e)   Since $AB \parallel DE$, $x = 50$.
   $y = x + 45$ **(Pr. 4)**
   $y = 50 + 45 = 95$

(c)   In △*ABC*, $x + 65 = 90$ **(Pr. 7)**
          $x = 25$
   In △I, $x + y = 90$ **(Pr. 7)**
      $25 + y = 90$
          $y = 65$

(f)   Since $AB \parallel CD$, $2x + 80 = 180$.
   $2x = 100$, $x = 50$
   $y = x + 80$ **(Pr. 4)**
   $y = 50 + 80 = 130$

3.2 APPLYING ANGLE SUM PRINCIPLES to ISOSCELES and EQUILATERAL TRIANGLES

Find x and y in each.

(a)

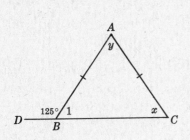

(b)

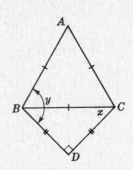

(c)

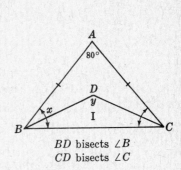

BD bisects $\angle B$
CD bisects $\angle C$

Solution:

(a) Since $AB = AC$, $\angle 1 = \angle x$.

$x = 180 - 125 = 55$
$2x + y = 180$ **(Pr. 1)**
$110 + y = 180$, $y = 70$

(b) By Pr. 8, $x = 45$.

Since $\angle ABC = 60°$ **(Pr. 6)**
and $\angle CBD = 45°$ **(Pr. 8)**
$y = 60 + 45 = 105$.

(c) Since $AB = AC$, $\angle x = \angle ACB$.

$2x + 80 = 180$ **(Pr. 1)**
$x = 50$
In $\triangle I$, $\frac{1}{2}x + \frac{1}{2}x + y = 180$ **(Pr. 1)**
$x + y = 180$
$50 + y = 180$
$y = 130$

3.3 APPLYING RATIOS to ANGLE SUMS

Find each angle

(a) of a triangle if its angles are in the ratio of $3 : 4 : 5$,

(b) of a quadrilateral if its angles are in the ratio of $3 : 4 : 5 : 6$,

(c) of a right triangle if the ratio of its acute angles is $2 : 3$.

Solution:

(a)

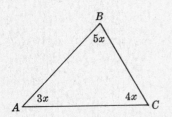

(b)

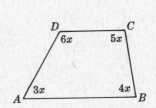

(c)

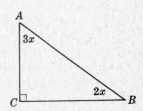

Let $3x$, $4x$ and $5x$ represent the angles.

$12x = 180$ **(Pr. 1)**
$x = 15$

$3x = 45$, $4x = 60$, $5x = 75$
Ans. $45°$, $60°$, $75°$

Let $3x$, $4x$, $5x$ and $6x$ represent the angles.

$18x = 360$ **(Pr. 3)**
$x = 20$

$3x = 60$, $4x = 80$, etc.
Ans. $60°$, $80°$, $100°$, $120°$

Let $2x$ and $3x$ represent the acute angles.

$5x = 90$ **(Pr. 7)**
$x = 18$

$2x = 36$, $3x = 54$
Ans. $36°$, $54°$, $90°$

3.4 USING ALGEBRA to PROVE ANGLE SUM PROBLEMS

(a) **Prove:** If one angle of a triangle equals the sum of the other two, then the triangle is a right triangle.

Given: $\triangle ABC$
$\angle C = \angle A + \angle B$

To Prove: $\triangle ABC$ is a right triangle.

Plan: Prove $\angle C = 90°$.

Algebraic Proof:

Let a = number of degrees in $\angle A$,
b = number of degrees in $\angle B$.
Then $a + b$ = number of degrees in $\angle C$.

$$a + b + (a + b) = 180 \qquad \textbf{(Pr. 1)}$$
$$2a + 2b = 180$$
$$a + b = 90$$

Since $\angle C = 90°$, $\triangle ABC$ is a rt. $\triangle$.

(b) **Prove:** If the opposite angles of a quadrilateral are equal, then its opposite sides are parallel.

Given: Quadrilateral $ABCD$
$\angle A = \angle C$
$\angle B = \angle D$

To Prove: $AB \parallel CD$
$BC \parallel AD$

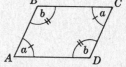

Plan: Prove int. $\angle$ on same side of transversal are supplementary.

Algebraic Proof:

Let a = number of degrees in $\angle A$ and $\angle C$,
b = number of degrees in $\angle B$ and $\angle D$.

$$2a + 2b = 360 \qquad \textbf{(Pr. 3)}$$
$$a + b = 180$$

Since $\angle A$ and $\angle B$ are supplementary, $BC \parallel AD$.
Since $\angle A$ and $\angle D$ are supplementary, $AB \parallel CD$.

4. Sum of the Angles of a Polygon

A *polygon* is a closed figure in a plane bounded by straight line segments as sides.

An *n-gon* is a polygon of n sides. Thus a polygon of 20 sides is a 20-gon.

A *regular polygon* is an equilateral and equiangular polygon. Thus a regular pentagon is a polygon having 5 equal angles and 5 equal sides. A square is a regular polygon of 4 sides.

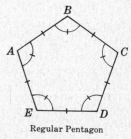

Regular Pentagon

NAMES OF POLYGONS ACCORDING TO THE NUMBER OF SIDES

Number of Sides	Polygon	Number of Sides	Polygon
3	Triangle	8	Octagon
4	Quadrilateral	9	Nonagon
5	Pentagon	10	Decagon
6	Hexagon	12	Dodecagon
7	Heptagon	n	n-gon

A. Sum of the Interior Angles of a Polygon

By drawing diagonals from any vertex to each of the other vertices, as in Fig. 1, a polygon of 7 sides is divisible into 5 triangles. Note that each triangle has one side of the polygon, except the first and last triangles which have two such sides.

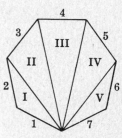

Fig. 1

In general, this process will divide a polygon of n sides into $(n-2)$ triangles; that is, the number of such triangles is always two less than the number of sides of the polygon.

The sum of the interior angles of the polygon equals the sum of the interior angles of the triangles. Hence:

Sum of interior angles of a polygon of n sides $= (n-2)180°$

B. *Sum of the Exterior Angles of a Polygon*

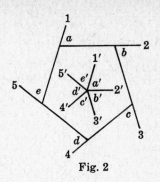

The exterior angles of a polygon can be brought together so that they will have the same vertex. To do this, from a point draw lines parallel to the sides of the polygon, as shown. If this is done, as in Fig. 2, it can be seen that regardless of the number of sides, the sum of the exterior angles equals 360°. Hence:

Sum of exterior angles of a polygon of *n* sides = 360°

Fig. 2

C. *Polygon Angle Principles*

(a) For any polygon

Pr. 1: If S is the sum of the interior angles of a polygon of n sides,

$$S \text{ (in st. } \angle) = n - 2 \quad \text{or} \quad S \text{ (in degrees)} = (n-2)180$$

Thus the sum of the interior angles of a polygon of 10 sides (decagon) equals 1440°, since $S = 8(180) = 1440$.

Pr. 2: The sum of the exterior angles of any polygon equals 360°.

Thus the sum of the exterior angles of a polygon of 23 sides equals 360°.

(b) For a regular polygon

Pr. 3: If a *regular polygon* of n sides has an interior angle i and an exterior angle e (in degrees), then

$$i = \frac{180(n-2)}{n}, \quad e = \frac{360}{n}, \quad \text{and} \quad i + e = 180$$

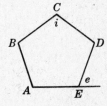

Regular Polygon

Thus for a regular polygon of 20 sides,

$$i = \frac{180(20-2)}{20} = 162, \quad e = \frac{360}{20} = 18,$$

and

$$i + e = 162 + 18 = 180$$

4.1   APPLYING ANGLE FORMULAS to a POLYGON

(a) Find the sum of the interior angles of a polygon of 9 sides and express your answer in straight angles and in degrees.

(b) Find the number of sides of a polygon if the sum of the interior angles is 3600°.

(c) Is it possible to have a polygon the sum of whose angles is 1890°?

Solution:

(a) S (in straight angles) $= n - 2 = 9 - 2 = 7$ straight angles
  S (in degrees) $= (n-2)180 = 7(180) = 1260°$

(b) S (in degrees) $= (n-2)180$. Then $3600 = (n-2)180$, from which $n = 22$.

(c) Since $1890 = (n-2)180$, then $n = 12\frac{1}{2}$. A polygon cannot have $12\frac{1}{2}$ sides.

4.2 APPLYING ANGLE FORMULAS to a REGULAR POLYGON

(a) Find each exterior angle of a regular polygon having 9 sides.
(b) Find each interior angle of a regular polygon having 9 sides.
(c) Find the number of sides of a regular polygon if each exterior angle is 5°.
(d) Find the number of sides of a regular polygon if each interior angle is 165°.

Solution:

(a) Since $n = 9$, $e = \dfrac{360}{n} = \dfrac{360}{9} = 40$.   *Ans.* 40°

(b) Since $n = 9$, $i = \dfrac{(n-2)180}{n} = \dfrac{(9-2)180}{9} = 140$.   *Ans.* 140°

   Another method: Since $i + e = 180$, $i = 180 - e = 180 - 40 = 140$.

(c) Substituting $e = 5$ in $e = \dfrac{360}{n}$, we have $5 = \dfrac{360}{n}$, $5n = 360$, $n = 72$.   *Ans.* 72 sides

(d) Substituting $i = 165$ in $i + e = 180$, we have $165 + e = 180$ or $e = 15$.

   Then using $e = \dfrac{360}{n}$ with $e = 15$, we have $15 = \dfrac{360}{n}$, $15n = 360$, $n = 24$.   *Ans.* 24 sides

4.3 APPLYING ALGEBRA to ANGLE SUMS of a POLYGON

   Find each interior angle of a quadrilateral

(a) if its interior angles are represented by $x + 10$, $2x + 20$, $3x - 50$, and $2x - 20$
(b) if its exterior angles are in the ratio $2:3:4:6$.

Solution:

(a) Since the sum of the interior $\angle = 360°$, $(x+10) + (2x+20) + (3x-50) + (2x-20) = 360$
$$8x - 40 = 360$$
$$x = 50.$$

   Then $x + 10 = 60$, $2x + 20 = 120$, $3x - 50 = 100$, $2x - 20 = 80$.   *Ans.* 60°, 120°, 100°, 80°

(b) Let the exterior angles be represented respectively by $2x$, $3x$, $4x$, $6x$.
   Then $2x + 3x + 4x + 6x = 360$. Solving, $15x = 360$ and $x = 24$.
   Hence the exterior angles are 48°, 72°, 96°, 144°.
   The interior angles are their supplements. *Ans.* 132°, 108°, 84°, 36°.

5. *Two New Congruency Theorems:*   s.a.a. = s.a.a.   and   hy. leg = hy. leg

   Three methods of proving triangles congruent have been previously studied. These are

   (1) s.a.s. = s.a.s.   (2) a.s.a. = a.s.a.   (3) s.s.s. = s.s.s.

   Two additional methods of proving that triangles are congruent are as follows:

   (4) s.a.a. = s.a.a.   (5) hy. leg = hy. leg

A. *Two New Congruency Principles*

Pr. 1: (s.a.a. = s.a.a.) *If two angles and a side opposite one of them of one triangle equal the corresponding parts of the other, the triangles are congruent.*

   Thus if $\angle A = \angle A'$, $\angle B = \angle B'$, and $BC = B'C'$, then $\triangle ABC \cong \triangle A'B'C'$.

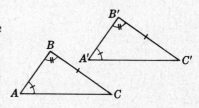

Pr. 2: (hy. leg = hy. leg) *If the hypotenuse and a leg of one right triangle equal the corresponding parts of another right triangle, the triangles are congruent.*

   Thus if hy. AB = hy. $A'B'$ and leg BC = leg $B'C'$, then rt. $\triangle ABC \cong$ rt. $\triangle A'B'C'$.

   A proof of this principle is given in Chapter 16.

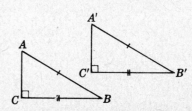

5.1 SELECTING CONGRUENT TRIANGLES USING s.a.a. = s.a.a. or hy. leg = hy. leg

From each of the following, select congruent triangles and state the reason for the congruency.

(a)

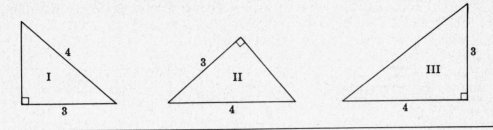

(b)

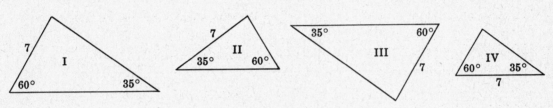

Solution:

(a) $\triangle$I $\cong$ $\triangle$II by hy. leg = hy. leg. In $\triangle$III, 4 is not a hypotenuse.

(b) $\triangle$I $\cong$ $\triangle$III by s.a.a. = s.a.a. In $\triangle$II, 7 is opposite 60° instead of 35°. In $\triangle$IV, 7 is included between 60° and 35°.

5.2 DETERMINING the REASON for CONGRUENCY of TRIANGLES

In each, $\triangle$I can be proved congruent to $\triangle$II. Make a diagram showing the equal parts of both triangles and state the reason for the congruency.

(a)

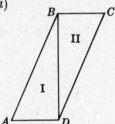

Given: $BD \perp BC$
$BD \perp AD$
$AB = CD$

To Prove: $\triangle$I $\cong$ $\triangle$II

(b)

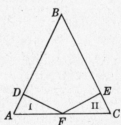

Given: $AB = BC$
$FD \perp AB$
$FE \perp BC$
F is midpoint of AC.

To Prove: $\triangle$I $\cong$ $\triangle$II

Solution:

(a)

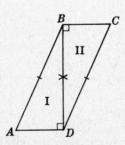

$\triangle$I $\cong$ $\triangle$II by hy. leg = hy. leg

(b)

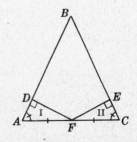

$\triangle$I $\cong$ $\triangle$II by s.a.a. = s.a.a.

5.3 PROVING a CONGRUENCY PROBLEM

Given: Quadrilateral $ABCD$
$DF \perp AC$, $BE \perp AC$
$AE = FC$, $BC = AD$

To Prove: $BE = FD$

Plan: Prove $\triangle I \cong \triangle II$

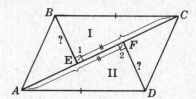

PROOF: Statements	Reasons
1. $BC = AD$	1. Given
2. $DF \perp AC$, $BE \perp AC$	2. Given
3. $\angle 1 = \angle 2$	3. Perpendiculars form rt. ∡, and all rt. ∡ are equal.
4. $AE = FC$	4. Given
5. $EF = EF$	5. Identity
6. $AF = EC$	6. If equals are added to equals, the sums are equal.
7. $\triangle I \cong \triangle II$	7. hy. leg = hy. leg
8. $BE = FD$	8. Corresponding parts of congruent ▲ are equal.

5.4 PROVING a CONGRUENCY PROBLEM STATED in WORDS

Prove: In an isosceles triangle, altitudes to the equal sides are equal.

Given: Isosceles $\triangle ABC$ ($AB = BC$)
AD is altitude to BC
CE is altitude to AB

To Prove: $AD = CE$

Plan: Prove $\triangle ACE \cong \triangle ACD$
or $\triangle I \cong \triangle II$

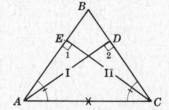

PROOF: Statements	Reasons
1. $AB = BC$	1. Given
2. $\angle A = \angle C$	2. In a △, angles opposite equal sides are equal.
3. AD is altitude to BC. CE is altitude to AB.	3. Given
4. $\angle 1 = \angle 2$	4. Altitudes form rt. ∡ and rt. ∡ are equal.
5. $AC = AC$	5. Identity
6. $\triangle I \cong \triangle II$	6. s.a.a. = s.a.a.
7. $AD = CE$	7. Corresponding parts of congruent ▲ are equal.

Supplementary Problems

1. In each, find x and y. (1.1)

(a)
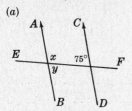

(b)
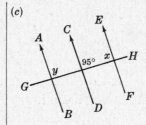

(c)
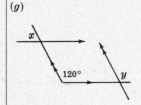

(d)
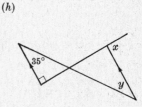

(e)
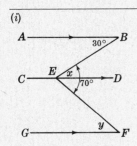

(f)
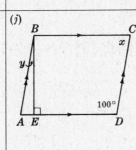

(g)
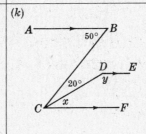

(h)
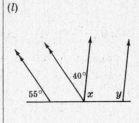

(i)
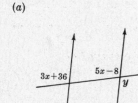

(j)
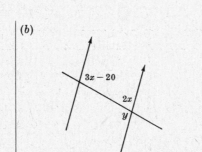

(k)
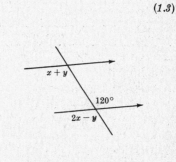

(l)
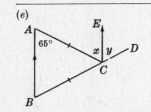

2. In each, find x and y. (1.3)

(a)

(b)

(c)

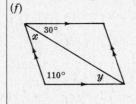

3. If two lines are parallel, find (1.3)

(a) two alternate interior angles represented by $3x$ and $5x - 70$,

(b) two corresponding angles represented by $2x + 10$ and $4x - 50$,

(c) two interior angles on the same side of the transversal represented by $2x$ and $3x$.

4. Prove each of the following. (1.4)

(a)
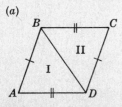

Given: Quad. $ABCD$
        $AB = CD$
        $BC = AD$

To Prove: $AB \parallel CD$
             $BC \parallel AD$

(b)
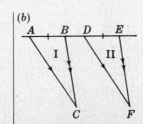

Given: $AB = DE$
        $AC \parallel DF$
        $BC \parallel EF$

To Prove: $AC = DF$

5. Prove each of the following.   (1.4)

(a)

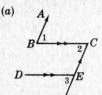

Given: $AB \parallel CF$
           $BC \parallel DE$
To Prove: $\angle 1 = \angle 3$

(b)

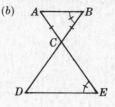

Given: $AC = BC$
           $\angle B = \angle E$
To Prove: $AB \parallel DE$

(c)

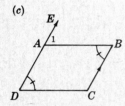

Given: Quad. $ABCD$
           $DE \parallel BC$
           $\angle B = \angle D$
To Prove: $AB \parallel CD$

(d)

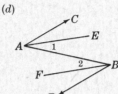

Given: $AC \parallel BD$
           AE bisects $\angle A$
           BF bisects $\angle B$
To Prove: $BF \parallel AE$

6. Prove each of the following.   (1.5)

  (a) If the opposite sides of a quadrilateral are parallel, then they are also equal.

  (b) If AB and CD bisect each other at E, then $AC \parallel BD$.

  (c) In quadrilateral $ABCD$, $BC \parallel AD$. If the diagonals AC and BD intersect at E and $AE = DE$, then $BE = CE$.

  (d) AB and CD are parallel lines cut by a transversal at E and F. If EG and FH bisect a pair of corresponding angles, then $EG \parallel FH$.

  (e) If a line through vertex B of $\triangle ABC$ is parallel to AC and bisects the angle formed by extending AB through B, then $\triangle ABC$ is isosceles.

7. In Fig. 1, find the distance from (a) A to B, (b) E to AC, (c) A to BC, (d) ED to BC.   (2.1)

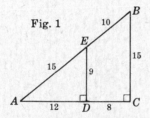

Fig. 1

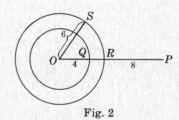

Fig. 2

8. In Fig. 2, find the distance (a) from P to the outer circle, (b) from P to the inner circle, (c) between the concentric circles, (d) from P to O.   (2.1)

9. (a) Locate P, a point on AD, equidistant from B and C.   (2.2)
          Locate Q, a point on AD, equidistant from AB and BC.

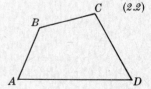

  (b) Locate R, a point equidistant from A, B and C.
          Locate S, a point equidistant from B, C and D.

  (c) Locate T, a point equidistant from BC, CD and AD.
          Locate U, a point equidistant from AB, BC and CD.

10. In each, describe P, Q and R as equidistant points and locate them on a bisector.   (2.3)

(a)

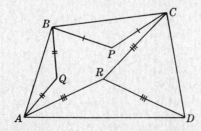

(b)

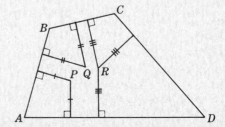

11. In each, describe P, Q and R as equidistant points.   *(2.4)*

(a)

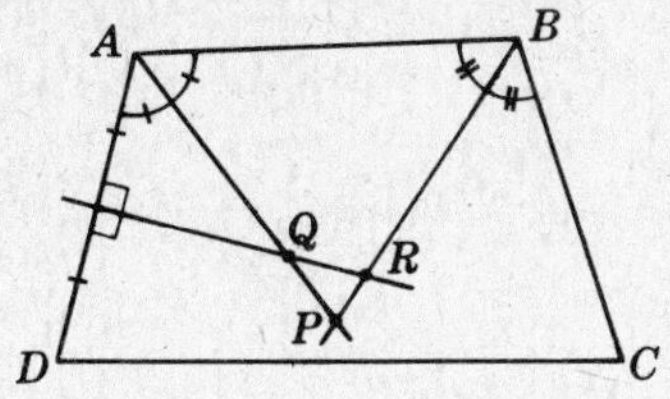

(b)

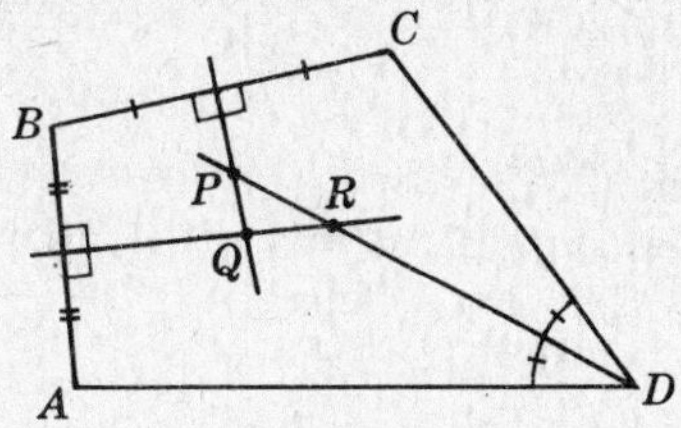

12. Find x and y in each.   *(3.1)*

(a)

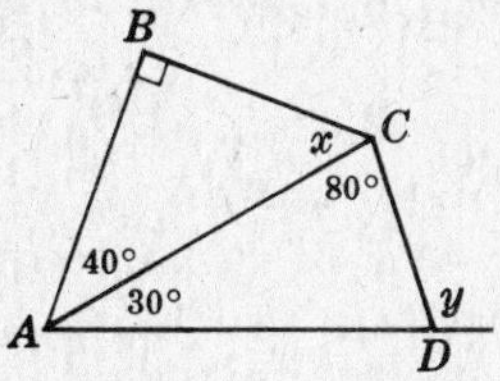

(b)

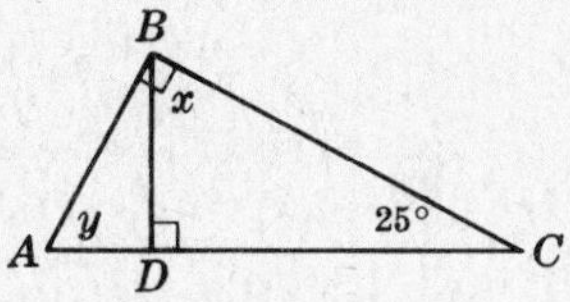

(c)

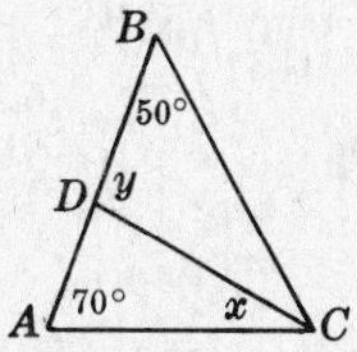

CD bisects $\angle C$

(d)

BD bisects $\angle B$
CD bisects $\angle C$

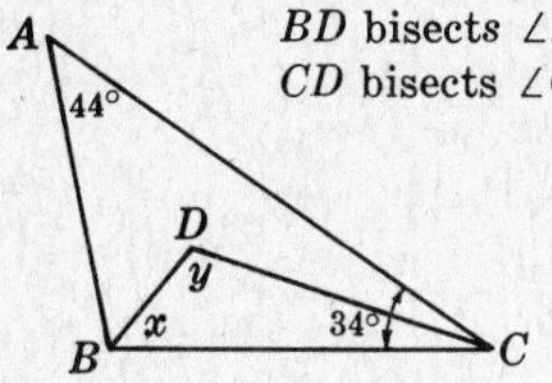

(e)

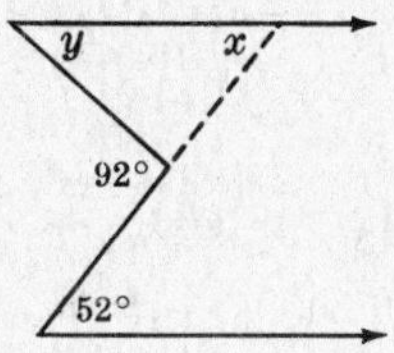

(f)

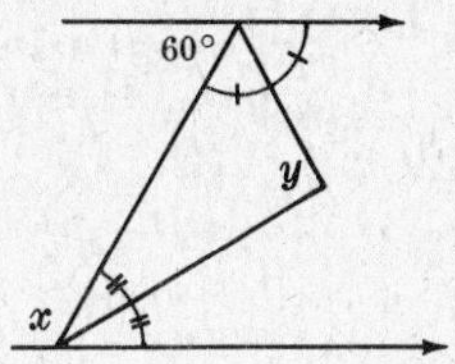

13. Find x and y in each.   *(3.2)*

(a)

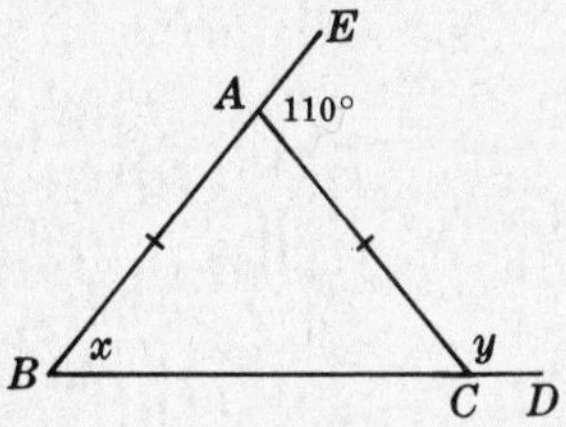

(b)

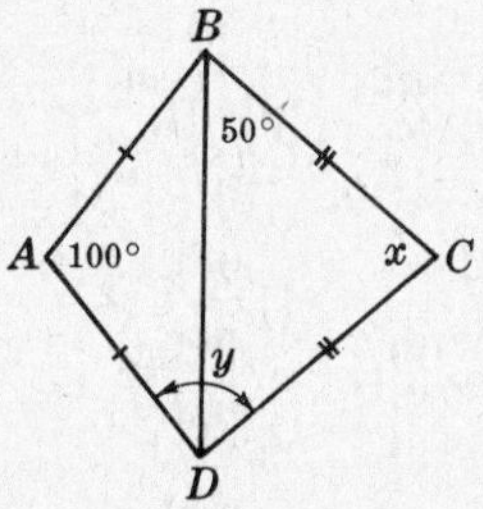

(c)

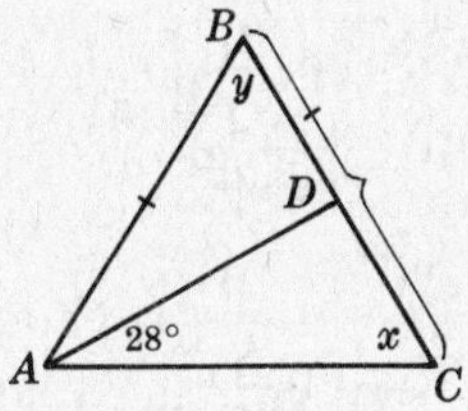

AD bisects $\angle A$.

(d)

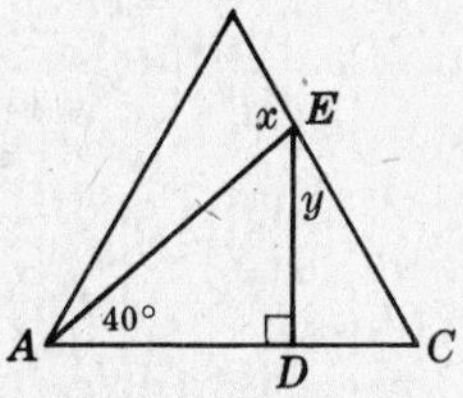

$\triangle ABC$ is equilateral.

(e)

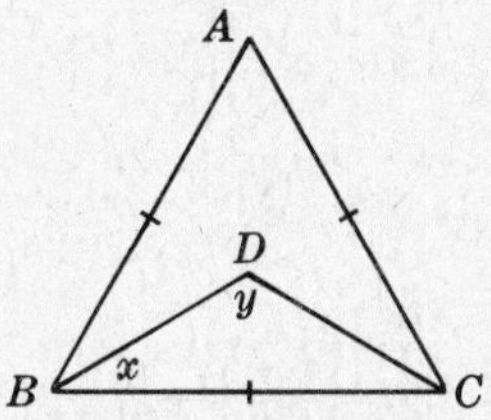

$\triangle ABC$ is equilateral.
BD bisects $\angle B$, *CD* bisects $\angle C$.

(f)

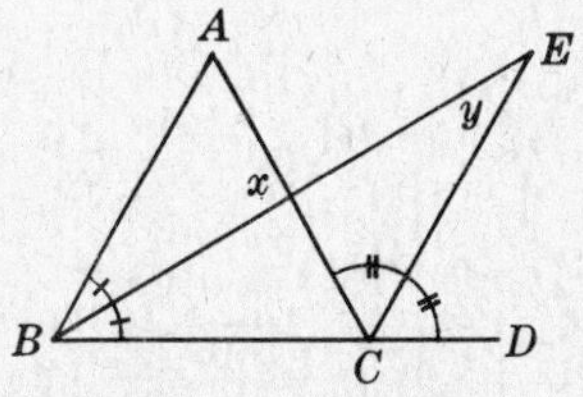

$\triangle ABC$ is equilateral.

14. Find each angle   *(3.3)*

  (a) of a triangle if its angles are in the ratio of $1:3:6$.

  (b) of a right triangle if its acute angles are in the ratio of $4:5$.

  (c) of an isosceles triangle if the ratio of its base angle to a vertex angle is $1:3$.

  (d) of a quadrilateral if its angles are in the ratio of $1:2:3:4$.

  (e) of a triangle, one of whose angles is $55°$ and the other two angles are in the ratio of $2:3$.

  (f) of a triangle if the ratio of its exterior angles is $2:3:4$.

15. Prove each of the following.   (3.4)

    (a) In quadrilateral $ABCD$ if $\angle A = \angle D$ and $\angle B = \angle C$, then $BC \parallel AD$.

    (b) Two parallel lines are cut by a transversal. Prove that the bisectors of two interior angles on the same side of the transversal are perpendicular to each other.

    (c) A triangle is a right triangle if a median to one of the sides equals one-half of that side.

16. Show that a triangle is   (3.4)

    (a) equilateral if its angles are represented by $x + 15$, $3x - 75$ and $2x - 30$.

    (b) isosceles if its angles are represented by $x + 15$, $3x - 35$ and $4x$.

    (c) a right triangle if its angles are in the ratio $2 : 3 : 5$.

    (d) an obtuse triangle if one angle is $64°$ and the larger of the other two is $10°$ less than five times the smaller.

17. (a) Find the sum of the interior angles in straight angles of a polygon of 9 sides; 32 sides.   (4.1)

    (b) Find the sum of the interior angles in degrees of a polygon of 11 sides; 32 sides; 1002 sides.

    (c) Find the number of sides of a polygon if the sum of the interior angles is 28 straight angles; 20 right angles; $4500°$; $36,000°$.

    (d) Is it possible to have a polygon the sum of whose angles is $13\frac{1}{4}$ straight angles? $18,050°$? $2700°$?

18. (a) Find each exterior angle of a regular polygon having 18 sides; 20 sides; 40 sides.   (4.2)

    (b) Find each interior angle of a regular polygon having 18 sides; 20 sides; 40 sides.

    (c) Find the number of sides of a regular polygon if each exterior angle is $120°$; $40°$; $18°$; $2°$.

    (d) Find the number of sides of a regular polygon if each interior angle is $60°$; $150°$; $170°$; $175°$; $179°$.

    (e) Find the number of sides of a regular polygon if each interior angle is three times each exterior angle.

19. (a) Find each interior angle of a quadrilateral if its interior angles are represented by $x - 5$, $x + 20$, $2x - 45$, and $2x - 30$.   (4.3)

    (b) Find each interior angle of a quadrilateral if its exterior angles are in the ratio of $1 : 2 : 3 : 3$.

    (c) Find each interior angle of a pentagon if its interior angles are represented by $x - 10$, $2x - 20$, $2x - 10$, $2x + 10$, and $3x - 30$.

20. From each of the following, select congruent triangles and state the reason for the congruency.   (5.1)

(a)

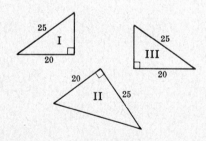

(b)

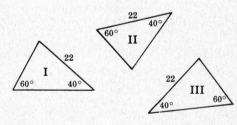

21. In each, two triangles can be proved congruent. Make a diagram showing the equal parts of both triangles and state the reason for the congruency.   (5.2)

(a)

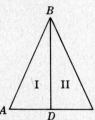

    Given: Isosceles $\triangle ABC$
            $(AB = BC)$
            BD is altitude to AC

    To Prove: $\triangle \text{I} \cong \triangle \text{II}$.

(b)

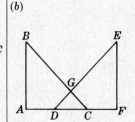

    Given: $AB = EF$
            $AB \perp AF$, $EF \perp AF$
            $DG = GC$

    To Prove: $\triangle ABC \cong \triangle DEF$

22. Prove each of the following.   (5.3)

(a)

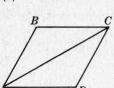

Given: $\angle B = \angle D$
$BC \parallel AD$

To Prove: $BC = AD$

(b)

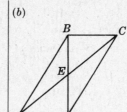

Given: $AE = EC$
$BD \perp BC$
$BD \perp AD$

To Prove: $BE = ED$

(c)

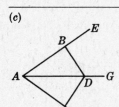

Given: $BD = DC$
$BD \perp AE$
$DC \perp AF$

To Prove: AG bisects $\angle A$.

(d)

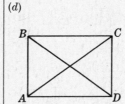

Given: $AC = BD$
$AB \perp AD$
$CD \perp AD$

To Prove: $AB = CD$

23. Prove each of the following.   (5.4)

   (a) If the perpendiculars to two sides of a triangle from the midpoint of the third side are equal, then the triangle is isosceles.

   (b) Perpendiculars from a point in the bisector of an angle to the sides of the angle are equal.

   (c) If the altitudes to two sides of a triangle are equal, then the triangle is isosceles.

   (d) Two right triangles are congruent if the hypotenuse and an acute angle of one equal the corresponding parts of the other.

Chapter 5

Parallelograms, Trapezoids, Medians and Midpoints

1. Trapezoids

(a) A *trapezoid* is a quadrilateral having two and only two parallel sides.

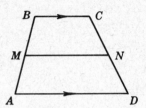

The *bases* of a trapezoid are its parallel sides. The *legs* are its non-parallel sides. The *median* of a trapezoid is the line joining the midpoints of its legs.

Thus in trapezoid $ABCD$, the bases are AD and BC, and the legs are AB and CD. If M and N are midpoints, then MN is the median of the trapezoid.

(b) An *isosceles trapezoid* is a trapezoid whose legs are equal. Thus in isosceles trapezoid $ABCD$, $AB = CD$.

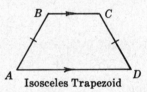

Isosceles Trapezoid

The base angles of a trapezoid are the angles at the ends of its longer base. Thus $\angle A$ and $\angle D$ are the base angles of isosceles trapezoid $ABCD$.

A. Trapezoid Principles

Pr. 1: *The base angles of an isosceles trapezoid are equal.*

Thus in trapezoid $ABCD$, if $AB = CD$, then $\angle A = \angle D$.

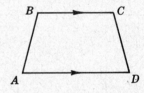

Pr. 2: *If the base angles of a trapezoid are equal, the trapezoid is isosceles.*

Thus in trapezoid $ABCD$, if $\angle A = \angle D$, then $AB = CD$.

1.1 APPLYING ALGEBRA to the TRAPEZOID

In each, find x and y.

(a)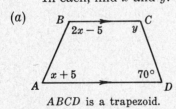

ABCD is a trapezoid.

(b)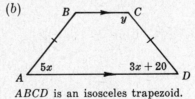

ABCD is an isosceles trapezoid.

(c)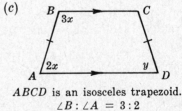

ABCD is an isosceles trapezoid.
$\angle B : \angle A = 3 : 2$

Solution:

(a) Since $AD \parallel BC$, $(2x - 5) + (x + 5) = 180$, $3x = 180$, $x = 60$.
Also, $y + 70 = 180$ or $y = 110$.

(b) Since $\angle A = \angle D$, $5x = 3x + 20$, $2x = 20$, $x = 10$.
Since $BC \parallel AD$, $y + (3x + 20) = 180$, $y + 50 = 180$, $y = 130$.

(c) Let the number of degrees in $\angle B$ and $\angle A$ be $3x$ and $2x$ respectively.
Since $BC \parallel AD$, $3x + 2x = 180$ or $x = 36$.
Since $\angle D = \angle A$, $y = 2x$ or $y = 72$.

1.2   PROOF of a TRAPEZOID PRINCIPLE STATED in WORDS (Pr. 1)

Prove: The base angles of an isosceles trapezoid are equal.

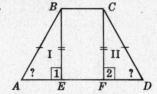

Given: Isosceles trapezoid $ABCD$ ($BC \parallel AD$, $AB = CD$)

To Prove: $\angle A = \angle D$

Plan: Draw $\perp$s to base from B and C.
Prove $\triangle I \cong \triangle II$.

PROOF: Statements	Reasons
1. Draw $BE \perp AD$ and $CF \perp AD$.	1. A $\perp$ may be drawn to a line from an outside point.
2. $BC \parallel AD$, $AB = CD$	2. Given
3. $BE = CF$	3. Parallel lines are everywhere equidistant.
4. $\angle 1 = \angle 2$	4. Perpendiculars form rt. $\angle$. Rt. $\angle$ are equal.
5. $\triangle I \cong \triangle II$	5. hy. leg = hy. leg
6. $\angle A = \angle D$	6. Corresponding parts of congruent $\triangle$ are equal.

2. *Parallelograms*

(*a*) A parallelogram is a quadrilateral whose opposite sides are parallel. The symbol for parallelogram is $\square$.

Thus in $\square ABCD$, $AB \parallel CD$ and $AD \parallel BC$.

(*b*) If the opposite sides of a quadrilateral are parallel, then it is a parallelogram. (This is the converse of the above definition.)

Thus if $AB \parallel CD$ and $AD \parallel BC$, then $ABCD$ is a $\square$.

A. *Properties of a Parallelogram*

Pr. 1: *The opposite sides of a parallelogram are parallel.* (Def.)

Pr. 2: *A diagonal of a parallelogram divides it into two congruent triangles.*

Thus if BD is a diagonal of $\square ABCD$, then $\triangle I = \triangle II$.

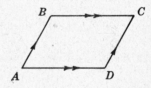

Pr. 3: *The opposite sides of a parallelogram are equal.*

Thus in $\square ABCD$, $AB = CD$ and $AD = BC$.

Pr. 4: *The opposite angles of a parallelogram are equal.*

Thus in $\square ABCD$, $\angle A = \angle C$ and $\angle B = \angle D$.

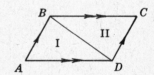

Pr. 5: *The consecutive angles of a parallelogram are supplementary.*

Thus in $\square ABCD$, $\angle A$ is the supplement of $\angle B$ or $\angle D$.

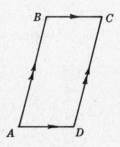

Pr. 6: *The diagonals of a parallelogram bisect each other.*

Thus in $\square ABCD$, $AE = EC$ and $BE = ED$.

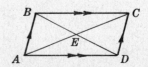

B. *Proving a Quadrilateral is a Parallelogram*

Pr. 7: *A quadrilateral is a parallelogram if its opposite sides are parallel.*

Thus if $AB \parallel CD$ and $AD \parallel BC$, then $ABCD$ is a $\square$.

Pr. 8: *A quadrilateral is a parallelogram if its opposite sides are equal.*

Thus if $AB = CD$ and $AD = BC$, then $ABCD$ is a $\square$.

Pr. 9: *A quadrilateral is a parallelogram if two sides are equal and parallel.*

Thus if $BC = AD$ and $BC \parallel AD$, then $ABCD$ is a $\square$.

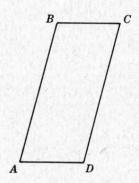

Pr. 10: *A quadrilateral is a parallelogram if its opposite angles are equal.*

Thus if $\angle A = \angle C$ and $\angle B = \angle D$, then $ABCD$ is a $\square$.

Pr. 11: *A quadrilateral is a parallelogram if its diagonals bisect each other.*

Thus if $AE = EC$ and $BE = ED$, then $ABCD$ is a $\square$.

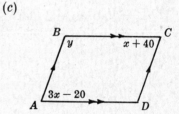

2.1 APPLYING PROPERTIES of PARALLELOGRAMS

If $ABCD$ is a parallelogram, find x and y in each.

(a)

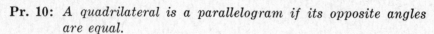

Perimeter = 40

(b)

(c)

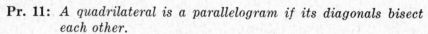

Solution:

(a) By Pr. 3, $BC = AD = 3x$ and $CD = AB = 2x$; then $2(2x + 3x) = 40$, $10x = 40$, $x = 4$.
By Pr. 3, $2y - 2 = 3x$; then $2y - 2 = 3(4)$, $2y = 14$, $y = 7$.

(b) By Pr. 6, $x + 2y = 15$ and $x = 3y$.
Substituting $3y$ for x in the first equation, $3y + 2y = 15$ or $y = 3$.   Then $x = 3y = 9$.

(c) By Pr. 4, $3x - 20 = x + 40$, $2x = 60$, $x = 30$.
By Pr. 5, $y + (x + 40) = 180$.   Then $y + (30 + 40) = 180$ or $y = 110$.

2.2 APPLYING PR. 7 to DETERMINE QUADRILATERALS which are PARALLELOGRAMS

State the parallelograms in each.

(a)   (b)   (c)

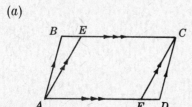

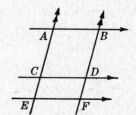

Ans.   (a) $ABCD, AECF$ (b) $ABFD, BCDE$ (c) $ABDC, CDFE, ABFE$

2.3 APPLYING PR. 9, 10, and 11

In each, state why $ABCD$ is a parallelogram.

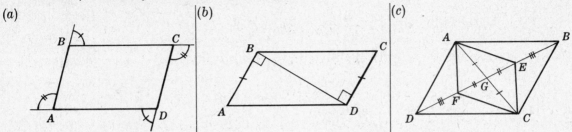

Solution:

(a) Since supplements of equal angles are equal, opposite angles of $ABCD$ are equal. Thus by Pr. 10, $ABCD$ is a parallelogram.

(b) Since perpendiculars to the same line are parallel, $AB \parallel CD$. Hence by Pr. 9, $ABCD$ is a parallelogram.

(c) Using the addition axiom, $DG = GB$. Hence by Pr. 11, $ABCD$ is a parallelogram.

2.4 PROVING a PARALLELOGRAM PROBLEM

Given: $\square ABCD$
  E is midpoint of BC.
  G is midpoint of AD.
  $EF \perp BD$, $GH \perp BD$

To Prove: $EF = GH$

Plan: Prove $\triangle BFE \cong \triangle GHD$

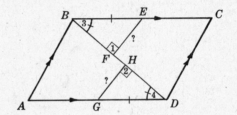

PROOF: Statements	Reasons
1. E is midpoint of BC. G is midpoint of AD.	1. Given
2. $BE = \frac{1}{2}BC$, $GD = \frac{1}{2}AD$	2. A midpoint cuts a line in half.
3. $ABCD$ is a $\square$.	3. Given
4. $BC = AD$	4. Opposite sides of a $\square$ are equal.
5. $BE = GD$	5. Halves of equals are equal.
6. $EF \perp BD$, $GH \perp BD$	6. Given
7. $\angle 1 = \angle 2$	7. Perpendiculars form rt. $\angle$. Rt. $\angle$ are $=$.
8. $BC \parallel AD$	8. Opposite sides of a $\square$ are $\parallel$.
9. $\angle 3 = \angle 4$	9. Alternate interior $\angle$ of $\parallel$ lines are $=$.
10. $\triangle BFE \cong \triangle GHD$	10. s.a.a. = s.a.a.
11. $EF = GH$	11. Corresponding parts of congruent $\triangle$ are $=$.

3. Special Parallelograms: Rectangle, Rhombus, Square

A. Definitions and Relationships among the Special Parallelograms

Rectangles, rhombuses and squares belong to the set of parallelograms. Each of these may be defined as a parallelogram, as follows:

1. A rectangle is an equiangular parallelogram.

2. A rhombus is an equilateral parallelogram.

3. A square is an equilateral and equiangular parallelogram. Thus a square is both a rectangle and a rhombus.

The relations among the special parallelograms can be pictured, using a circle to represent each set. Note the following:

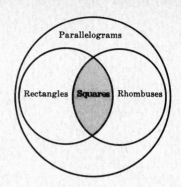

1. Since every rectangle and every rhombus must be a parallelogram, the circle for the set of rectangles and the circle for the set of rhombuses must be inside the circle for the set of parallelograms.

2. Since every square is both a rectangle and a rhombus, the overlapping shaded section represents the set of squares.

B. Properties of the Special Parallelograms

Pr. 1: *A rectangle, rhombus, or a square has all the properties of a parallelogram.*

Pr. 2: *Each angle of a rectangle is a right angle.*

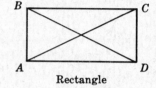
Rectangle

Pr. 3: *The diagonals of a rectangle are equal.*
  Thus in rectangle $ABCD$, $AC = BD$.

Pr. 4: *All the sides of a rhombus are equal.*

Pr. 5: *The diagonals of a rhombus are perpendicular bisectors of each other.*
  Thus in rhombus $ABCD$, AC and BD are $\perp$ bisectors of each other.

Pr. 6: *The diagonals of a rhombus bisect the vertex angles.*
  Thus in rhombus $ABCD$, AC bisects $\angle A$ and $\angle C$.

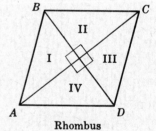
Rhombus

Pr. 7: *The diagonals of a rhombus form four congruent triangles.*
  Thus in rhombus $ABCD$, $\triangle I \cong \triangle II \cong \triangle III \cong \triangle IV$.

Pr. 8: *A square has all the properties of both the rhombus and the rectangle.*
  By definition, a square is both a rectangle and a rhombus.

C. Diagonal Properties of Parallelograms, Rectangles, Rhombuses and Squares
Each check in the following table indicates a diagonal property of the figure.

Diagonal Properties	Parallelo-gram	Rectangle	Rhombus	Square
Diagonals bisect each other.	√	√	√	√
Diagonals are equal.		√		√
Diagonals are perpendicular.			√	√
Diagonals bisect vertex angles.			√	√
Diagonals form 2 pairs of congruent triangles.	√	√	√	√
Diagonals form 4 congruent triangles.			√	√

D. Proving that a Parallelogram is a Rectangle, Rhombus or a Square

1. PROVING THAT A PARALLELOGRAM IS A RECTANGLE

The basic or minimum definition of a rectangle is as follows: *A rectangle is a parallelogram having one right angle.* Since the consecutive angles of a parallelogram are supplementary, if one angle is a right angle, the remaining angles must be right angles.

The converse of this basic definition of a rectangle provides a useful method of proving that a parallelogram is a rectangle, as follows:

Pr. 9:   *If a parallelogram has one right angle, then it is a rectangle.*

Thus if *ABCD* is a ▱ and ∠*A* = 90°, then *ABCD* is a rectangle.

Pr. 10:   *If a parallelogram has equal diagonals, then it is a rectangle.*

Thus if *ABCD* is a ▱ and *AC* = *BD*, then *ABCD* is a rectangle.

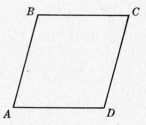

2. PROVING THAT A PARALLELOGRAM IS A RHOMBUS

The basic or minimum definition of a rhombus is as follows: *A rhombus is a parallelogram having two equal adjacent sides.*

The converse of this basic definition of a rhombus provides a useful method of proving that a parallelogram is a rhombus, as follows:

Pr. 11:   *If a parallelogram has equal adjacent sides, then it is a rhombus.*

Thus if *ABCD* is a ▱ and *AB* = *BC*, then *ABCD* is a rhombus.

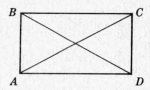

3. PROVING THAT A PARALLELOGRAM IS A SQUARE

Pr. 12:   *If a parallelogram has a right angle and two equal adjacent sides, then it is a square.*

This follows from the fact that a square is both a rectangle and a rhombus.

3.1   APPLYING ALGEBRA to the RHOMBUS

If *ABCD* is a rhombus, find x and y in each.

(a) 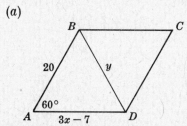   (b) 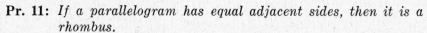   (c)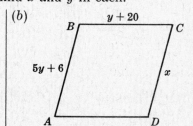

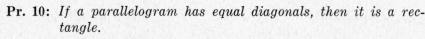
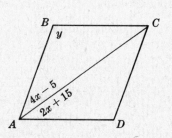

Solution:

(a) Since $AB = AD$, $3x - 7 = 20$ or $x = 9$. Since $\triangle ABD$ is equilateral, $y = 20$.

(b) Since $BC = AB$, $5y + 6 = y + 20$ or $y = 3\frac{1}{2}$. Since $CD = BC$, $x = y + 20$ or $x = 23\frac{1}{2}$.

(c) Since AC bisects $\angle A$, $4x - 5 = 2x + 15$ or $x = 10$. Hence, $2x + 15 = 35$ and $\angle A = 2(35°) = 70°$.

Since $\angle B$ and $\angle A$ are supplementary, $y + 70 = 180$ or $y = 110$.

3.2   PROVING a SPECIAL PARALLELOGRAM PROBLEM

Given: Rectangle $ABCD$
   E is midpoint of BC.

To Prove: $AE = ED$

Plan: Prove $\triangle AEB \cong \triangle CED$.

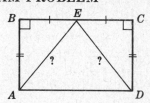

PROOF: **Statements**	**Reasons**
1. $ABCD$ is a rectangle.	1. Given
2. E is midpoint of BC.	2. Given
3. $BE = EC$	3. A midpoint divides a line into two equal parts.
4. $\angle B = \angle C$	4. A rectangle is equiangular.
5. $AB = CD$	5. Opposite sides of a $\square$ are equal.
6. $\triangle AEB \cong \triangle CED$	6. s.a.s. = s.a.s.
7. $AE = ED$	7. Corresponding parts of congruent $\triangle$ are equal.

3.3   PROVING a SPECIAL PARALLELOGRAM PROBLEM STATED in WORDS

Prove: A diagonal of a rhombus bisects each vertex angle through which it passes.

Given: Rhombus $ABCD$
   AC is a diagonal.

To Prove: AC bisects $\angle A$ and $\angle C$.

Plan: Prove
   (1) $\angle 1$ and $\angle 2$ each equals $\angle 3$.
   (2) $\angle 3$ and $\angle 4$ each equals $\angle 1$.

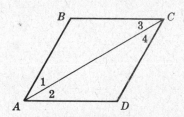

PROOF: **Statements**	**Reasons**
1. $ABCD$ is a rhombus.	1. Given
2. $AB = BC$	2. A rhombus is equilateral.
3. $\angle 1 = \angle 3$	3. In a $\triangle$, angles opposite equal sides are equal.
4. $BC \parallel AD$, $AB \parallel CD$	4. Opposite sides of a $\square$ are $\parallel$.
5. $\angle 2 = \angle 3$, $\angle 1 = \angle 4$	5. Alternate interior $\angle$ of $\parallel$ lines are equal.
6. $\angle 1 = \angle 2$, $\angle 3 = \angle 4$	6. Things equal to the same thing are equal to each other.
7. AC bisects $\angle A$ and $\angle C$.	7. To divide into two equal parts is to bisect.

4. *Three or more Parallels.   Medians and Midpoints*

A. *Three or More Parallels*

Pr. 1: *If three or more parallels cut off equal segments on one transversal, then they cut off equal segments on any other transversal.*

   Thus if $l_1 \parallel l_2 \parallel l_3$ and segments a and b of transversal AB are equal, then segments c and d of transversal CD are equal.

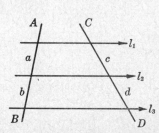

B. Midpoint and Median Principles of Triangles and Trapezoids

Pr. 2: *If a line is drawn from the midpoint of one side of a tri-*
angle and parallel to a second side, then it passes through
the midpoint of the third side.

Thus in $\triangle ABC$, if M is the midpoint of AB and $MN \parallel AC$, then
N is the midpoint of BC.

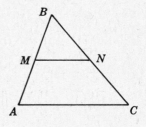

Pr. 3: *If a line joins the midpoints of two sides of a triangle, then*
it is parallel to the third side and equal to one-half of it.

Thus in $\triangle ABC$, if M and N are the midpoints of AB and BC, then
$MN \parallel AC$ and $MN = \frac{1}{2}AC$.

Pr. 4: *The median of a trapezoid is parallel to its bases and equal*
to one-half of their sum.

Thus if m is the median of trapezoid $ABCD$, then $m \parallel b$, $m \parallel b'$ and
$m = \frac{1}{2}(b + b')$.

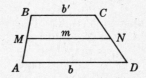

Pr. 5: *The median to the hypotenuse of a right triangle equals one-*
half of the hypotenuse.

Thus in rt. $\triangle ABC$, if CM is the median to hypotenuse AB, then
$CM = \frac{1}{2}AB$; that is, $CM = AM = MB$.

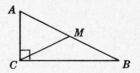

Pr. 6: *The medians of a triangle meet in a point which is two-*
thirds of the distance from any vertex to the midpoint of
the opposite side.

Thus if AN, BP and CM are medians of $\triangle ABC$, then they meet in
a point G which is two-thirds of the distance from A to N, B to P, and
C to M.

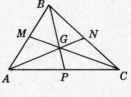

Note. G is the center of gravity or the centroid of $\triangle ABC$. If the
triangle is made of a firm substance such as cardboard, it can
be balanced at the centroid on the point of a needle.

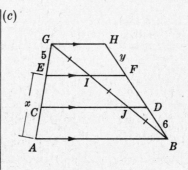

4.1 APPLYING PRINCIPLE 1 to THREE or MORE PARALLELS

In each, find x and y.

(a)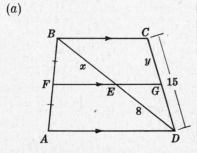

(b)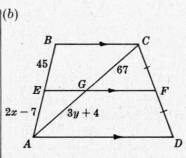

(c)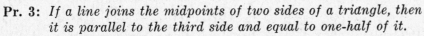

Solution:

(a) Since $BE = ED$ and $CG = \frac{1}{2}CD$, $x = 8$ and $y = 7\frac{1}{2}$.

(b) Since $BE = EA$ and $CG = AG$, $2x - 7 = 45$ and $3y + 4 = 67$. Hence $x = 26$ and $y = 21$.

(c) Since $AC = CE = EG$ and $HF = FD = DB$, $x = 10$ and $y = 6$.

4.2   APPLYING PRINCIPLES 2 and 3

In each, find x and y.

(a)

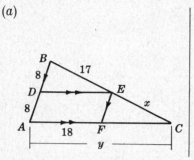

(b)

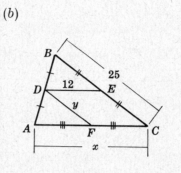

(c)

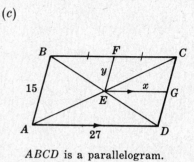

ABCD is a parallelogram.

Solution:

(a) By Pr. 2, E is midpoint of BC and F is midpoint of AC.   Hence $x = 17$ and $y = 36$.

(b) By Pr. 3, $DE = \frac{1}{2}AC$ and $DF = \frac{1}{2}BC$. Hence $x = 24$ and $y = 12\frac{1}{2}$.

(c) Since $ABCD$ is a parallelogram, E is midpoint of AC.

  By Pr. 2, G is midpoint of CD.

  By Pr. 3, $x = \frac{1}{2}(27) = 13\frac{1}{2}$ and $y = \frac{1}{2}(15) = 7\frac{1}{2}$.

4.3   APPLYING PRINCIPLE 4 to the MEDIAN of a TRAPEZOID

If MP is the median of trapezoid $ABCD$,

(a) find m if $b = 20$ and $b' = 28$,

(b) find b' if $b = 30$ and $m = 26$,

(c) find b if $b' = 35$ and $m = 40$.

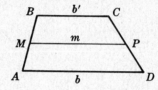

Solution:

  In each, apply the formula $m = \frac{1}{2}(b + b')$.

(a) $m = \frac{1}{2}(20 + 28)$, $m = 24$ (b) $26 = \frac{1}{2}(30 + b')$, $b' = 22$ (c) $40 = \frac{1}{2}(b + 35)$, $b = 45$

4.4   APPLYING PRINCIPLES 5 and 6 to the MEDIANS of a TRIANGLE

In each, find x and y.

(a)

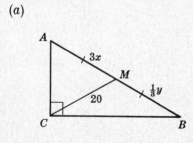

(b)

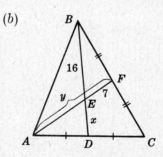

(c)

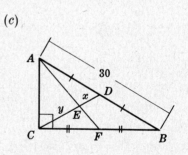

Solution:

(a) Since $AM = MB$, CM is the median to hypotenuse AB.

  Hence by Pr. 5, $3x = 20$ and $\frac{1}{3}y = 20$. Thus $x = 6\frac{2}{3}$ and $y = 60$.

(b) BD and AF are medians of $\triangle ABC$.

  Hence by Pr. 6, $x = \frac{1}{2}(16) = 8$ and $y = 3(7) = 21$.

(c) CD is the median to hypotenuse AB; hence by Pr. 5, $CD = 15$.

  CD and AF are medians of $\triangle ABC$; hence by Pr. 6, $x = \frac{1}{3}(15) = 5$ and $y = \frac{2}{3}(15) = 10$.

4.5 PROVING a MIDPOINT PROBLEM

Given: Quadrilateral $ABCD$
E, F, G and H are midpoints of AD, AB, BC and CD, respectively.

To Prove: $EFGH$ is a $\square$.

Plan: Prove EF and GH are equal and parallel.

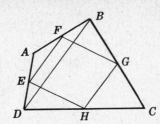

PROOF: Statements	Reasons
1. Draw BD.	1. A straight line may be drawn between any two points.
2. E, F, G and H are midpoints.	2. Given
3. EF and $GH \parallel BD$. EF and $GH = \frac{1}{2}BD$.	3. A line joining the midpoints of two sides of a $\triangle$, is parallel and equal to half of the third side.
4. $EF \parallel GH$	4. Two lines parallel to a third line are parallel to each other.
5. $EF = GH$	5. Things equal to the same thing are equal to each other.
6. $EFGH$ is a $\square$.	6. If two sides of a quadrilateral are $=$ and $\parallel$, the quadrilateral is a $\square$.

Supplementary Problems

1. Find x and y in each.   (1.1)

(a)

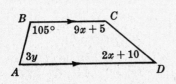

(b)

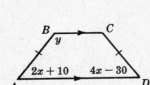

(c)

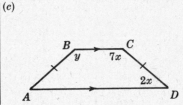

ABCD is a trapezoid.   ABCD is an isosceles trapezoid.   ABCD is an isosceles trapezoid.

2. Prove: If the base angles of a trapezoid are equal, the trapezoid is isosceles.   (1.2)

3. Prove: (a) The diagonals of an isosceles trapezoid are equal.   (1.2)
(b) If the non-parallel sides AB and CD of an isosceles trapezoid are extended until they meet at E, triangle ADE thus formed is isosceles.

4. State the parallelograms in each.   (2.2)

(a)

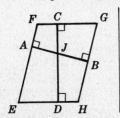

(b)

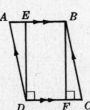

(c)

(d)

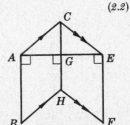

5. In each, state why $ABCD$ is a parallelogram.   (2.3)

(a)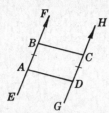

(b)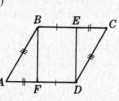

(c)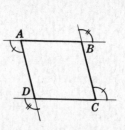

(d)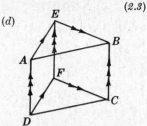

6. If $ABCD$ is a parallelogram, find x and y in each.   (2.1)
 (a) $AD = 5x$, $AB = 2x$, $CD = y$, perimeter $= 84$.
 (b) $AB = 2x$, $BC = 3y + 8$, $CD = 7x - 25$, $AD = 5y - 10$.
 (c) $\angle A = 4y - 60$, $\angle C = 2y$, $\angle D = x$.
 (d) $\angle A = 3x$, $\angle B = 10x - 15$, $\angle C = y$.

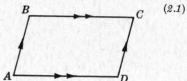

7. If $ABCD$ is a parallelogram, find x and y in each.   (2.1)
 (a) $AE = x + y$, $EC = 20$, $BE = x - y$, $ED = 8$.
 (b) $AE = x$, $EC = 4y$, $BE = x - 2y$, $ED = 9$.
 (c) $AE = 3x - 4$, $EC = x + 12$, $BE = 2y - 7$, $ED = x - y$.
 (d) $AE = 2x + y$, $AC = 30$, $BE = x + y$, $BD = 24$.

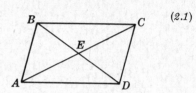

8. Prove each of the following.   (2.4)

(a)
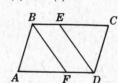

 Given: $\square ABCD$
 G is midpoint of AB.
 F is midpoint of CD.
 $HG \perp AB$, $EF \perp CD$
 To Prove: $EF = GH$

(b) and (c)
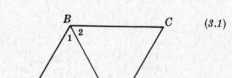

(b) **Given:** $\square ABCD$
 $BE = FD$
 To Prove: $BF = ED$

(c) **Given:** $\square ABCD$
 BF bisects $\angle B$,
 ED bisects $\angle D$.
 To Prove: $BF = ED$

9. Prove each of the following.   (2.4)
 (a) The opposite sides of a parallelogram are equal.   (Pr. 3)
 (b) If the opposite sides of a quadrilateral are equal, the quadrilateral is a parallelogram.   (Pr. 8)
 (c) If two sides of a quadrilateral are equal and parallel, the quadrilateral is a parallelogram. (Pr. 9)
 (d) The diagonals of a parallelogram bisect each other.   (Pr. 6)
 (e) If the diagonals of a quadrilateral bisect each other, the quadrilateral is a parallelogram.   (Pr. 11)

10. If $ABCD$ is a rhombus, find x and y in each.   (3.1)
 (a) $BC = 35$, $CD = 8x - 5$, $BD = 5y$, $\angle C = 60°$.
 (b) $AB = 43$, $AD = 4x + 3$, $BD = y + 8$, $\angle B = 120°$.
 (c) $AB = 7x$, $AD = 3x + 10$, $BC = y$.
 (d) $AB = x + y$, $AD = 2x - y$, $BC = 12$.
 (e) $\angle B = 130°$, $\angle 1 = 3x - 10$, $\angle A = 2y$.
 (f) $\angle 1 = 8x - 29$, $\angle 2 = 5x + 4$, $\angle D = y$.

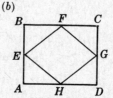

11. Prove each of the following.   (3.2)

(a)
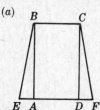

 Given: Rectangle $ABCD$
 $EA = DF$
 To Prove: $BE = CF$

(b)
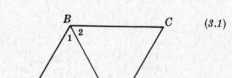

 Given: Rectangle $ABCD$
 E, F, G and H are the midpoints
 of the sides of the rectangle.
 To Prove: $EFGH$ is a rhombus.

12. Prove each of the following.　　　　　　　　　　　　　　　　(3.3)

　(a) If the diagonals of a parallelogram are equal, the parallelogram is a rectangle.

　(b) If the diagonals of a parallelogram are perpendicular to each other, the parallelogram is a rhombus.

　(c) If a diagonal of a parallelogram bisects a vertex angle, then the parallelogram is a rhombus.

　(d) The diagonals of a rhombus divide it into four congruent triangles.

　(e) The diagonals of a rectangle are equal.

13. In each, find x and y.　　　　　　　　　　　　　　　　　(4.1)

(a)　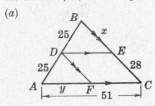　(b)　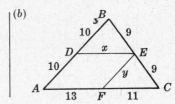　(c)　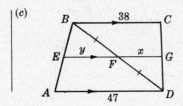

14. In each, find x and y.　　　　　　　　　　　　　　　　　(4.2)

(a)　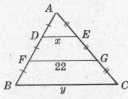　(b)　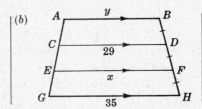　(c)　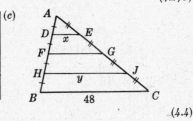

15. If MP is the median of trapezoid $ABCD$,　　　　　　　　　(4.3)

　(a) find m if $b = 23$ and $b' = 15$,

　(b) find b' if $b = 46$ and $m = 41$,

　(c) find b if $b' = 51$ and $m = 62$.

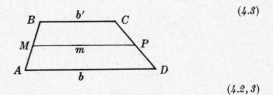

16. In each, find x and y.　　　　　　　　　　　　　　　　(4.2, 3)

(a)　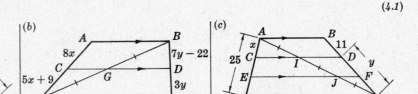

17. In a right triangle,　　　　　　　　　　　　　　　　　　　(4.4)

　(a) find the length of the median to a hypotenuse whose length is 45.

　(b) find the length of the hypotenuse, if the length of its median is 35.

18. If the medians of $\triangle ABC$ meet in D,　　　　　　　　　(4.4)

　(a) find the length of the median whose shorter segment is 7,

　(b) find the length of the median whose longer segment is 20,

　(c) find the length of the shorter segment of the median of length 42,

　(d) find the length of the longer segment of the median of length 39.

19. Prove each of the following.　　　　　　　　　　　　　　　(4.5)

　(a) If the midpoints of the sides of a rhombus are joined in order, the quadrilateral formed is a rectangle.

　(b) If the midpoints of the sides of a square are joined in order, the quadrilateral formed is a square.

　(c) In $\triangle ABC$, M, P and Q are the midpoints of AB, BC and AC respectively. Prove that $QMPC$ is a parallelogram.

　(d) In right $\triangle ABC$, $\angle C = 90°$. If Q, M and P are the midpoints of AC, AB and BC respectively, prove that $QMPC$ is a rectangle.

Chapter 6

Circles

1. The Circle, and Circle Relationships

The following terms are associated with the circle. Although a few have been previously mentioned, they are repeated here for ready reference.

1. A *circle* is a closed curve all of whose points lie in the same plane and are at the same distance from the center.

   The symbol for circle is ⊙, for circles, ⊚.

2. The *circumference* of a circle is the distance around the circle. It contains 360°.

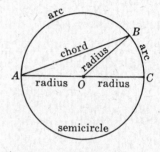

3. A *radius* of a circle is a line joining the center to a point on the circumference.

   Thus *AO*, *BO* and *CO* are radii.

4. A *central angle* is an angle formed by two radii.

   Thus ∠*AOB* and ∠*BOC* are central angles.

5. An *arc* is a part of the circumference of a circle. The symbol for arc is ⌒.

   Thus $\overarc{AB}$ stands for arc *AB*.

Fig. 1

6. A *semicircle* is an arc equal to one-half of the circumference of a circle.

   Thus $\overarc{ABC}$ is a semicircle.

7. A *minor arc* is an arc less than a semicircle. A *major arc* is an arc greater than a semicircle.

   Thus in the figure above, $\overarc{BC}$ is a minor arc and $\overarc{BAC}$ is a major arc. Three letters are needed to indicate a major arc.

8. *To intercept* an arc is to cut off the arc.

   Thus in the figure above, ∠*BAC* and ∠*BOC* intercept $\overarc{BC}$.

9. A *chord* of a circle is a line joining two points of the circumference.

   Thus in the adjacent figure, *AB* is a chord.

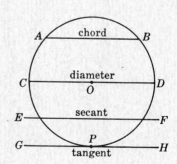

10. A *diameter* of a circle is a chord through the center.

    Thus *CD* is a diameter of circle *O*.

11. A *secant* of a circle is a line that intersects the circle at two points.

    Thus *EF* is a secant.

12. A *tangent* of a circle is a line that touches the circle at one and only one point no matter how far produced.

    Thus *GH* is a tangent to the circle at *P*. *P* is the point of contact or the point of tangency.

Fig. 2

13. An *inscribed polygon* is a polygon all of whose sides are chords of a circle.

   Thus △*ABD*, △*BCD* and quadrilateral *ABCD* are inscribed polygons of circle *O*.

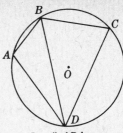

Inscribed Polygons
Circumscribed Circle

14. A *circumscribed circle* is a circle passing through each vertex of a polygon.

   Thus circle *O* is a circumscribed circle of quadrilateral *ABCD*.

15. A *circumscribed polygon* is a polygon all of whose sides are tangents to a circle.

   Thus △*ABC* is a circumscribed polygon of circle *O*.

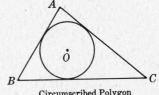

16. An *inscribed circle* is a circle to which all the sides of a polygon are tangents.

   Thus circle *O* is an inscribed circle of △*ABC*.

Circumscribed Polygon
Inscribed Circle

17. *Concentric circles* are circles that have the same center.

   Thus the two circles shown are concentric circles.

   AB is a tangent of the inner circle and a chord of the outer one. *CD* is a secant of the inner circle and a chord of the outer one.

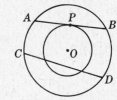

Concentric Circles

A. Circle Principles

Pr. 1: *A diameter divides a circle into two equal parts.*

   Thus diameter *AB* divides circle *O* into two equal semicircles, $\overarc{ACB}$ and $\overarc{ADB}$.

Pr. 2: *If a chord divides a circle into two equal parts, then it is a diameter.* (Converse of Pr. 1)

   Thus if $\overarc{ACB} = \overarc{ADB}$, then *AB* is a diameter.

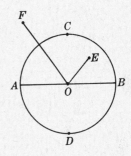

Pr. 3: *A point is outside, on, or inside a circle according to whether its distance from the center is greater than, equal to, or less than the radius.*

   Thus, *F* is outside circle *O* since *FO* is greater than a radius.

   E is inside circle *O* since *EO* is less than a radius.

   A is on circle *O* since *AO* is a radius.

Pr. 4: *Radii of the same or equal circles are equal.*

   Thus in circle *O*, *OA* = *OC*.

Pr. 5: *Diameters of the same or equal circles are equal.*

   Thus in circle *O*, *AB* = *CD*.

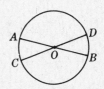

Pr. 6: *In the same or equal circles, equal central angles have equal arcs.*

   Thus in circle *O*, if ∠1 = ∠2, then $\overarc{AC} = \overarc{CB}$.

Pr. 7: *In the same or equal circles, equal arcs have equal central angles.*

   Thus in circle *O*, if $\overarc{AC} = \overarc{CB}$, then ∠1 = ∠2.

   (Principles 6 and 7 are converses.)

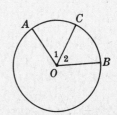

Pr. 8: *In the same or equal circles, equal chords have equal arcs.*

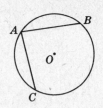

Thus in circle O, if $AB = AC$, then $\overarc{AB} = \overarc{AC}$.

Pr. 9: *In the same or equal circles, equal arcs have equal chords.*

Thus in circle O, if $\overarc{AB} = \overarc{AC}$, then $AB = AC$.
(Principles 8 and 9 are converses.)

Pr. 10: *A diameter perpendicular to a chord bisects the chord and its arcs.*

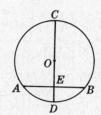

Thus in circle O, if $CD \perp AB$, then CD bisects AB, $\overarc{AB}$ and $\overarc{ACB}$.

(A proof of this principle is given in Chapter 16.)

Pr. 11: *A perpendicular bisector of a chord passes through the center of the circle.*

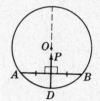

Thus in circle O, if PD is the perpendicular bisector of AB, then PD passes through center O.

Pr. 12: *In the same or equal circles, equal chords are equally distant from the center.*

Thus in circle O, if $AB = CD$, $OE \perp AB$ and $OF \perp CD$, then $OE = OF$.

Pr. 13: *In the same or equal circles, chords that are equally distant from the center are equal.*

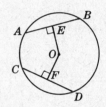

Thus in circle O, if $OE = OF$, $OE \perp AB$ and $OF \perp CD$, then $AB = CD$.

1.1 MATCHING TEST of CIRCLE VOCABULARY

Match each figure on the left with one of the names on the right:

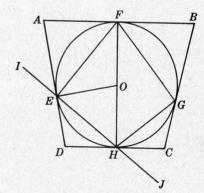

(a) OE

(b) FG

(c) FH

(d) CD

(e) IJ

(f) $\overarc{EF}$

(g) $\overarc{FGH}$

(h) $\overarc{FEG}$

(i) $\angle EOF$

(j) circle O about $EFGH$

(k) circle O in $ABCD$

(l) quadrilateral $EFGH$

(m) quadrilateral $ABCD$

1. radius
2. central angle
3. semicircle
4. minor arc
5. major arc
6. chord
7. diameter
8. secant
9. tangent
10. inscribed polygon
11. circumscribed polygon
12. inscribed circle
13. circumscribed circle

Ans.   (a) 1,   (b) 6,   (c) 7,   (d) 9,   (e) 8,   (f) 4,   (g) 3,   (h) 5,   (i) 2,   (j) 13,   (k) 12,   (l) 10,   (m) 11

1.2   APPLYING PRINCIPLES 4 and 5: To a circle and to equal circles

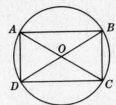

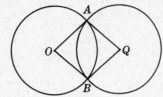

(a) What kind of triangle is *OCD*?

(b) What kind of quadrilateral is *ABCD*?

(c) If circle *O* = circle *Q*, what kind of quad-rilateral is *OAQB*?

Solution:

Radii or diameters of the same or equal circles are equal.

(a) Since $OC = OD$, $\triangle OCD$ is isosceles.

(b) Since diagonals AC and BD are equal and bisect each other, $ABCD$ is a rectangle.

(c) Since the circles are equal, $OA = AQ = QB = BO$ and $OAQB$ is a rhombus.

1.3   PROVING a CIRCLE PROBLEM

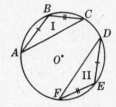

Given: $AB = DE$
$BC = EF$

To Prove: $\angle B = \angle E$

Plan: Prove $\triangle I \cong \triangle II$.

PROOF: Statements	Reasons
1. $AB = DE$, $BC = EF$	1. Given
2. $\overarc{AB} = \overarc{DE}$, $\overarc{BC} = \overarc{EF}$	2. In a circle, equal chords have equal arcs.
3. $\overarc{ABC} = \overarc{DEF}$	3. If equals are added to equals, the sums are equal.
4. $AC = DF$	4. In a circle, equal arcs have equal chords.
5. $\triangle I \cong \triangle II$	5. s.s.s. = s.s.s.
6. $\angle B = \angle E$	6. Corresponding parts of congruent $\triangle$ are equal.

1.4   PROVING a CIRCLE PROBLEM STATED in WORDS

Prove: If a radius bisects a chord, then it is perpendicular to the chord.

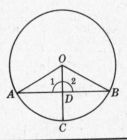

Given: Circle *O*
OC bisects AB.

To Prove: $OC \perp AB$

Plan: Prove $\triangle AOD \cong \triangle BOD$ to show $\angle 1 = \angle 2$.
Also, $\angle 1$ and $\angle 2$ are supplementary.

PROOF: Statements	Reasons
1. Draw OA and OB.	1. A straight line may be drawn between any two points.
2. $OA = OB$	2. Radii of a circle are equal.
3. OC bisects AB.	3. Given
4. $AD = DB$	4. To bisect is to divide into two equal parts.
5. $OD = OD$	5. Identity
6. $\triangle AOD \cong \triangle BOD$	6. s.s.s. = s.s.s.
7. $\angle 1 = \angle 2$	7. Corresponding parts of congruent $\triangle$ are equal.
8. $\angle 1$ is the supplement of $\angle 2$.	8. Adjacent $\angle$ are supplementary if exterior sides lie in a straight line.
9. $\angle 1$ and $\angle 2$ are right angles.	9. Equal supplementary angles are right angles.
10. $OC \perp AB$.	10. Rt. $\angle$ are formed by perpendiculars.

2. Tangents

The *length of a tangent* from a point to a circle is the length of the segment of the tangent from the given point to the point of tangency.

Thus *PA* is the length of the tangent from *P* to circle *O*.

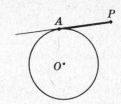

A. Tangent Principles

Pr. 1: *A tangent is perpendicular to the radius drawn to the point of contact.*

Thus if *AB* is a tangent to circle *O* at *P*, and *OP* is drawn, then *AB* ⊥ *OP*.

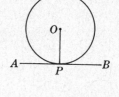

Pr. 2: *A line is tangent to a circle if it is perpendicular to a radius at its outer end.*

Thus if *AB* ⊥ radius *OP* at *P*, then *AB* is tangent to circle *O*.

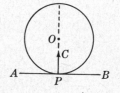

Pr. 3: *A line passes through the center of a circle if it is per- pendicular to a tangent at its point of contact.*

Thus if *AB* is tangent to circle *O* at *P*, and *CP* ⊥ *AB* at *P*, then *CP* extended will pass through the center *O*.

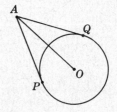

Pr. 4: *Tangents to a circle from an outside point are equal.*

Thus if *AP* and *AQ* are tangent to circle *O* at *P* and *Q*, then *AP* = *AQ*.

Pr. 5: *The line from the center of a circle to an outside point bisects the angle between the tangents from the point to the circle.*

Thus *OA* bisects ∠*PAQ* if *AP* and *AQ* are tangents to circle. *O*.

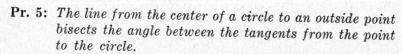

B. Two Circles in Varying Relative Positions

The *line of centers of two circles* is the line joining their centers. Thus *OO'* is the line of centers of circles *O* and *O'*.

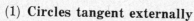

(1) Circles tangent externally

As shown, circles *O* and *O'* are tangent externally at *P*. *AB* is the common internal tangent of both circles. The line of centers *OO'* passes through *P*, is perpendicular to *AB*, and equals the sum of the radii, $R + r$. Also, *AB* bisects each of the common external tangents, *CD* and *EF*.

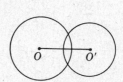

(2) Circles tangent internally

As shown, circles *O* and *O'* are tangent internally at *P*. *AB* is the common external tangent of both circles. The line of centers *OO'* if extended, passes through *P*, is perpendicular to *AB*, and equals the difference of the radii, $R - r$.

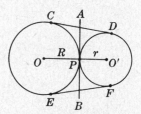

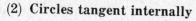
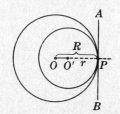

(3) Overlapping circles

As shown, circles O and O' overlap. Their common chord is AB. If the circles are unequal, their (equal) common external tangents CD and EF meet at P. The line of centers OO' is the perpendicular bisector of AB and, if extended, passes through P.

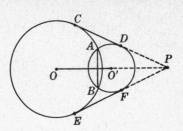

(4) Circles outside each other

As shown, circles O and O' are entirely outside of each other. The common internal tangents, AB and CD, meet at P. If the circles are unequal, their common external tangents, EF and GH, if extended meet at P'. The line of centers OO' passes through P and P'. Also, $AB = CD$ and $EF = GH$.

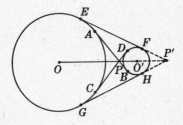
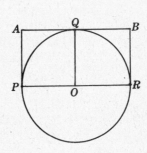

2.1   TRIANGLES and QUADRILATERALS HAVING TANGENT SIDES

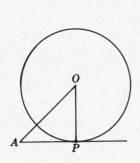

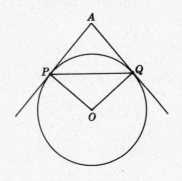

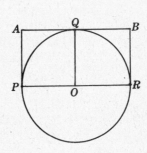

AP is a tangent.

(a) If $AP = OP$, what kind of triangle is OPA?

AP and AQ are tangents.

(b) If $AP = PQ$, what kind of triangle is APQ?

(c) If $AP = OP$, what kind of quadrilateral is $OPAQ$?

AP, AB and BR are tangents.

(d) If $OQ \perp PR$, what kind of quadrilateral is $PABR$?

Solution:

(a) AP is tangent to the circle at P; then by Pr. 1, $\angle OPA$ is a right angle. Also, $AP = OP$.

Hence $\triangle OAP$ is an isosceles right triangle.

(b) AP and AQ are tangents from a point to the circle; hence by Pr. 4, $AP = AQ$. Also, $AP = PQ$.

Then $\triangle APQ$ is an equilateral triangle.

(c) By Pr. 4, $AP = AQ$. Also, OP and OQ are equal radii and $AP = OP$.

Then $AP = AQ = OP = OQ$; hence $OPAQ$ is a rhombus with a right angle, or a square.

(d) By Pr. 1, $AP \perp PR$ and $BR \perp PR$. Then $AP \parallel BR$, since both are $\perp$ to PR.

By Pr. 1, $AB \perp OQ$; also, $PR \perp OQ$ (Given). Then $AB \parallel PR$, since both are $\perp$ to OQ.

Hence $PABR$ is a parallelogram with a right angle, or a rectangle.

2.2 APPLYING PRINCIPLE 1: To angles having a tangent side

(a)
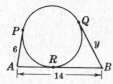

AP is a tangent.
Find ∠A if ∠A : ∠O = 2 : 3

(b)
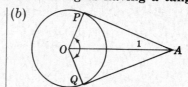

AP and AQ are tangents.
Find ∠1 if ∠O = 140°.

(c)
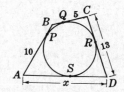

DP and CQ are tangents.
Find ∠2 and ∠3 if ∠OPD is
trisected and PQ is a diameter.

Solution:

   Pr. 1: A tangent is perpendicular to the radius drawn to the point of contact.

(a) By Pr. 1, ∠P = 90°. Then ∠A + ∠O = 90°.

   If ∠A = 2x and ∠O = 3x, then 5x = 90 and x = 18. Hence ∠A = 36°.

(b) By Pr. 1, ∠P = ∠Q = 90°.

   Since ∠P + ∠Q + ∠O + ∠A = 360°, ∠A + ∠O = 180°. Since ∠O = 140°, ∠A = 40°.

   By Pr. 5, ∠1 = ½∠A = 20°.

(c) By Pr. 1, ∠DPQ = ∠PQC = 90°. Since ∠1 = 30°, ∠2 = 60°.

   Since ∠3 is an exterior angle of △PQB, ∠3 = 90° + 60° = 150°.

2.3 APPLYING PRINCIPLE 4: To the lengths of tangents

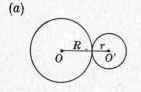

(a) AP, BQ and AB are tangents.
   Find y.

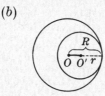

(b) △ABC is circumscribed.
   Find x.

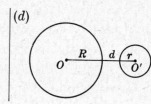

(c) Quadrilateral ABCD is
   circumscribed.
   Find x.

Solution:

   Pr. 4: Tangents to a circle from an outside point are equal.

(a) By Pr. 4, AR = 6 and RB = y. Then RB = AB − AR = 14 − 6 = 8. Hence, y = RB = 8.

(b) By Pr. 4, PC = 8, QB = 4 and AP = AQ. Then AQ = AB − QB = 11. Hence, x = AP + PC = 11 + 8 = 19.

(c) By Pr. 4, AS = 10, CR = 5 and RD = SD. Then RD = CD − CR = 8. Hence, x = AS + SD = 10 + 8 = 18.

2.4 FINDING the LINE of CENTERS

   If two circles have radii of 9 and 4 respectively, find their line of centers (a) if the circles are tangent externally, (b) if the circles are tangent internally, (c) if the circles are concentric, (d) if the circles are 5 units apart.

(a)   (b)   (c)   (d)

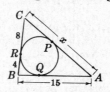

Solution:

   Let R = radius of larger circle, r = radius of smaller circle.

(a) Since R = 9 and r = 4, OO′ = R + r = 9 + 4 = 13.

(b) Since R = 9 and r = 4, OO′ = R − r = 9 − 4 = 5.

(c) Since the circles have the same center, their line of centers has zero length.

(d) Since R = 9, r = 4 and d = 5, OO′ = R + d + r = 9 + 5 + 4 = 18.

2.5 PROVING a TANGENT PROBLEM STATED in WORDS

Prove: Tangents to a circle from an outside point
are equal. (Pr. 4)

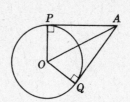

Given: Circle O
AP is tangent at P.
AQ is tangent at Q.

To Prove: $AP = AQ$

Plan: Draw OP, OQ and OA and prove $\triangle AOP \cong \triangle AOQ$.

PROOF: Statements	Reasons
1. Draw OP, OQ and OA.	1. A straight line may be drawn between any two points.
2. $OP = OQ$	2. Radii of a circle are equal.
3. $\angle P$ and $\angle Q$ are right angles.	3. A tangent is $\perp$ to radius drawn to point of contact.
4. $OA = OA$	4. Identity
5. $\triangle AOP \cong \triangle AOQ$	5. hy. leg = hy. leg
6. $AP = AQ$	6. Corresponding parts of congruent $\triangle$ are equal.

3. *Measurement of Angles and Arcs in a Circle*

(a) A *central angle* has the same number of degrees as the
arc it intercepts.

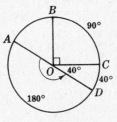

Thus a central angle which is a right angle intercepts a $90°$ arc;
a $40°$ central angle intercepts a $40°$ arc, and a central angle which
is a straight angle intercepts a semicircle of $180°$.

Since the numerical measures in degrees of both the central angle
and its intercepted arc is the same, we may restate the above principle as follows: A central angle is measured by its intercepted arc.
The symbol $\stackrel{\circ}{=}$ may be used to mean "is measured by." (Do not say
that a central angle equals its intercepted arc. An angle cannot
equal an arc.)

(b) An *inscribed angle* is an angle whose vertex is on the circle
and whose sides are chords. *An angle inscribed in an arc
has its vertex on the arc and its sides passing through
the ends of the arc.*

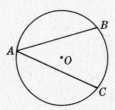

Thus $\angle A$ is an inscribed angle whose sides are the chords AB
and AC. Note that $\angle A$ intercepts $\overset{\frown}{BC}$ and is inscribed in $\overset{\frown}{BAC}$.

A. *Angle Measurement Principles*

Pr. 1: *A central angle is measured by its intercepted arc.*

Pr. 2: *An inscribed angle is measured by one-half of its intercepted arc.*

A proof of this principle is given in Chapter 16.

Pr. 3: *In the same or equal circles, equal inscribed angles
have equal intercepted arcs.*

Thus in Fig. 1, if $\angle 1 = \angle 2$, then $\overset{\frown}{BC} = \overset{\frown}{DE}$.

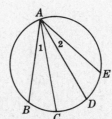

Pr. 4: *In the same or equal circles, inscribed angles having
equal intercepted arcs are equal.*

Thus in Fig. 1, if $\overset{\frown}{BC} = \overset{\frown}{DE}$, then $\angle 1 = \angle 2$.
(Converse of Pr. 3.)

Fig. 1

Pr. 5: *Angles inscribed in the same or equal arcs are equal.*

     Thus in Fig. 2, if $\angle C$ and $\angle D$ are inscribed in $\overset{\frown}{ACB}$, then $\angle C = \angle D$.

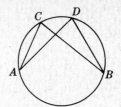

Fig. 2

Pr. 6: *An angle inscribed in a semicircle is a right angle.*

     Thus in Fig. 3, since $\angle C$ is inscribed in semicircle ACD, $\angle C = 90°$.

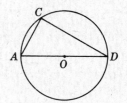

Fig. 3

Pr. 7: *Opposite angles of an inscribed quadrilateral are supplementary.*

     Thus in Fig. 4, if $ABCD$ is an inscribed quadrilateral, $\angle A$ is the supplement of $\angle C$.

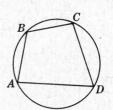

Fig. 4

Pr. 8: *Parallel lines intercept equal arcs on a circle.*

     Thus in Fig. 5, if $AB \parallel CD$, then $\overset{\frown}{AC} = \overset{\frown}{BD}$. If tangent FG is parallel to CD, then $\overset{\frown}{PC} = \overset{\frown}{PD}$.

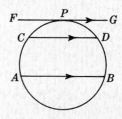

Fig. 5

Pr. 9: *An angle formed by a tangent and a chord is measured by one-half of its intercepted arc.*

Pr. 10: *An angle formed by two intersecting chords is measured by one-half the sum of the intercepted arcs.*

Pr. 11: *An angle formed by two secants intersecting outside a circle is measured by one-half the difference of the intercepted arcs.*

Pr. 12: *An angle formed by a tangent and a secant intersecting outside a circle is measured by one-half the difference of the intercepted arcs.*

Pr. 13: *An angle formed by two tangents intersecting outside a circle is measured by one-half the difference of the intercepted arcs.*

     Proofs of Principles 10-13 are given in Chapter 16.

B. Table of Angle Measurement Principles

Position of Vertex	Kind of Angle	Diagram	Measurement Formula	Method of Measurement
CENTER OF CIRCLE	Central Angle (Apply Pr. 1.)		$\angle O \triangleq \overset{\frown}{AB}$ $\angle O = a°$	By intercepted arc
ON CIRCLE	Inscribed Angle (Apply Pr. 2.)		$\angle A \triangleq \frac{1}{2}\overset{\frown}{BC}$ $\angle A = \frac{1}{2}a°$	By one-half of intercepted arc
	Angle formed by a tangent and a chord (Apply Pr. 9.)			
INSIDE CIRCLE	Angle formed by two intersecting chords (Apply Pr. 10.)		$\angle 1 \triangleq \frac{1}{2}(\overset{\frown}{AC} + \overset{\frown}{BD})$ $\angle 1 = \frac{1}{2}(a° + b°)$	By one-half sum of intercepted arcs
OUTSIDE CIRCLE	Angle formed by two secants (Apply Pr. 11.)		$\angle A \triangleq \frac{1}{2}(\overset{\frown}{BC} - \overset{\frown}{DE})$ $\angle A = \frac{1}{2}(a° - b°)$	
	Angle formed by a secant and a tangent (Apply Pr. 12.)		$\angle A \triangleq \frac{1}{2}(\overset{\frown}{BC} - \overset{\frown}{BD})$ $\angle A = \frac{1}{2}(a° - b°)$	By one-half difference of intercepted arcs
	Angle formed by two tangents (Apply Pr. 13.)		$\angle A \triangleq \frac{1}{2}(\overset{\frown}{BDC} - \overset{\frown}{BC})$ $\angle A = \frac{1}{2}(a° - b°)$ Also, $\angle A = (180 - b)°$	

Note. In the case of an angle formed by a secant and a chord meeting on the circle, first find the inscribed angle adjacent to it and then subtract from 180°.

Thus if secant AB meets chord CD at C on the circle, to find $\angle y$, first find inscribed $\angle x$. Obtain $\angle y$ by subtracting $\angle x$ from 180°.

3.1   APPLYING PRINCIPLES 1-2: To measure central and inscribed angles

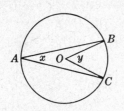

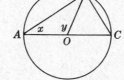

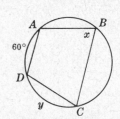

(a) If $\angle y = 46°$, find $\angle x$.   (b) If $\angle y = 112°$, find $\angle x$.   (c) If $\angle x = 75°$, find $\widehat{y}$.

Solution:

(a) $\angle y \stackrel{\circ}{=} \widehat{BC}$; then $\widehat{BC} = 46°$. $\angle x \stackrel{\circ}{=} \frac{1}{2}\widehat{BC} = \frac{1}{2}(46°) = 23°$.

(b) $\angle y \stackrel{\circ}{=} \widehat{AB}$; then $\widehat{AB} = 112°$.
    $\widehat{BC} = \widehat{ABC} - \widehat{AB} = 180° - 112° = 68°$. $\angle x \stackrel{\circ}{=} \frac{1}{2}\widehat{BC} = \frac{1}{2}(68°) = 34°$.

(c) $\angle x \stackrel{\circ}{=} \frac{1}{2}\widehat{ADC}$; then $\widehat{ADC} = 150°$. $\widehat{y} = \widehat{ADC} - \widehat{AD} = 150° - 60° = 90°$.

3.2   APPLYING PRINCIPLES 3-8: To measure angles and arcs
    In each, find x and y.

(a)
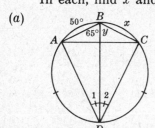

(b)
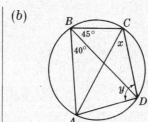

(c)
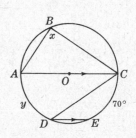

Solution:

(a) Since $\angle 1 = \angle 2$, $\widehat{x} = \widehat{AB} = 50°$.   Since $\widehat{AD} = \widehat{CD}$, $\angle y = \angle ABD = 65°$.

(b) $\angle ABD$ and $\angle x$ are inscribed in $\widehat{ABD}$; hence $\angle x = \angle ABD = 40°$.
    $ABCD$ is an inscribed quadrilateral; hence $\angle y = 180° - \angle B = 95°$.

(c) Since $\angle x$ is inscribed in a semicircle, $\angle x = 90°$.   Since $AC \parallel DE$, $\widehat{y} = \widehat{CE} = 70°$.

3.3   APPLYING PRINCIPLE 9: To measure an angle formed by a tangent and a chord
    In each of the following, CD is a tangent at P.

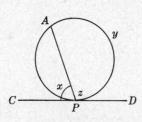

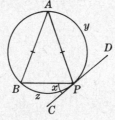

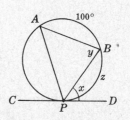

(a) If $\widehat{y} = 220°$, find $\angle x$.   (b) If $\widehat{y} = 140°$, find $\angle x$.   (c) If $\angle y = 75°$, find $\angle x$.

Solution:

(a) $\angle z \stackrel{\circ}{=} \frac{1}{2}\widehat{y} = \frac{1}{2}(220°) = 110°$. $\angle x = 180° - 110° = 70°$.

(b) Since $AB = AP$, $\widehat{AB} = \widehat{y} = 140°$ and $\widehat{z} = 360° - 140° - 140° = 80°$. $\angle x \stackrel{\circ}{=} \frac{1}{2}\widehat{z} = 40°$.

(c) $\angle y \stackrel{\circ}{=} \frac{1}{2}\widehat{AP}$; then $\widehat{AP} = 150°$ and $\widehat{z} = 360° - 100° - 150° = 110°$. $\angle x \stackrel{\circ}{=} \frac{1}{2}\widehat{z} = 55°$.

3.4 APPLYING PRINCIPLE 10

Pr. 10: *An angle formed by two intersecting chords is measured by one-half the sum of the intercepted arcs.*

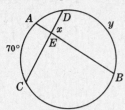
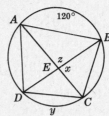
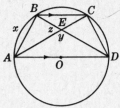

(a) If $\angle x = 95°$, find $\widehat{y}$. (b) If $\widehat{y} = 80°$, find $\angle x$. (c) If $\widehat{x} = 78°$, find $\angle y$.

Solution:

(a) $\angle x \triangleq \frac{1}{2}(\widehat{AC} + \widehat{y})$; then $95° = \frac{1}{2}(70° + \widehat{y})$, $\widehat{y} = 120°$.

(b) $\angle z \triangleq \frac{1}{2}(\widehat{y} + \widehat{AB}) = \frac{1}{2}(80° + 120) = 100°$. $\angle x = 180° - \angle z = 80°$.

(c) $BC \parallel AD$; then $\widehat{CD} = \widehat{x} = 78°$. $\angle z \triangleq \frac{1}{2}(\widehat{x} + \widehat{CD}) = 78°$. $\angle y = 180° - \angle z = 102°$.

3.5 APPLYING PRINCIPLES 11 to 13

Pr. 11 to 13: *An angle formed by two secants, by a secant and a tangent or by two tangents is measured by one-half the difference of the intercepted arcs.*

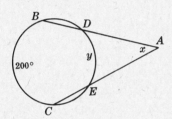
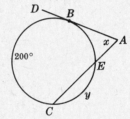
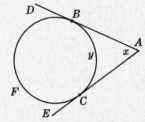

(a) If $\angle x = 40°$, find $\widehat{y}$. (b) If $\angle x = 67°$, find $\widehat{y}$. (c) If $\angle x = 61°$, find $\widehat{y}$.

Solution:

(a) $\angle x \triangleq \frac{1}{2}(\widehat{BC} - \widehat{y})$; then $40° = \frac{1}{2}(200° - \widehat{y})$, $\widehat{y} = 120°$.

(b) $\angle x \triangleq \frac{1}{2}(\widehat{BC} - \widehat{BE})$; then $67° = \frac{1}{2}(200° - \widehat{BE})$, $\widehat{BE} = 66°$. $\widehat{y} = 360° - 266° = 94°$.

(c) $\angle x \triangleq \frac{1}{2}(\widehat{BFC} - \widehat{y})$; then $61° = \frac{1}{2}[(360° - \widehat{y}) - \widehat{y}]$, $\widehat{y} = 119°$.

3.6 USING EQUATIONS IN TWO UNKNOWNS

In each, find x and y, using equations in two unknowns.

(a)

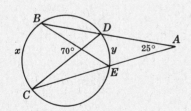

(b)

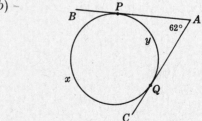

Solution:

(a) By Pr. 10, (1) $70° = \frac{1}{2}(\widehat{x} + \widehat{y})$
By Pr. 11, (2) $25° = \frac{1}{2}(\widehat{x} - \widehat{y})$
Add (1) and (2) to get $\widehat{x} = 95°$.
Subtract (1) from (2) to get $\widehat{y} = 45°$.

(b) Since $\widehat{x} + \widehat{y} = 360°$, (1) $180° = \frac{1}{2}(\widehat{x} + \widehat{y})$
By Pr. 13, (2) $62° = \frac{1}{2}(\widehat{x} - \widehat{y})$
Add (1) and (2) to find $\widehat{x} = 242°$.
Subtract (1) from (2) to find $\widehat{y} = 118°$.

3.7 MEASURING ANGLES and ARCS in GENERAL

In each, find x and y.

(a)

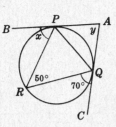

(b)

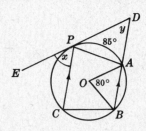

Solution:

(a) By Pr. 2, $50° = \frac{1}{2}\overset{\frown}{PQ}$ or $\overset{\frown}{PQ} = 100°$.

By Pr. 9, $70° = \frac{1}{2}\overset{\frown}{QR}$ or $\overset{\frown}{QR} = 140°$.

Then $\overset{\frown}{PR} = 360° - \overset{\frown}{PQ} - \overset{\frown}{QR} = 120°$.

By Pr. 9, $\angle x \overset{\circ}{=} \frac{1}{2}\overset{\frown}{PR} = 60°$.

By Pr. 13, $\angle y \overset{\circ}{=} \frac{1}{2}(\overset{\frown}{PRQ} - \overset{\frown}{PQ})$
$= \frac{1}{2}(260° - 100°) = 80°$.

(b) By Pr. 1, $\overset{\frown}{AB} = 80°$.

By Pr. 8, $\overset{\frown}{BC} = \overset{\frown}{PA} = 85°$.

Then $\overset{\frown}{PC} = 360° - \overset{\frown}{PA} - \overset{\frown}{AB} - \overset{\frown}{BC} = 110°$.

By Pr. 9, $\angle x \overset{\circ}{=} \frac{1}{2}\overset{\frown}{PC} = 55°$.

By Pr. 12, $\angle y \overset{\circ}{=} \frac{1}{2}(\overset{\frown}{PCB} - \overset{\frown}{PA})$
$= \frac{1}{2}(195° - 85°) = 55°$.

3.8 PROVING an ANGLE MEASUREMENT PROBLEM

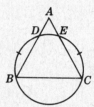

Given: $\overset{\frown}{BD} = \overset{\frown}{CE}$

To Prove: $AB = AC$

Plan: First prove $\overset{\frown}{CD} = \overset{\frown}{BE}$.
Use this to show that $\angle B = \angle C$.

PROOF: Statements	Reasons
1. $\overset{\frown}{BD} = \overset{\frown}{CE}$	1. Given
2. $\overset{\frown}{DE} = \overset{\frown}{DE}$	2. Identity
3. $\overset{\frown}{BE} = \overset{\frown}{CD}$	3. If equals are added to equals, the sums are equal.
4. $\angle B = \angle C$	4. In a circle, inscribed angles having equal intercepted arcs are equal.
5. $AB = AC$	5. In a triangle, sides opposite equal angles are equal.

3.9 PROVING an ANGLE MEASUREMENT PROBLEM STATED in WORDS

Prove: Parallel chords drawn at the ends of a diameter are equal.

Given: Circle O
AB is a diameter.
$AC \parallel BD$

To Prove: $AC = BD$

Plan: Prove $\overset{\frown}{AC} = \overset{\frown}{BD}$

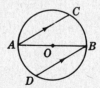

PROOF: Statements	Reasons
1. AB is a diameter.	1. Given
2. $\overset{\frown}{ACB} = \overset{\frown}{ADB}$	2. A diameter cuts a circle into two equal semicircles.
3. $AC \parallel BD$	3. Given
4 $\overset{\frown}{AD} = \overset{\frown}{BC}$	4. Parallel lines intercept equal arcs, on a circle.
5. $\overset{\frown}{AC} = \overset{\frown}{BD}$	5. If equals are subtracted from equals, the differences are equal.
6. $AC = BD$	6. In a circle, equal arcs have equal chords.

Supplementary Problems

1. (a) **Given:** $AB = DE$
          $AC = DF$
     To Prove: $\angle B = \angle E$

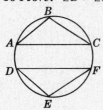

(b) **Given:** Circle O, $AB = BC$
          Diameter BD
     To Prove: BD bisects $\angle AOC$.

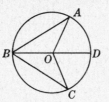

(c) **Given:** Circle O (1.3)
          $\overset{\frown}{AB} = \overset{\frown}{CD}$
     To Prove: $\angle AOC = \angle BOD$

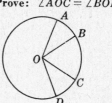

2. (a) **Given:** $AB = AC$
     To Prove: $\overset{\frown}{ABC} = \overset{\frown}{ACB}$

(b) **Given:** $\overset{\frown}{ABC} = \overset{\frown}{ACB}$
     To Prove: $AB = AC$

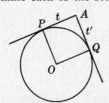

(c) **Given:** Circle O, $AB = AD$
          Diameter AC
     To Prove: $BC = CD$

(d) **Given:** Circle O
          $AB = AD$, $BC = CD$
     To Prove: AC is a diameter.

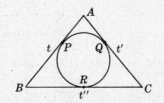

(e) **Given:** $AD = BC$ (1.3)
     To Prove: $AC = BD$

(f) **Given:** $AC = BD$
     To Prove: $AD = BC$

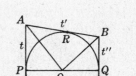

3. Prove each of the following.   (1.4)

(a) If a radius bisects a chord, then it bisects its arcs.

(b) If a diameter bisects the major arc of a chord, then it is perpendicular to the chord.

(c) If a diameter is perpendicular to a chord, it bisects the chord and its arcs.

4. Prove each of the following.   (1.4)

(a) A radius through the point of intersection of two equal chords bisects an angle formed by them.

(b) If chords drawn from the ends of a diameter make equal angles with the diameter, the chords are equal.

(c) In a circle, equal chords are equally distant from the center of the circle.

(d) In a circle, chords that are equally distant from the center are equal.

5. Determine each of the following. (t, t' and t'' are tangents.)   (2.1)

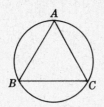

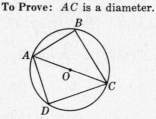

(a) If $\angle A = 90°$, what kind of quadrilateral is $PAQO$?

(b) If $BR = RC$, what kind of triangle is ABC?

(c) What kind of quadrilateral is $PABQ$ if PQ is a diameter?

(d) What kind of triangle is AOB?

6. In circle O, radii OA and OB are drawn to the points of tangency of PA and PB. Find $\angle AOB$ if $\angle APB$ equals (a) $40°$, (b) $120°$, (c) $90°$, (d) $x°$, (e) $(180-x)°$, (f) $(90-x)°$.   (2.2)

7. Find each of the following. (t and t' are tangents.)   (2.2)

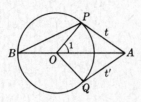   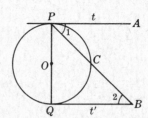

(a) If $\angle POQ = 80°$, find $\angle PAQ$.   (d) If PB bisects $\angle APQ$, find $\angle 2$.
(b) If $\angle PBO = 25°$, find $\angle 1$ and $\angle PAQ$.   (e) If $\angle 1 = 35°$, find $\angle 2$.
(c) If $\angle PAQ = 72°$, find $\angle 1$ and $\angle PBO$.   (f) If $PQ = QB$, find $\angle 1$.

8. $\triangle ABC$ is circumscribed.   Quadrilateral $ABCD$ Quadrilateral $ABCD$ (2.3)
                                             is circumscribed.   is circumscribed.

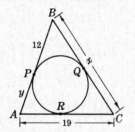   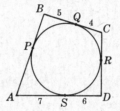   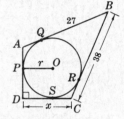

(a) If $y = 9$, find x.   (c) Find $AB + CD$.   (e) If $r = 10$, find x.
(b) If $x = 25$, find y.   (d) Find perimeter of $ABCD$. (f) If $x = 25$, find r.

9. If two circles have radii of 20 and 13 respectively, find their line of centers   (2.4)
   (a) if the circles are concentric,   (c) if the circles are tangent externally,
   (b) if the circles are 7 units apart,   (d) if the circles are tangent internally.

10. If the line of centers of two circles is 30, what is the relation between the two circles   (2.4)
   (a) if their radii are 25 and 5,   (c) if their radii are 20 and 5,
   (b) if their radii are 35 and 5,   (d) if their radii are 25 and 10.

11. What is the relation between two circles if their line of centers is   (2.4)
   (a) 0, (b) equal to the difference of their radii, (c) equal to the sum of their radii, (d) greater than the sum of their radii, (e) less than the difference of their radii and greater than 0, (f) greater than the difference and less than the sum of their radii.

12. Prove each of the following   (2.5)
   (a) The line from the center of a circle to an outside point bisects the angle between the tangents from the point to the circle.
   (b) If two circles are tangent externally, their common internal tangent bisects a common external tangent.
   (c) If two circles are outside each other, their common internal tangents are equal.
   (d) In a circumscribed quadrilateral, the sum of two opposite sides equals the sum of the other two.

13. Find the number of degrees in a central angle which intercepts an arc of   (3.1)
   (a) $40°$, (b) $90°$, (c) $170°$, (d) $180°$, (e) $2x°$, (f) $(180-x)°$, (g) $(2x-2y)°$.

14. Find the number of degrees in an inscribed angle which intercepts an arc of   *(3.1)*
    (a) 40°, (b) 90°, (c) 170°, (d) 180°, (e) 260°, (f) 348°, (g) 2x°, (h) (180 − x)°, (i) (2x − 2y)°.

15. Find the number of degrees in the arc intercepted by   *(3.1)*

    (a) a central angle of 85°,
    (b) an inscribed angle of 85°,
    (c) a central angle of c°,
    (d) an inscribed angle of i°,

    (e) the central angle of a triangle formed by two radii and a chord equal to a radius,
    (f) the smallest angle of an inscribed triangle whose angles intercept arcs in the ratio of 1 : 2 : 3.

16. Find the number of degrees in each of the arcs intercepted by the angles of an inscribed triangle if these angles are in the ratio of (a) 1 : 2 : 3, (b) 2 : 3 : 4, (c) 5 : 6 : 7, (d) 1 : 4 : 5.   *(3.1)*

17.   *(3.1)*

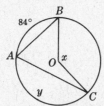

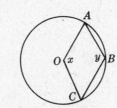

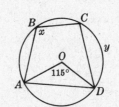

    (a) If $\widehat{y} = 140°$, find ∠x.
    (b) If ∠x = 165°, find $\widehat{y}$.

    (c) If ∠y = 115°, find ∠x.
    (d) If ∠x = 108°, find ∠y.

    (e) If $\widehat{y} = 105°$, find ∠x.
    (f) If ∠x = 96°, find $\widehat{y}$.

18. If quadrilateral ABCD is inscribed in a circle, find   *(3.2)*

    (a) ∠A if ∠C = 45°,
    (b) ∠B if ∠D = 90,
    (c) ∠C if ∠A = x°,
    (d) ∠D if ∠B = (90 − x)°,

    (e) ∠A if $\widehat{BAD} = 160°$,
    (f) ∠B if $\widehat{ABC} = 200°$,
    (g) ∠C if $\widehat{BC} = 140°$ and $\widehat{CD} = 110°$,
    (h) ∠D if ∠D : ∠B = 2 : 3.

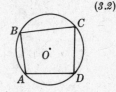

19. If BC and AD are the parallel sides of inscribed trapezoid ABCD, as shown, find   *(3.2)*

    (a) $\widehat{AB}$ if $\widehat{CD} = 85°$,
    (b) $\widehat{CD}$ if $\widehat{AB} = y°$,
    (c) $\widehat{AB}$ if $\widehat{BC} = 60°$ and $\widehat{AD} = 80°$,
    (d) $\widehat{CD}$ if $\widehat{AD} + \widehat{BC} = 170°$,

    (e) ∠A if ∠D = 72°,
    (f) ∠A if ∠C = 130°,
    (g) ∠B if ∠C = 145°,
    (h) ∠B if $\widehat{AD} = 90°$ and $\widehat{AB} = 84°$.

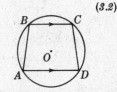

20. A diameter is parallel to a chord. Find the number of degrees in an arc between the diameter and chord if the chord intercepts (a) a minor arc of 80°, (b) a major arc of 300°.   *(3.2)*

21. In each, find x and y.   *(3.2)*

(a)
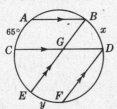

(b)
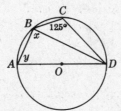

(c)
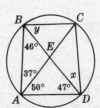

22. Find the number of degrees in the angle formed by a tangent and a chord drawn to the point of tangency if the intercepted arc is   *(3.3)*

    (a) 38°, (b) 90°, (c) 138°, (d) 180°, (e) 250°, (f) 334°, (g) x°, (h) (360 − x)°, (i) (2x + 2y)°.

23. Find the number of degrees in the arc intercepted by an angle formed by a tangent and a chord drawn to the point of tangency if the angle equals   *(3.3)*

    (a) 55°, (b) $67\frac{1}{2}$°, (c) 90°, (d) 135°, (e) (90 − x)°, (f) (180 − x)°, (g) (x − y)°, (h) $3\frac{1}{2}x$°

24. Find the number of degrees in the acute angle formed by a tangent through the vertex and an adjacent side of an inscribed (a) square, (b) equilateral triangle, (c) regular hexagon, (d) regular decagon.   *(3.3)*

25. In each, find x and y. (t and t' are tangents.)   *(3.3)*

(a) 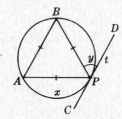   (b) 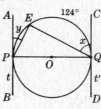   (c)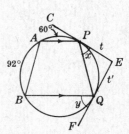

26. If AC and BD are chords intersecting in a circle as shown, find   *(3.4)*

(a) $\angle x$ if $\widehat{AB} = 90°$ and $\widehat{CD} = 60°$ (e) $\widehat{AB} + \widehat{CD}$ if $\angle x = 70°$,

(b) $\angle x$ if $\widehat{AB}$ and $\widehat{CD}$ each equals 75°   (f) $\widehat{BC} + \widehat{AD}$ if $\angle x = 65°$,

(c) $\angle x$ if $\widehat{AB} + \widehat{CD} = 230°$,   (g) $\widehat{BC}$ if $\angle x = 60°$ and $\widehat{AD} = 160°$,

(d) $\angle x$ if $\widehat{BC} + \widehat{AD} = 160°$,   (h) $\widehat{BC}$ if $\angle y = 72°$ and $\widehat{AD} = 2\widehat{BC}$.

27. If AC and BD are diagonals of an inscribed quadrilateral $ABCD$ as shown, find   *(3.4)*

(a) $\angle 1$ if $\widehat{a} = 95°$ and $\widehat{c} = 75°$,   (e) $\angle 2$ if $\widehat{b} + \widehat{d} = \widehat{a} + \widehat{c}$,

(b) $\angle 1$ if $\widehat{b} = 88°$ and $\widehat{d} = 66°$,   (f) $\angle 2$ if $BC \parallel AD$ and $\widehat{a} = 70°$,

(c) $\angle 1$ if $\widehat{b}$ and $\widehat{d}$ each equals 100°,   (g) $\angle 2$ if AD is a diameter and $b = 80°$,

(d) $\angle 1$ if $\widehat{a} : \widehat{b} : \widehat{c} : \widehat{d} = 1 : 2 : 3 : 4$,   (h) $\angle 2$ if $ABCD$ is a rectangle and $\widehat{a} = 70°$.

28. In each, find x and y.   *(3.4)*

(a) 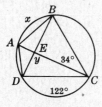   (b) 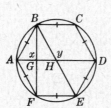   (c)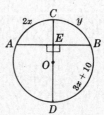

29. If AB and AC are intersecting secants as shown, find   *(3.5)*

(a) $\angle A$ if $\widehat{c} = 100°$ and $\widehat{a} = 40°$,   (e) $\widehat{a}$ if $\widehat{c} = 160°$ and $\angle A = 20°$,

(b) $\angle A$ if $\widehat{c} - \widehat{a} = 74°$,   (f) $\widehat{c}$ if $\widehat{a} = 60°$ and $\angle A = 35°$,

(c) $\angle A$ if $\widehat{c} = \widehat{a} + 40°$,   (g) $\widehat{c} - \widehat{a}$ if $\angle A = 47°$,

(d) $\angle A$ if $\widehat{a} : \widehat{b} : \widehat{c} : \widehat{d} = 1 : 4 : 3 : 2$,   (h) $\widehat{a}$ if $\widehat{c} = 3\widehat{a}$ and $\angle A = 25°$.

30. If tangent AP and secant AB intersect as shown, find   *(3.5)*

(a) $\angle A$ if $\widehat{c} = 150°$ and $\widehat{a} = 60°$,   (f) $\widehat{a}$ if $\widehat{c} = 220°$ and $\angle A = 40°$,

(b) $\angle A$ if $\widehat{c} = 200°$ and $\widehat{b} = 110°$,   (g) $\widehat{c}$ if $\widehat{a} = 55°$ and $\angle A = 30°$,

(c) $\angle A$ if $\widehat{b} = 120°$ and $\widehat{a} = 70°$,   (h) $\widehat{a}$ if $\widehat{c} = 3\widehat{a}$ and $\angle A = 45°$,

(d) $\angle A$ if $\widehat{c} - \widehat{a} = 73°$,   (i) $\widehat{a}$ if $\widehat{b} = 100°$ and $\angle A = 50°$.

(e) $\angle A$ if $\widehat{a} : \widehat{b} : \widehat{c} = 1 : 4 : 7$,

31. If AP and AQ are intersecting tangents as shown, find   *(3.5)*

(a) $\angle A$ if $\widehat{b} = 200°$,   (f) $\angle A$ if $\widehat{b} = \widehat{a} + 50°$,   (k) $\widehat{a}$ if $\angle A = 35°$,

(b) $\angle A$ if $\widehat{a} = 95°$,   (g) $\angle A$ if $\widehat{b} - \widehat{a} = 84°$,   (l) $\widehat{a}$ if $\angle A = y°$,

(c) $\angle A$ if $\widehat{a} = x°$,   (h) $\angle A$ if $\widehat{b} : \widehat{a} = 5 : 1$,   (m) $\widehat{b}$ if $\angle A = 60°$,

(d) $\angle A$ if $\widehat{a} = (90 - x)°$,   (i) $\angle A$ if $\widehat{b} : \widehat{a} = 7 : 3$,   (n) $\widehat{b}$ if $\angle A = x°$,

(e) $\angle A$ if $\widehat{b} = 3\widehat{a}$,   (j) $\angle A$ if $\widehat{b} = 5\widehat{a} - 60°$,   (o) $\widehat{b}$ if $AP \perp AQ$.

32. In each, find *x and y*. (*t* and *t'* are tangents.)   *(3.5)*

(a)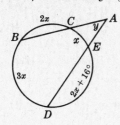

(b)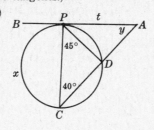

(c)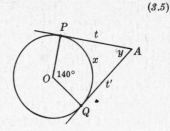

33. If *AB* and *AC* are intersecting secants as shown, find   *(3.6)*

(a) $\widehat{x}$ if $\angle 1 = 80°$ and $\angle A = 40°$,

(b) $\widehat{x}$ if $\angle 1 + \angle A = 150°$,

(c) $\widehat{x}$ if $\angle 1$ and $\angle A$ are supplementary,

(d) $\widehat{y}$ if $\angle 1 = 95°$ and $\angle A = 45°$,

(e) $\widehat{y}$ if $\angle 1 - \angle A = 22\frac{1}{2}°$,

(f) $\widehat{y}$ if $\widehat{x} + \widehat{y} = 190°$ and $\angle A = 50°$.

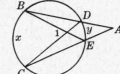

34. In each, find *x and y*. (*t* and *t'* are tangents.)   *(3.6)*

(a)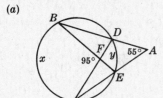

(b)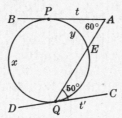

(c)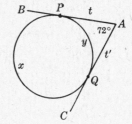

35. If *ABC* is an inscribed triangle as shown, find   *(3.7)*

(a) $\angle A$ if $\widehat{a} = 110°$ and $\widehat{c} = 200°$,

(b) $\angle A$ if $AB \perp BC$ and $\widehat{a} = 102°$,

(c) $\angle A$ if AC is a diameter and $\widehat{a} = 80°$,

(d) $\angle A$ if $\widehat{a}:\widehat{b}:\widehat{c} = 3:1:2$,

(e) $\angle A$ if AC is a diameter and $\widehat{a}:\widehat{b} = 5:4$,

(f) $\angle B$ if $\widehat{ABC} = 208°$,

(g) $\angle B$ if $\widehat{a} + \widehat{b} = 3\widehat{c}$,

(h) $\angle B$ if $\widehat{a} = 75°$ and $\widehat{c} = 2\widehat{b}$,

(i) $\angle C$ if $AB \perp BC$ and $\widehat{a} = 5\widehat{b}$,

(j) $\widehat{c}$ if $\angle A:\angle B:\angle C = 5:4:3$.

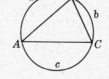

36. If *ABCP* is an inscribed quadrilateral, *PD* a tangent and *AF* a secant, find   *(3.7)*

(a) $\angle 1$ if $\widehat{a} = 94°$ and $\widehat{c} = 54°$,

(b) $\angle 2$ if AP is a diameter,

(c) $\angle 3$ if $\widehat{CPA} = 250°$,

(d) $\angle 3$ if $\widehat{ABC} = 120°$,

(e) $\angle 4$ if $\widehat{BCP} = 130°$ and $\widehat{b} = 50°$,

(f) $\angle 4$ if $BC \parallel AP$ and $\widehat{a} = 74°$,

(g) $\widehat{a}$ if $BC \parallel AP$ and $\angle 6 = 42°$,

(h) $\widehat{a}$ if AC is a diameter and $\angle 5 = 35°$,

(i) $\widehat{b}$ if $AC \perp BP$ and $\angle 2 = 57°$,

(j) $\widehat{c}$ if AC and BP are diameters and $\angle 5 = 41°$,

(k) $\widehat{d}$ if $\angle 1 = 95°$ and $\widehat{b} = 95°$,

(l) $\angle CPA$ if $\angle 3 = 79°$.

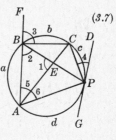

37. Find *x* and *y* in each. (*t* and *t'* are tangents.)   *(3.7)*

(a)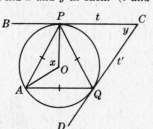

(b)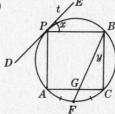

PBCA is an inscribed square

(c)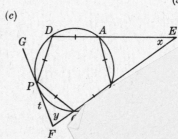

38. In each, find x and y.   (3.7)

(a)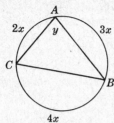

(b)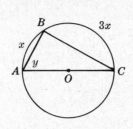

(c)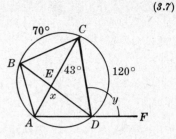

39. Prove each of the following.   (3.8)

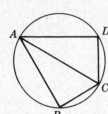

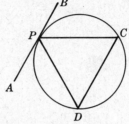

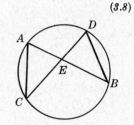

(a) **Given:** AC bisects $\angle A$.
   To Prove: $BC = CD$.

(b) **Given:** $BC = CD$.
   To Prove: AC bisects $\angle A$.

(c) **Given:** $AB \parallel CD$
           AB is a tangent.
   To Prove: $PC = PD$.

(d) **Given:** $PC = PD$
           AB is a tangent.
   To Prove: $AB \parallel CD$.

(e) **Given:** $\overset{\frown}{AC} = \overset{\frown}{BD}$
   To Prove: $CE = EB$.

(f) **Given:** $CE = EB$
   To Prove: $\overset{\frown}{AC} = \overset{\frown}{BD}$.

40. In each, prove the indicated triangles are mutually equiangular; that is, the three angles of one triangle are equal to the corresponding angles of the other.   (3.8)

(a) $\triangle AEC$ and $\triangle DEB$.

(b) $\triangle APC$ and $\triangle ABP$.

(c) $\triangle ABE$ and $\triangle ACD$.

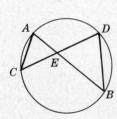

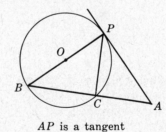

AP is a tangent

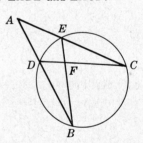

41. Prove each of the following.   (3.9)

(a) The base angles of an inscribed trapezoid are equal.

(b) A parallelogram inscribed in a circle is a rectangle.

(c) In a circle, parallel chords intercept equal arcs.

(d) Diagonals drawn from a vertex of a regular inscribed pentagon trisect the vertex angle.

(e) If a tangent through a vertex of an inscribed triangle is parallel to its opposite side, the triangle is isosceles.

Chapter 7

Similarity

1. Ratios

Ratios are used to compare quantities by division. The ratio of two quantities is the first divided by the second. A ratio is an abstract number, that is, a number without a unit of measure. Thus the ratio of 10 ft. to 5 ft. is 10 ft. ÷ 5 ft., which equals 2.

A ratio can be expressed in the following ways: (a) using a colon, $3:4$; (b) using "to", 3 to 4; (c) as a common fraction, $\frac{3}{4}$; (d) as a decimal, .75; (e) as a percent, 75%.

To find ratios, the quantities involved must have the same unit. A ratio should be simplified by reducing to lowest terms and eliminating fractions contained in the ratio. Thus to find the ratio of 1 ft. to 4 in., first change the foot to 12 inches, then take the ratio of 12 inches to 4 inches; the result is a ratio of 3 to 1, or 3. Also, the ratio of $2\frac{1}{2}:\frac{1}{2}$ equals $5:1$ or 5.

The ratio of three or more quantities may be expressed as a *continued ratio*. Thus the ratio of \$2 to \$3 to \$5 is the continued ratio $2:3:5$. This enlarged ratio is a combination of three separate ratios; these are $2:3$, $3:5$ and $2:5$.

1.1 RATIO of TWO QUANTITIES with the SAME UNIT

Express each ratio in lowest terms: (a) 15° to 3°, (b) \$1.25 to \$5, (c) $2\frac{1}{2}$ yr. to 2 yr.

Solution:

(a) $\frac{15}{3} = 5$ (b) $\frac{1.25}{5} = \frac{1}{4}$ (c) $\frac{2\frac{1}{2}}{2} = \frac{5}{4}$

1.2 RATIO of TWO QUANTITIES with DIFFERENT UNITS

Express each ratio in lowest terms: (a) 2 yr. to 3 mo., (b) 80¢ to \$3.20.

Solution:

(a) 2 yr. to 3 mo. $= 24$ mo. to 3 mo. $= \frac{24}{3} = 8$ (b) 80¢ to \$3.20 $= 80$¢ to 320¢ $= \frac{80}{320} = \frac{1}{4}$

1.3 CONTINUED RATIO of THREE QUANTITIES

Express each ratio in lowest terms: (a) 1 gal. to 2 qt. to 2 pt., (b) 1 ton to 1 lb. to 8 oz.

Solution:

(a) 1 gal. to 2 qt. to 2 pt. $= 4$ qt. to 2 qt. to 1 qt. $= 4:2:1$

(b) 1 ton to 1 lb. to 8 oz. $= 2000$ lb. to 1 lb. to $\frac{1}{2}$ lb. $= 2000:1:\frac{1}{2} = 4000:2:1$

1.4 NUMERICAL and ALGEBRAIC RATIOS

Express each ratio in lowest terms:

(a) 50 to 60, (b) 6.3 to .9, (c) 12 to $\frac{3}{8}$, (d) $2x$ to $5x$, (e) $5s^2$ to s^3, (f) x to $5x$ to $7x$.

Solution:

(a) $\frac{50}{60} = \frac{5}{6}$ (b) $\frac{6.3}{9} = 7$ (c) $12 \div \frac{3}{8} = 32$ (d) $\frac{2x}{5x} = \frac{2}{5}$ (e) $\frac{5s^2}{s^3} = \frac{5}{s}$ (f) $x:5x:7x = 1:5:7$

1.5 USING RATIOS in ANGLE PROBLEMS

If two angles are in the ratio of $3:2$, find the angles if (a) they are adjacent and form an angle of $40°$, (b) they are the acute angles of a right triangle, (c) they are two angles of a triangle whose third angle is $70°$.

Solution:

Let the number of degrees in the angles be $3x$ and $2x$.

(a) $3x + 2x = 40$, $5x = 40$, $x = 8$; hence the angles are $24°$ and $16°$.
(b) $3x + 2x = 90$, $5x = 90$, $x = 18$; hence the angles are $54°$ and $36°$.
(c) $3x + 2x + 70 = 180$, $5x = 110$, $x = 22$; hence the angles are $66°$ and $44°$.

1.6 THREE ANGLES HAVING a FIXED RATIO

If three angles are in the ratio of $4:3:2$, find the angles if (a) the first and the third are supplementary, (b) the angles are the three angles of a triangle.

Solution:

Let the number of degrees in the angles be $4x$, $3x$ and $2x$.

(a) $4x + 2x = 180$, $6x = 180$, $x = 30$; hence the angles are $120°$, $90°$ and $60°$.
(b) $4x + 3x + 2x = 180$, $9x = 180$, $x = 20$; hence the angles are $80°$, $60°$ and $40°$.

2. *Proportions*

A *proportion* is an equality of two ratios. Thus $2:5 = 4:10$ or $\frac{2}{5} = \frac{4}{10}$ is a proportion.

The fourth term of a proportion is the *fourth proportional* to the other three taken in order. Thus in $2:3 = 4:x$, x is the fourth proportional to 2, 3 and 4.

The *means* of a proportion are its middle terms, that is, its second and third terms. The *extremes* of a proportion are its outside terms, that is, its first and fourth terms. Thus in $a:b = c:d$, the means are b and c, and the extremes are a and d.

If the two means of a proportion are the same, either mean is the *mean proportional* between the first and fourth terms. Thus in $9:3 = 3:1$, 3 is the mean proportional between 9 and 1.

A. *Proportion Principles*

Pr. 1: *In any proportion, the product of the means equals the product of the extremes.*
   Thus if $a:b = c:d$, then $ad = bc$.

Pr. 2: *If the product of two numbers equals the product of two other numbers, either pair may be made the means of a proportion and the other pair may be made the extremes.*
   Thus if $3x = 5y$, then $x:y = 5:3$ or $y:x = 3:5$ or $3:y = 5:x$ or $5:x = 3:y$.

B. *Methods of Changing a Proportion Into a New Proportion*

Pr. 3: **Inversion Method.** *A proportion may be changed into a new proportion by inverting each ratio.*
   Thus if $\frac{1}{x} = \frac{4}{5}$, then $\frac{x}{1} = \frac{5}{4}$.

Pr. 4: **Alternation Method.** *A proportion may be changed into a new proportion by interchanging the means or by interchanging the extremes.*
   Thus if $\frac{x}{3} = \frac{y}{2}$, then $\frac{x}{y} = \frac{3}{2}$ or $\frac{2}{3} = \frac{y}{x}$.

Pr. 5: **Addition Method.** *A proportion may be changed into a new proportion by adding the terms of each ratio to obtain new first and third terms.*
   Thus if $\frac{a}{b} = \frac{c}{d}$, then $\frac{a+b}{b} = \frac{c+d}{d}$. If $\frac{x-2}{2} = \frac{9}{1}$, then $\frac{x}{2} = \frac{10}{1}$.

Pr. 6: **Subtraction Method.** *A proportion may be changed into a new proportion by subtracting the terms of each ratio to obtain new first and third terms.*

Thus if $\dfrac{a}{b} = \dfrac{c}{d}$, then $\dfrac{a-b}{b} = \dfrac{c-d}{d}$. If $\dfrac{x+3}{3} = \dfrac{9}{1}$, then $\dfrac{x}{3} = \dfrac{8}{1}$.

C. Other Proportion Principles

Pr. 7: *If any three terms of one proportion equal the corresponding three terms of another proportion, the remaining terms are equal.*

Thus if $\dfrac{x}{y} = \dfrac{3}{5}$ and $\dfrac{x}{4} = \dfrac{3}{5}$, then $y = 4$.

Pr. 8: *In a series of equal ratios, the sum of any of the numerators is to the sum of the corresponding denominators as any numerator is to its denominator.*

Thus if $\dfrac{a}{b} = \dfrac{c}{d} = \dfrac{e}{f}$, then $\dfrac{a+c+e}{b+d+f} = \dfrac{a}{b}$. If $\dfrac{x-y}{4} = \dfrac{y-3}{5} = \dfrac{3}{1}$, then $\dfrac{x-y+y-3+3}{4+5+1} = \dfrac{3}{1}$
or $\dfrac{x}{10} = \dfrac{3}{1}$.

2.1 FINDING UNKNOWNS in PROPORTIONS

Solve for x: (a) $x : 4 = 6 : 8$ (c) $x : 5 = 2x : (x+3)$ (e) $\dfrac{x}{2x-3} = \dfrac{3}{5}$

(b) $3 : x = x : 27$ (d) $\dfrac{3}{x} = \dfrac{2}{5}$ (f) $\dfrac{x-2}{4} = \dfrac{7}{x+2}$

Solution:

(a) $4(6) = 8x$, $8x = 24$, $x = 3$
(b) $x^2 = 3(27)$, $x^2 = 81$, $x = \pm 9$
(c) $5(2x) = x(x+3)$, $10x = x^2 + 3x$, $x^2 - 7x = 0$, $x = 0$ or 7
(d) $2x = 3(5)$, $2x = 15$, $x = 7\frac{1}{2}$
(e) $3(2x-3) = 5x$, $6x - 9 = 5x$, $x = 9$
(f) $4(7) = (x-2)(x+2)$, $28 = x^2 - 4$, $x^2 = 32$, $x = \pm 4\sqrt{2}$

2.2 FINDING FOURTH PROPORTIONALS to THREE GIVEN NUMBERS

Find the fourth proportional to (a) $2, 4, 6$; (b) $4, 2, 6$; (c) $\frac{1}{2}, 3, 4$; (d) b, d, c.

Solution:

(a) $2 : 4 = 6 : x$, $2x = 24$, $x = 12$ (c) $\frac{1}{2} : 3 = 4 : x$, $\frac{1}{2}x = 12$, $x = 24$
(b) $4 : 2 = 6 : x$, $4x = 12$, $x = 3$ (d) $b : d = c : x$, $bx = cd$, $x = cd/b$

2.3 FINDING MEAN PROPORTIONAL to TWO GIVEN NUMBERS

Find the positive mean proportional (x) between (a) 5 and 20, (b) $\frac{1}{2}$ and $\frac{8}{9}$.

Solution:

(a) $5 : x = x : 20$, $x^2 = 100$, $x = 10$ (b) $\frac{1}{2} : x = x : \frac{8}{9}$, $x^2 = \frac{4}{9}$, $x = \frac{2}{3}$

2.4 CHANGING EQUAL PRODUCTS into PROPORTIONS

(a) Form a proportion whose fourth term is x if $2bx = 3s^2$.

(b) Find the ratio of x to y if $ay = bx$.

Solution:

(a) $2b : 3s = s : x$ or $2b : 3 = s^2 : x$ or $2b : s^2 = 3 : x$ (b) $x : y = a : b$

2.5 CHANGING PROPORTIONS into NEW PROPORTIONS

In each, form a new proportion whose first term is x.

(a) $\dfrac{15}{x} = \dfrac{3}{4}$ (b) $\dfrac{x-6}{6} = \dfrac{5}{3}$ (c) $\dfrac{x+8}{8} = \dfrac{4}{3}$ (d) $\dfrac{5}{2} = \dfrac{15}{x}$

Solution:

(a) By Pr. 3, $\dfrac{x}{15} = \dfrac{4}{3}$.   (b) By Pr. 5, $\dfrac{x}{6} = \dfrac{8}{3}$.   (c) By Pr. 6, $\dfrac{x}{8} = \dfrac{1}{3}$.   (d) By Pr. 4, $\dfrac{x}{2} = \dfrac{15}{5}$.

2.6 COMBINING NUMERATORS and DENOMINATORS of PROPORTIONS

Use Pr. 8 to find x: (a) $\dfrac{x-2}{9} = \dfrac{2}{3}$, (b) $\dfrac{x+y}{8} = \dfrac{x-y}{4} = \dfrac{2}{3}$, (c) $\dfrac{3x-y}{15} = \dfrac{y-3}{10} = \dfrac{3}{5}$.

Solution:

(a) $\dfrac{x-2+2}{9+3} = \dfrac{2}{3}$, $\dfrac{x}{12} = \dfrac{2}{3}$, $x = 8$.   (b) $\dfrac{(x+y)+(x-y)}{8+4} = \dfrac{2}{3}$, $\dfrac{2x}{12} = \dfrac{2}{3}$, $x = 4$.

(c) $\dfrac{(3x-y)+(y-3)+3}{15+10+5} = \dfrac{3}{5}$, $\dfrac{3x}{30} = \dfrac{3}{5}$, $x = 6$.

3. Proportional Lines

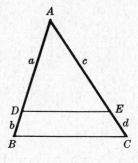

If two lines are divided proportionately,

(1) the corresponding segments are in proportion and

(2) the two lines and either pair of corresponding segments are in proportion.

Thus if AB and AC are divided proportionately by DE, a proportion such as $\dfrac{a}{b} = \dfrac{c}{d}$ may be obtained using the four segments, or a proportion such as $\dfrac{a}{AB} = \dfrac{c}{AC}$ may be obtained using the two lines and two of their segments.

Obtaining the Eight Arrangements of Any Proportion

A proportion such as $\dfrac{a}{b} = \dfrac{c}{d}$ can be arranged in eight ways. To obtain the eight variations, simply let each term of the proportion represent a segment of the above diagram. Each of the possible proportions is then obtained by using the same direction, as follows:

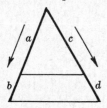 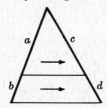 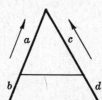 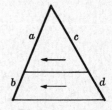

Direction: Down	Direction: Up	Direction: Right	Direction: Left
$\dfrac{a}{b} = \dfrac{c}{d}$ or $\dfrac{c}{d} = \dfrac{a}{b}$	$\dfrac{b}{a} = \dfrac{d}{c}$ or $\dfrac{d}{c} = \dfrac{b}{a}$	$\dfrac{a}{c} = \dfrac{b}{d}$ or $\dfrac{b}{d} = \dfrac{a}{c}$	$\dfrac{c}{a} = \dfrac{d}{b}$ or $\dfrac{d}{b} = \dfrac{c}{a}$

Pr. 1: *If a line is parallel to one side of a triangle, then it divides the other two sides proportionately.*

Thus in $\triangle ABC$, if $DE \parallel BC$, then $\dfrac{a}{b} = \dfrac{c}{d}$.

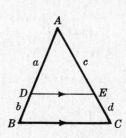

Pr. 2: *If a line divides two sides of a triangle proportionately, it is parallel to the third side.* (Principles 1 and 2 are converses.)

Thus in $\triangle ABC$ if $\dfrac{a}{b} = \dfrac{c}{d}$, then $DE \parallel BC$.

Pr. 3: *Three or more parallel lines divide any two transversals proportionately.*

Thus if $AB \parallel EF \parallel CD$, then $\dfrac{a}{b} = \dfrac{c}{d}$.

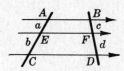

Pr. 4: *A bisector of an angle of a triangle divides the opposite side into segments which are proportional to the adjacent sides.*

Thus in $\triangle ABC$, if CD bisects $\angle C$, then $\dfrac{a}{b} = \dfrac{c}{d}$.

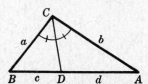

3.1 APPLYING PRINCIPLE 1

Find x in each.

(a)

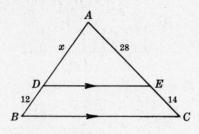

(b)

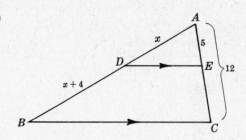

Solution:

(a) $DE \parallel BC$; hence $\dfrac{x}{12} = \dfrac{28}{14}$, $x = 24$.

(b) $EC = 7$. $DE \parallel BC$; hence $\dfrac{x}{x+4} = \dfrac{5}{7}$, $7x = 5x + 20$, $x = 10$.

3.2 APPLYING PRINCIPLE 3

Find x in each.

(a)

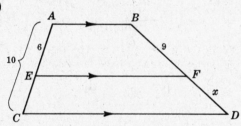

(b)

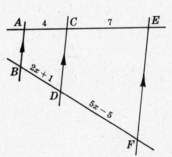

Solution:

(a) $EC = 4$. $AB \parallel EF \parallel CD$; hence $\dfrac{x}{9} = \dfrac{4}{6}$ and $x = 6$.

(b) $AB \parallel CD \parallel EF$; hence $\dfrac{5x-5}{2x+1} = \dfrac{7}{4}$, $20x - 20 = 14x + 7$, $6x = 27$ and $x = 4\frac{1}{2}$.

3.3 APPLYING PRINCIPLE 4

Find x in each.

(a)

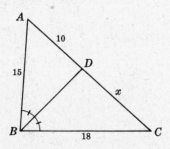

(b)

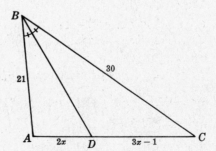

Solution:

(a) BD bisects $\angle B$; hence $\dfrac{x}{10} = \dfrac{18}{15}$ and $x = 12$.

(b) BD bisects $\angle B$; hence $\dfrac{3x-1}{2x} = \dfrac{30}{21} = \dfrac{10}{7}$, $21x - 7 = 20x$ and $x = 7$.

3.4 PROVING a PROPORTIONAL SEGMENTS PROBLEM

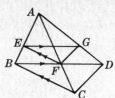

Given: $EG \parallel BD$, $EF \parallel BC$

To Prove: $FG \parallel CD$

Plan: Prove that FG divides AC and AD proportionately.

PROOF: Statements	Reasons
1. $EG \parallel BD$, $EF \parallel BC$	1. Given
2. $\dfrac{AE}{EB} = \dfrac{AG}{GD}$, $\dfrac{AE}{EB} = \dfrac{AF}{FC}$	2. If a line is parallel to a side of a triangle, it divides the other two sides proportionately.
3. $\dfrac{AF}{FC} = \dfrac{AG}{GD}$	3. Things equal to the same thing are equal to each other.
4. $FG \parallel CD$	4. If a line divides two sides of a triangle proportionately, it is parallel to the third side.

4. Similar Triangles

Similar polygons are polygons whose corresponding angles are equal and whose corresponding sides are in proportion. Similar polygons have the same shape although not necessarily the same size.

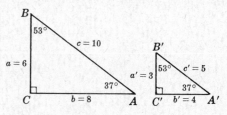

The symbol for "similar" is $\sim$. $\triangle ABC \sim \triangle A'B'C'$ is read "triangle ABC is similar to triangle A-prime B-prime C-prime." As in the case of congruent triangles, *corresponding sides of similar triangles are opposite equal angles.* (Note that corresponding sides and angles are usually designated by the same letters and primes.)

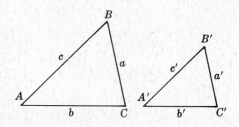

Thus $\triangle ABC \sim \triangle A'B'C'$, since

$$\angle A = \angle A' = 37°, \quad \angle B = \angle B' = 53°, \quad \angle C = \angle C' = 90°$$

and

$$\frac{a}{a'} = \frac{b}{b'} = \frac{c}{c'} \quad \text{or} \quad \frac{6}{3} = \frac{8}{4} = \frac{10}{5}$$

Selecting Similar Triangles in Order to Prove a Proportion

In Solved Problem 4.9, it is given that $ABCD$ is a parallelogram and it is required to prove that $\dfrac{AE}{BC} = \dfrac{AF}{FB}$. To prove this proportion, it is necessary to select similar triangles whose sides are in the proportion. This can be done by simply using the letters A, E and F in the numerators for one of the triangles and the letters B, C and F in the denominators for the other. Hence, prove $\triangle AEF \sim \triangle BCF$.

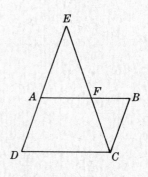

Suppose that the proportion to be proved is $\dfrac{AE}{AF} = \dfrac{BC}{FB}$. In such a case, interchanging the means leads to $\dfrac{AE}{BC} = \dfrac{AF}{FB}$. The needed triangles can then be selected from the numerators and the denominators.

Suppose that the proportion to be proved is $\dfrac{AE}{AD} = \dfrac{AF}{FB}$. The method of selecting triangles cannot be used until the term AD is replaced by BC. This can be done since AD and BC are opposite sides of the parallelogram $ABCD$ and therefore are equal.

A. *Principles of Similar Triangles*

Pr. 1: *Corresponding angles of similar triangles are equal.* (By definition.)

Pr. 2: *Corresponding sides of similar triangles are in proportion.* (By definition.)

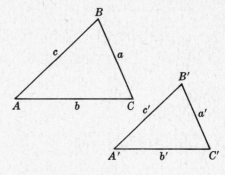

Pr. 3: *Two triangles are similar if two angles of one triangle equal respectively two angles of the other.*

Thus in Fig. 1, if $\angle A = \angle A'$ and $\angle B = \angle B'$, then $\triangle ABC \sim \triangle A'B'C'$.

Fig. 1

Pr. 4: *Two triangles are similar if an angle of one triangle equals an angle of the other and the sides including these angles are in proportion.*

Thus in Fig. 1, if $\angle C = \angle C'$ and $\dfrac{a}{a'} = \dfrac{b}{b'}$, then $\triangle ABC \sim \triangle A'B'C'$.

Pr. 5: *Two triangles are similar if their corresponding sides are in proportion.*

Thus in Fig. 1, if $\dfrac{a}{a'} = \dfrac{b}{b'} = \dfrac{c}{c'}$, then $\triangle ABC \sim \triangle A'B'C'$.

Pr. 6: *Two right triangles are similar if an acute angle of one equals an acute angle of the other.* (Corollary of Pr. 3.)

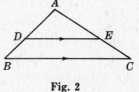

Pr. 7: *A line parallel to a side of a triangle cuts off a triangle similar to the given triangle.*

Thus in Fig. 2, if $DE \parallel BC$, then $\triangle ADE \sim \triangle ABC$.

Fig. 2

Pr. 8: *Triangles similar to the same triangle are similar to each other.*

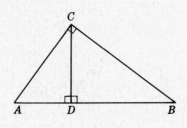

Fig. 3

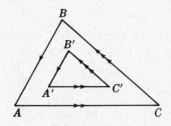

Fig. 4

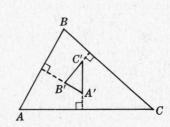

Fig. 5

Pr. 9: *The altitude to the hypotenuse of a right triangle divides it into two triangles which are similar to the given triangle and to each other.*

Thus in Fig. 3, $\triangle CDA \sim \triangle CDB \sim \triangle ABC$.

Pr. 10: *Triangles are similar if their sides are respectively parallel to each other.*

Thus in Fig. 4, $\triangle ABC \sim \triangle A'B'C'$.

Pr. 11: *Triangles are similar if their sides are respectively perpendicular to each other.*

Thus in Fig. 5, $\triangle ABC \sim \triangle A'B'C'$.

4.1 APPLYING PRINCIPLE 2

In similar triangles ABC and $A'B'C'$, find x and y if $\angle A = \angle A'$ and $\angle B = \angle B'$.

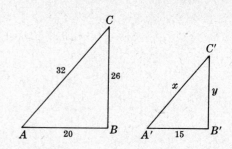

Solution:

Since $\angle A = \angle A'$ and $\angle B = \angle B'$, x and y correspond to 32 and 26 respectively.

Hence $\dfrac{x}{32} = \dfrac{15}{20}$ and $x = 24$; $\dfrac{y}{26} = \dfrac{15}{20}$ and $y = 19\frac{1}{2}$.

4.2 APPLYING PRINCIPLE 3

In each, two pairs of equal angles can be used to prove the indicated triangles similar. Determine the equal angles and state the reason.

(a) $\triangle BEC \sim \triangle AED$

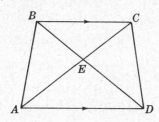

ABCD is a trapezoid.

(b) $\triangle AED \sim \triangle CEB$

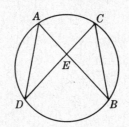

(c) $\triangle ADE \sim \triangle ABC$

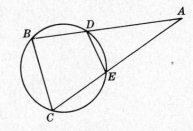

Solution:

(a) $\angle CBD = \angle BDA$ and $\angle BCA = \angle CAD$, since alternate interior angles of parallel lines are equal ($BC \parallel AD$). Also, $\angle BEC$ and $\angle AED$ are equal vertical angles.

(b) $\angle A = \angle C$ and $\angle B = \angle D$, since angles inscribed in the same arc are equal. Also, $\angle AED$ and $\angle CEB$ are equal vertical angles.

(c) $\angle ABC = \angle AED$, since each is a supplement of $\angle DEC$. $\angle ACB = \angle ADE$, since each is a supplement of $\angle BDE$. Also, $\angle A = \angle A$.

4.3 APPLYING PRINCIPLE 6

In each, determine the angles that can be used to prove the indicated triangles similar.

(a) $\triangle ACD \sim \triangle ACB$

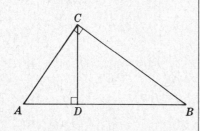

(b) $\triangle AEC \sim \triangle CDB$

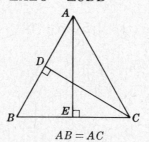

$AB = AC$

(c) $\triangle ADE \sim \triangle ABC$

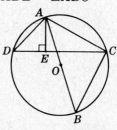

AB is a diameter.

Solution:

(a) $\angle ACB$ and $\angle ADC$ are right angles. $\angle A = \angle A$.

(b) $\angle AEC$ and $\angle BDC$ are right angles.

$\angle B = \angle ACE$, since angles in a triangle opposite equal sides are equal.

(c) $\angle ACB$ is a right angle, since it is inscribed in a semicircle. Hence, $\angle AED = \angle ACB$.

$\angle D = \angle B$, since angles inscribed in the same arc are equal.

4.4 APPLYING PRINCIPLE 4

In each, determine the pair of equal angles and the proportion needed to prove the indicated triangles similar.

(a) $\triangle AEB \sim \triangle DEC$

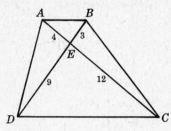

(b) $\triangle AED \sim \triangle ABC$

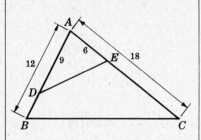

(c) $\triangle ABC \sim \triangle ADC$

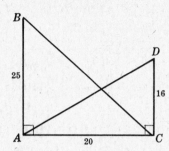

Solution:

(a) $\angle AEB = \angle DEC$, $\dfrac{3}{9} = \dfrac{4}{12}$

(b) $\angle A = \angle A$, $\dfrac{6}{12} = \dfrac{9}{18}$

(c) $\angle BAC = \angle ACD$, $\dfrac{20}{16} = \dfrac{25}{20}$

4.5 APPLYING PRINCIPLE 5

In each, determine the proportion needed to prove the indicated triangles similar.

(a) $\triangle ABC \sim \triangle DEF$

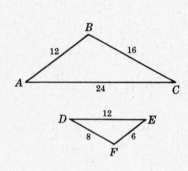

(b) $\triangle ABD \sim \triangle BDC$

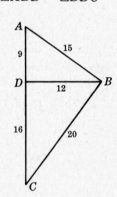

(c) $\triangle ABD \sim \triangle BEC$

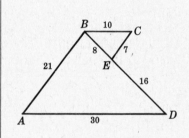

Solution:

(a) $\dfrac{6}{12} = \dfrac{8}{16} = \dfrac{12}{24}$

(b) $\dfrac{9}{12} = \dfrac{12}{16} = \dfrac{15}{20}$

(c) $\dfrac{7}{21} = \dfrac{8}{24} = \dfrac{10}{30}$

4.6 PROPORTIONS OBTAINED from SIMILAR TRIANGLES

Find x in each.

(a)

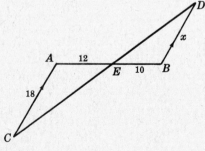

(b)

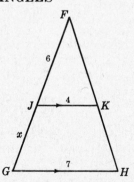

Solution:

(a) Since $BD \parallel AC$, $\angle A = \angle B$ and $\angle C = \angle D$; hence $\triangle AEC \sim \triangle DEB$. Then $\dfrac{x}{18} = \dfrac{10}{12}$ and $x = 15$.

(b) Since $JK \parallel GH$, $\triangle FJK \sim \triangle FGH$ by Pr. 7. Hence $\dfrac{6}{x+6} = \dfrac{4}{7}$ and $x = 4\frac{1}{2}$.

4.7   FINDING HEIGHTS USING GROUND SHADOWS

A tree casts a 15 ft. shadow at a time when a nearby upright pole of 6 ft. casts a shadow of 2 ft. Find the height of the tree if both tree and pole make right angles with the ground.

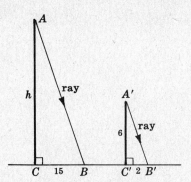

Solution:

At the same time in localities near each other, the rays of the sun strike the ground at equal angles; hence $\angle B = \angle B'$. Since the tree and the pole make right angles with the ground, $\angle C = \angle C'$. Hence $\triangle ABC \sim \triangle A'B'C'$, $\dfrac{h}{6} = \dfrac{15}{2}$ and $h = 45$.

4.8   PROVING a SIMILAR TRIANGLE PROBLEM STATED in WORDS

Prove: Isosceles triangles are similar if a base angle of one equals a base angle of the other.

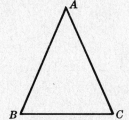

Given: Isosceles $\triangle ABC$ $(AB = AC)$
Isosceles $\triangle A'B'C'$ $(A'B' = A'C')$
$\angle B = \angle B'$

To Prove: $\triangle ABC \sim \triangle A'B'C'$

Plan: Prove $\angle C = \angle C'$ and use Pr. 3.

PROOF: Statements	Reasons
1. $\angle B = \angle B'$	1. Given
2. $\angle B = \angle C$, $\angle B' = \angle C'$	2. Base angles of an isosceles triangle are equal.
3. $\angle C = \angle C'$	3. Things equal to equal things are equal to each other.
4. $\triangle ABC \sim \triangle A'B'C'$	4. Two triangles are similar if two angles of one triangle equal two angles of the other.

4.9   PROVING a PROPORTION PROBLEM INVOLVING SIMILAR TRIANGLES

Given: Parallelogram $ABCD$

To Prove: $\dfrac{AE}{BC} = \dfrac{AF}{BF}$

Plan: Prove $\triangle AEF \sim \triangle BFC$

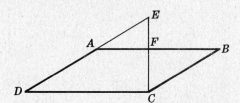

PROOF: Statements	Reasons
1. $ABCD$ is a parallelogram.	1. Given
2. $ED \parallel BC$	2. Opposite sides of a parallelogram are parallel.
3. $\angle DEC = \angle ECB$	3. Alternate interior angles of parallel lines are equal.
4. $\angle EFA = \angle BFC$	4. Vertical angles are equal.
5. $\triangle AEF \sim \triangle BFC$	5. Two triangles are similar if two angles of one triangle equal respectively two angles of the other.
6. $\dfrac{AE}{BC} = \dfrac{AF}{BF}$	6. Corresponding sides of similar triangles are in proportion.

5. *Extending a Basic Proportion Principle*

Pr. 1:   Corresponding *sides* of similar *triangles* are in proportion.

Pr. 2:   Corresponding *lines* of similar *triangles* are in proportion.

Pr. 3:   Corresponding *lines* of similar *polygons* are in proportion.

When "lines" replaces "sides", Pr. 1 becomes the more general Pr. 2.   When "polygons" replaces "triangles", Pr. 2 becomes the even more general Pr. 3.

By "lines" is meant straight, broken or curved lines, such as: altitudes, medians, angle bisectors, perimeters, radii of inscribed or circumscribed circles, circumferences of inscribed or circumscribed circles.

The *ratio of similitude* of similar polygons is the ratio of any pair of corresponding lines.

Corollaries of Pr. 2 and 3, such as the following, can be devised for any combination of corresponding lines:

(a) Corresponding *altitudes* of similar triangles have the same ratio as any two corresponding *medians*.

   Thus in Fig. 1, if $\triangle ABC \sim \triangle A'B'C'$, then $\dfrac{h}{h'} = \dfrac{m}{m'}$.

(b) *Perimeters* of similar polygons have the same ratio as any two corresponding *sides*.

   Thus in Fig. 2, if   quad. I $\sim$ quad. I', then $\dfrac{34}{17} = \dfrac{4}{2} = \dfrac{6}{3} = \dfrac{10}{5} = \dfrac{14}{7}$.

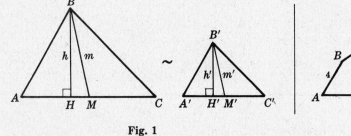

<table>
<tr><td>Fig. 1</td><td>Fig. 2</td></tr>
</table>

5.1 LINE RATIOS from SIMILAR TRIANGLES

(a) In similar triangles, if corresponding sides are in the ratio of $3:2$, find the ratio of corresponding medians.

(b) The sides of a triangle are 4, 6 and 7.   If the perimeter of a similar triangle is 51, find its longest side.

(c) In $\triangle ABC$, $BC = 25$ and the altitude to $BC = 10$.   A line segment terminating in the sides of the triangle is parallel to BC and 3 units from A.   Find its length.

(a)

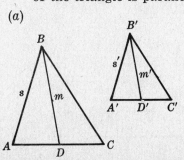

(b)

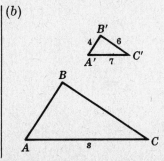

(c)

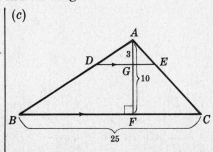

Solution:

(a) If $\triangle ABC \sim \triangle A'B'C'$ and $\dfrac{s}{s'} = \dfrac{3}{2}$, then $\dfrac{m}{m'} = \dfrac{3}{2}$.

(b) Perimeter of $\triangle A'B'C' = 4 + 6 + 7 = 17$. Since $\triangle ABC \sim \triangle A'B'C'$, $\dfrac{s}{7} = \dfrac{51}{17}$ and $s = 21$.

(c) Since $\triangle ADE \sim \triangle ABC$, $\dfrac{DE}{25} = \dfrac{3}{10}$ and $DE = 7\frac{1}{2}$.

5.2 LINE RATIOS from SIMILAR POLYGONS

Complete each statement.

(a) If corresponding sides of two similar polygons are in the ratio of $4:3$, then the ratio of their perimeters is (   ).

(b) The perimeters of two similar quadrilaterals are 30 and 24. If a side of the smaller quadrilateral is 8, the corresponding side of the larger is (   ).

(c) If each side of a pentagon is tripled and the angles remain the same, then each diagonal is (   ).

Solution:

(a) If the polygons are similar, then $\dfrac{p}{p'} = \dfrac{s}{s'} = \dfrac{4}{3}$.

(b) If the quadrilaterals are similar, then $\dfrac{s}{s'} = \dfrac{p}{p'}$, $\dfrac{s}{8} = \dfrac{30}{24}$ and $s = 10$.

(c) Tripled, since polygons are similar if their corresponding angles are equal and their corresponding sides are in proportion.

6. *Proving Equal Products of Lines*

In a problem, to prove that the product of two lines equals the product of another pair of lines, it is necessary first to set up the proportion which will lead to the two equal products.

6.1 PROVING an EQUAL PRODUCTS PROBLEM

Prove: If two secants intersect outside a circle, the product of one of the secants and its external segment equals the product of the other secant and its external segment.

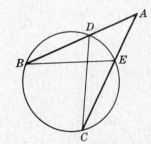

Given: Secants AB and AC.

To Prove: $AB \times AD = AC \times AE$

Plan: Prove $\triangle ABE \sim \triangle ACD$ to obtain $\dfrac{AB}{AC} = \dfrac{AE}{AD}$.

PROOF: Statements	Reasons
1. Draw BE and CD.	1. A straight line may be drawn between any two points.
2. $\angle A = \angle A$	2. Identity.
3. $\angle B = \angle C$	3. Angles inscribed in the same arc are equal.
4. $\triangle AEB \sim \triangle ADC$	4. Two triangles are similar if two angles of one triangle equal respectively two angles of the other.
5. $\dfrac{AB}{AC} = \dfrac{AE}{AD}$	5. Corresponding sides of similar triangles are in proportion.
6. $AB \times AD = AC \times AE$	6. In a proportion, the product of the means equals the product of the extremes.

7. *Lines Intersecting Inside and Outside a Circle*

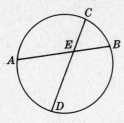

Fig. 1

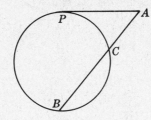

Fig. 2

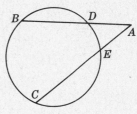

Fig. 3

Pr. 1: *If two chords intersect in a circle the product of the segments of one chord equals the product of the segments of the other.*

Thus in Fig. 1, $AE \times EB = CE \times ED$.

Pr. 2: *If a tangent and a secant intersect outside a circle, the tangent is the mean proportional between the secant and its external segment.*

Thus in Fig. 2, if PA is a tangent, $\dfrac{AB}{AP} = \dfrac{AP}{AC}$.

Pr. 3: *If two secants intersect outside a circle, the product of one of the secants and its external segment equals the product of the other secant and its external segment.*

Thus in Fig. 3, $AB \times AD = AC \times AE$.

7.1 APPLYING PRINCIPLE 1

Find x in each, if chords AB and CD intersect in E.

(a)

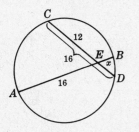

(b)

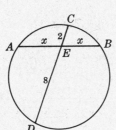

(c)

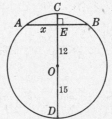

Diameter $CD \perp AB$

Solution:

(a) $ED = 4$.   Then $16x = 4(12)$, $16x = 48$, $x = 3$.

(b) $AE = EB = x$.   Then $x^2 = 8(2)$, $x^2 = 16$ and $x = 4$.

(c) $CE = 3$, $AE = EB = x$.   Then $x^2 = 27(3)$, $x^2 = 81$ and $x = 9$.

7.2 APPLYING PRINCIPLE 2

Find x in each, if tangent AP and secant AB intersect at A.

(a)

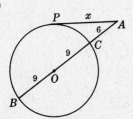

(b)

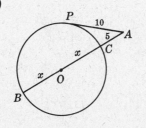

(c)

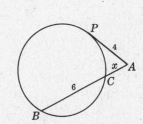

Solution:

(a) $AB = 9+9+6 = 24$. Then $x^2 = 24(6)$, $x^2 = 144$ and $x = 12$.

(b) $AB = 2x+5$. Then $5(2x+5) = 100$ and $x = 7\frac{1}{2}$.

(c) $AB = x+6$. Then $x(x+6) = 16$, $x^2+6x-16 = 0$, $(x+8)(x-2) = 0$ and $x = 2$.

7.3 APPLYING PRINCIPLE 3

Find x in each, if secants AB and AC intersect in A.

(a)

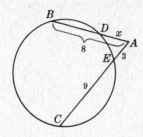

(b)

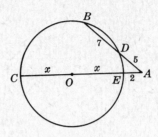

Solution:

(a) $AC = 12$. Then $8x = 12(3)$ and $x = 4\frac{1}{2}$.

(b) $AC = 2x+2$, $AB = 12$. Then $2(2x+2) = 12(5)$ and $x = 14$.

8. Mean Proportionals in a Right Triangle

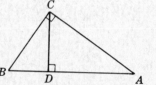

Pr. 1: *The altitude to the hypotenuse of a right triangle is the mean proportional between the segments of the hypotenuse.*

Thus in right $\triangle ABC$, $\dfrac{BD}{CD} = \dfrac{CD}{DA}$.

Pr. 2: *In a right triangle, either leg is the mean proportional between the hypotenuse and the projection of that leg on the hypotenuse.*

Thus in right $\triangle ABC$, $\dfrac{AB}{BC} = \dfrac{BC}{BD}$ and $\dfrac{AB}{AC} = \dfrac{AC}{AD}$.

A proof of this principle is given in Chapter 16.

8.1 FINDING MEAN PROPORTIONALS in a RIGHT TRIANGLE

In each, find x and y.

(a)

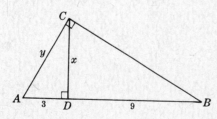

(b)

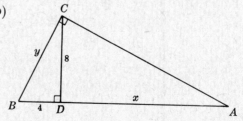

Solution:

(a) By Pr. 1, $\dfrac{3}{x} = \dfrac{x}{9}$, $x^2 = 27$ and $x = 3\sqrt{3}$. By Pr. 2, $\dfrac{12}{y} = \dfrac{y}{3}$, $y^2 = 36$ and $y = 6$.

(b) By Pr. 1, $\dfrac{x}{8} = \dfrac{8}{4}$ and $x = 16$. By Pr. 2, $\dfrac{20}{y} = \dfrac{y}{4}$, $y^2 = 80$ and $y = 4\sqrt{5}$.

9. Law of Pythagoras

In a right triangle, the square of the hypotenuse equals the sum of the squares of the legs.

Thus in Fig. 1, $c^2 = a^2 + b^2$.

A proof of the Law of Pythagoras is given in Chapter 16.

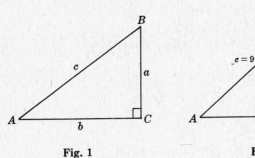

Fig. 1　　　　　　　　　　Fig. 2　　　　　　　　　　Fig. 3

A. Tests for Right, Acute, and Obtuse Triangles:

1. If $c^2 = a^2 + b^2$ applies to the three sides of a triangle, then the triangle is a right triangle; but if $c^2 \neq a^2 + b^2$, then the triangle is not a right triangle.

2. In $\triangle ABC$, if $c^2 < a^2 + b^2$ where c is the longest side of the triangle, then the triangle is an acute triangle.

Thus in Fig. 2, $9^2 < 6^2 + 8^2$ or $81 < 100$; hence $\triangle ABC$ is an acute triangle.

3. In $\triangle ABC$, if $c^2 > a^2 + b^2$ where c is the longest side of the triangle, then the triangle is an obtuse triangle.

Thus in Fig. 3, $11^2 > 6^2 + 8^2$ or $121 > 100$; hence $\triangle ABC$ is an obtuse triangle.

9.1　FINDING the SIDES of a RIGHT TRIANGLE

In the right triangle shown (a) find hypotenuse c if $a = 12$ and $b = 9$, (b) find arm a if $b \doteq 6$ and $c = 8$, (c) find arm b if $a = 4\sqrt{3}$ and $c = 8$.

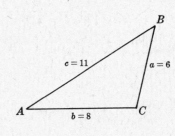

Solution:

(a) $c^2 = a^2 + b^2 = 12^2 + 9^2 = 225$ and $c = 15$.

(b) $a^2 = c^2 - b^2 = 8^2 - 6^2 = 28$ and $a = 2\sqrt{7}$.

(c) $b^2 = c^2 - a^2 = 8^2 - (4\sqrt{3})^2 = 64 - 48 = 16$ and $b = 4$.

9.2　RATIOS in a RIGHT TRIANGLE

In a right triangle, the hypotenuse is 20 and the ratio of the two arms is $3:4$. Find each arm.

Solution:

Let the two arms be denoted by $3x$ and $4x$.

Then $20^2 = (3x)^2 + (4x)^2$, $400 = 9x^2 + 16x^2$, $400 = 25x^2$ and $x = 4$; hence the arms are 12 and 16.

9.3 APPLYING the LAW of PYTHAGORAS to an ISOSCELES TRIANGLE

Find the altitude to the base of an isosceles triangle if the base is 8 and the equal sides are 12.

Solution:

The altitude h of an isosceles triangle bisects the base.

Then $h^2 = a^2 - (\tfrac{1}{2}b)^2 = 12^2 - 4^2 = 128$ and $h = 8\sqrt{2}$.

9.4 APPLYING the LAW of PYTHAGORAS to a RHOMBUS

In a rhombus, find (a) a side s if the diagonals are 30 and 40, (b) a diagonal d if a side is 26 and the other diagonal is 20.

Solution:

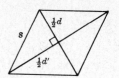

The diagonals of a rhombus are perpendicular bisectors of each other; hence $s^2 = (\tfrac{1}{2}d)^2 + (\tfrac{1}{2}d')^2$.

(a) $d = 30,\ d' = 40$. Then $s^2 = 15^2 + 20^2 = 625,\ s = 25$.

(b) $s = 26,\ d' = 20$. Then $26^2 = (\tfrac{1}{2}d)^2 + 10^2,\ 576 = (\tfrac{1}{2}d)^2,\ \tfrac{1}{2}d = 24,\ d = 48$.

9.5 APPLYING the LAW of PYTHAGORAS to a TRAPEZOID

Find x in each, if $ABCD$ is a trapezoid.

(a)

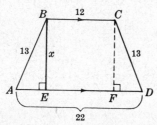

(b)

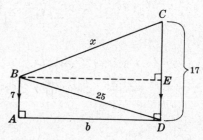

Solution:

The dotted perpendicular lines shown in each diagram are additional lines needed only for solution. Note how rectangles are formed by these added lines.

(a) $EF = BC = 12,\ AE = \tfrac{1}{2}(22 - 12) = 5$. Then $x^2 = 13^2 - 5^2 = 144,\ x = 12$.

(b) $b^2 = 25^2 - 7^2 = 576,\ b = 24;\quad BE = b = 24,\ CE = 17 - 7 = 10$. Then $x^2 = 24^2 + 10^2,\ x = 26$.

9.6 APPLYING the LAW of PYTHAGORAS to a CIRCLE

(a) Find the distance d from the center of a circle of radius 17 to a chord whose length is 30.

(b) Find the length of a common external tangent to two externally tangent circles with radii 4 and 9.

(a)

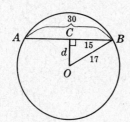

(b)

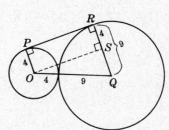

Solution:

(a) $BC = \tfrac{1}{2}(30) = 15$. Then $d^2 = 17^2 - 15^2 = 64$ and $d = 8$.

(b) $OS = PR,\ RS = 4,\ OQ = 13,\ SQ = 9 - 4 = 5$.

In right $\triangle OSQ$, $(OS)^2 = 13^2 - 5^2 = 144$ and $OS = 12$; hence $PR = 12$.

10. *Special Right Triangles*

1. 30°-60°-90° TRIANGLE

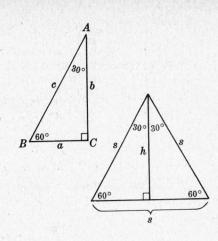

A 30°-60°-90° triangle is one-half an equilateral triangle.

Thus in right $\triangle ABC$, $a = \frac{1}{2}c$.

Consider that $c = 2$; then $a = 1$, and applying the Law of Pythagoras,

$$b^2 = c^2 - a^2 = 2^2 - 1^2 = 3, \quad b = \sqrt{3}$$

and the ratio of the sides is

$$a:b:c = 1:\sqrt{3}:2$$

A. *Principles of the 30°-60°-90° Triangle*

Pr. 1: *The leg opposite the 30° angle equals one-half the hypotenuse, i.e. $a = \frac{1}{2}c$.*

Pr. 2: *The leg opposite the 60° angle equals one-half the hypotenuse times the square root of 3, i.e. $b = \frac{1}{2}c\sqrt{3}$.*

Pr. 3: *The leg opposite the 60° angle equals the leg opposite the 30° angle times the square root of 3, i.e. $b = a\sqrt{3}$.*

B. *Equilateral Triangle Principle*

Pr. 4: *The altitude of an equilateral triangle equals one-half a side times the square root of 3, i.e. $h = \frac{1}{2}s\sqrt{3}$.* (Pr. 4 is a corollary of Pr. 2.)

2. 45°-45°-90° TRIANGLE

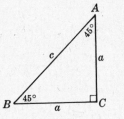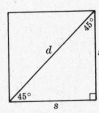

A 45°-45°-90° triangle is one-half a square.

In right triangle ABC, $c^2 = a^2 + a^2$ or $c = a\sqrt{2}$. Hence the ratio of the sides is

$$a:a:c = 1:1:\sqrt{2}$$

C. *Principles of the 45°-45°-90° Triangle*

Pr. 5: *The leg opposite a 45° angle equals one-half the hypotenuse times the square root of 2, i.e. $a = \frac{1}{2}c\sqrt{2}$.*

Pr. 6: *The hypotenuse equals a side times the square root of 2, i.e. $c = a\sqrt{2}$.*

D. *Square Principle*

Pr. 7: *In a square, a diagonal equals a side times the square root of 2, i.e. $d = s\sqrt{2}$.*

10.1 APPLYING PRINCIPLES 1 to 4

(a) If the hypotenuse of a 30°-60°-90° triangle is 12, find its legs.

(b) Each leg of an isosceles trapezoid is 18. If the base angles are 60° and the upper base is 10, find the altitude and the lower base.

(a) 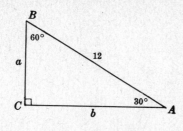   (b)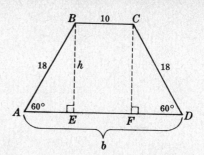

Solution:

(a) By Pr. 1, $a = \frac{1}{2}(12) = 6$. By Pr. 2, $b = \frac{1}{2}(12)\sqrt{3} = 6\sqrt{3}$.

(b) By Pr. 2, $h = \frac{1}{2}(18)\sqrt{3} = 9\sqrt{3}$. By Pr. 1, $AE = FD = \frac{1}{2}(18) = 9$; hence $b = 9 + 10 + 9 = 28$.

10.2 APPLYING PRINCIPLES 5 and 6

(a) Find the leg of an isosceles right triangle whose hypotenuse is 28.

(b) An isosceles trapezoid has base angles of 45°. If the upper base is 12 and the altitude is 3, find the lower base and each leg.

(a) 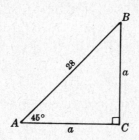   (b)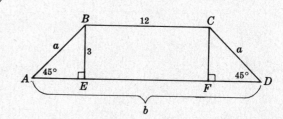

Solution:

(a) By Pr. 5, $a = \frac{1}{2}(28)\sqrt{2} = 14\sqrt{2}$.

(b) By Pr. 6, $a = 3\sqrt{2}$. $AE = BE = 3$ and $EF = 12$; hence $b = 3 + 12 + 3 = 18$.

Supplementary Problems

1. Express each ratio in lowest terms: (1.1)

   (a) 20¢ to 5¢   (f) 50% to 25%   (k) $\frac{1}{2}$ lb. to $\frac{1}{4}$ lb.

   (b) 5 dimes to 15 dimes   (g) 15° to 75°   (l) $2\frac{1}{2}$ da. to $3\frac{1}{2}$ da.

   (c) 30 lb. to 25 lb.   (h) 33% to 77%   (m) 5 ft. to $\frac{1}{4}$ ft.

   (d) 20° to 14°   (i) $2.20 to $3.30   (n) $\frac{1}{2}$ yd. to $1\frac{1}{2}$ yd.

   (e) 27 min. to 21 min.   (j) $.84 to $.96   (o) $16\frac{1}{2}$ ft. to $5\frac{1}{2}$ ft.

2. Express each ratio in lowest terms: (1.2)

   (a) 1 yr. to 2 mo.   (e) 2 yd. to 2 ft.   (i) 100 lb. to 1 ton

   (b) 2 wk. to 5 da.   (f) $2\frac{1}{3}$ yd. to 2 ft.   (j) $2 to 25¢

   (c) 3 da. to 3 wk.   (g) $1\frac{1}{2}$ ft. to 9 in.   (k) 2 quarters to 3 dimes

   (d) $\frac{1}{2}$ hr. to 20 min.   (h) 2 lb. to 8 oz.   (l) 1 sq. yd. to 2 sq. ft.

3. Express each ratio in lowest terms: *(1.3)*
 (a) 20¢ to 30¢ to $1
 (b) $3 to $1.50 to 25¢
 (c) 1 quarter to 1 dime to 1 nickel
 (d) 1 da. to 4 da. to 1 wk.
 (e) $\frac{1}{2}$ da. to 9 hr. to 3 hr.
 (f) 2 hr. to $\frac{1}{2}$ hr. to 15 min.
 (g) 1 ton to 200 lb. to 40 lb.
 (h) 3 lb. to 1 lb. to 8 oz.
 (i) 1 gal. to 1 qt. to 1 pt.

4. Express each ratio in lowest terms: *(1.4)*
 (a) 60 to 70
 (b) 84 to 7
 (c) 65 to 15
 (d) 125 to 500
 (e) 630 to 105
 (f) 1760 to 990
 (g) .7 to 2.1
 (h) .36 to .24
 (i) .002 to .007
 (j) .055 to .005
 (k) 6.4 to 8
 (l) 144 to 2.4
 (m) $7\frac{1}{2}$ to $2\frac{1}{2}$
 (n) $1\frac{1}{2}$ to 10
 (o) $\frac{5}{6}$ to $1\frac{2}{3}$
 (p) $\frac{7}{4}$ to $\frac{1}{8}$

5. Express each ratio in lowest terms: *(1.4)*
 (a) x to $8x$
 (b) $15c$ to 5
 (c) $11d$ to 22
 (d) $2\pi r$ to πD
 (e) πab to πa^2
 (f) $4S$ to S^2
 (g) S^3 to $6S^2$
 (h) $9r^2t$ to $6rt^2$
 (i) x to $4x$ to $10x$
 (j) $15y$ to $10y$ to $5y$
 (k) x^3 to x^2 to x
 (l) $12w$ to $10w$ to $8w$ to $2w$

6. Using x as their common factor, represent the numbers and their sum if   *(1.4)*
 (a) two numbers have a ratio of $5:4$,
 (b) two numbers have a ratio of 9 to 1,
 (c) three numbers have a ratio of $2:5:11$,
 (d) five numbers have a ratio of $1:2:2:3:7$.

7. If two angles in the ratio of $5:4$ are represented by $5x$ and $4x$, express each statement as an equation; then find x and the angles: *(1.5)*
 (a) The angles are adjacent and form an angle of 45°.
 (b) The angles are complementary.
 (c) The angles are supplementary.
 (d) The angles are two angles of a triangle whose third angle is their difference.

8. If three angles in the ratio of $7:6:5$ are represented by $7x$, $6x$ and $5x$, express each statement as an equation; then find x and the angles: *(1.6)*
 (a) The first and second are adjacent and form an angle of 91°.
 (b) The first and third are supplementary.
 (c) The first and one-half the second are complementary.
 (d) The angles are the three angles of a triangle.

9. Solve for x: *(2.1)*
 (a) $x:6 = 8:3$
 (b) $5:4 = 20:x$
 (c) $9:x = x:4$
 (d) $x:2 = 10:x$
 (e) $(x+4):3 = 3:(x-4)$
 (f) $(2x+8):(x+2) = (2x+5):(x+1)$
 (g) $a:b = c:x$
 (h) $x:2y = 18y:x$

10. Solve for x: *(2.1)*
 (a) $\dfrac{5}{7} = \dfrac{15}{x}$
 (b) $\dfrac{7}{x} = \dfrac{3}{2}$
 (c) $\dfrac{3}{x} = \dfrac{x}{12}$
 (d) $\dfrac{x}{5} = \dfrac{15}{x}$
 (e) $\dfrac{x+2}{5} = \dfrac{6}{3}$
 (f) $\dfrac{x-1}{3} = \dfrac{5}{x+1}$
 (g) $\dfrac{2x}{x+7} = \dfrac{3}{5}$
 (h) $\dfrac{a}{x} = \dfrac{x}{b}$

11. Find the fourth proportional to each set of numbers: *(2.2)*
 (a) 1, 3, 5
 (b) 8, 6, 4
 (c) 2, 3, 4
 (d) 3, 4, 2
 (e) 3, 2, 5
 (f) $\frac{1}{3}$, 2, 5
 (g) 2, 8, 8
 (h) $b, 2a, 3b$

12. Find the positive mean proportional between each pair of numbers: *(2.3)*
 (a) 4 and 9, (b) 12 and 3, (c) $\frac{1}{3}$ and 27, (d) $2b$ and $8b$, (e) 2 and 5, (f) 3 and 9, (g) p and q.

13. In each, form a proportion whose fourth term is x: (2.4)

(a) $cx = bd$,   (b) $pq = ax$,   (c) $hx = a^2$,   (d) $3x = 7$,   (e) $x = ab/c$.

14. In each, find the ratio of x to y: (2.4)

(a) $2x = y$,   (b) $3y = 4x$,   (c) $x = \frac{1}{2}y$,   (d) $ax = hy$,   (e) $x = by$.

15. Which of the following is not a proportion?

(a) $\frac{4}{3} \overset{?}{=} \frac{24}{18}$,   (b) $\frac{3}{5} \overset{?}{=} \frac{7}{12}$,   (c) $\frac{25}{45} \overset{?}{=} \frac{10}{18}$,   (d) $\frac{.2}{.3} \overset{?}{=} \frac{6}{9}$,   (e) $\frac{x}{8} \overset{?}{=} \frac{3}{4}$ when $x = 6$.

16. In each, form a new proportion whose first term is x. Then find x: (2.5)

(a) $\frac{3}{2} = \frac{9}{x}$ (b) $\frac{1}{x} = \frac{5}{4}$ (c) $\frac{a}{x} = \frac{2}{b}$ (d) $\frac{x+5}{5} = \frac{11}{10}$ (e) $\frac{x-20}{20} = \frac{1}{4}$

17. Find x in each:

(a) $a:b = c:x$ and $a:b = c:d$ (c) $2:3x = 4:5y$ and $2:15 = 4:5y$

(b) $5:7 = x:42$ and $5:7 = 35:42$ (d) $7:5x-2 = 14:3y$ and $7:18 = 14:3y$

18. Find x in each: (2.6)

(a) $\frac{x-7}{8} = \frac{7}{4}$ (b) $\frac{x+y}{6} = \frac{x-y}{3} = \frac{1}{3}$ (c) $\frac{2x-y}{8} = \frac{y-1}{10} = \frac{1}{2}$

19. Find x in each: (3.1)

(a)   (b)   (c)

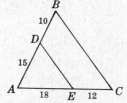 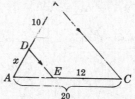 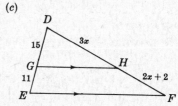

20. In which of the following is a line parallel to one side of the triangle?

(a)   (b)   (c)

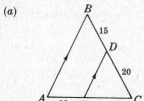 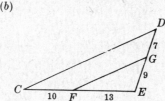 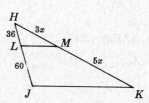

21. Find x in each: (3.2)

(a)   (b)   (c)

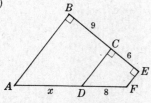 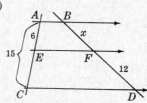 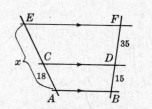

22. Find x in each: (3.3)

(a)   (b)   (c)

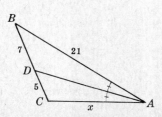 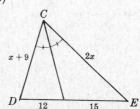 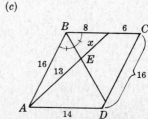

23. Prove: Three or more parallel lines divide any two transversals proportionately.   (3.4)

24. In similar triangles ABC and $A'B'C'$, $\angle B$ and
    $\angle B'$ are corresponding angles. Find $\angle B$ if
    (a) $\angle A' = 120°$ and $\angle C' = 25°$
    (b) $\angle A' + \angle C' = 127°$.

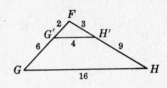

25. In similar triangles ABC and $A'B'C'$, $\angle A =$
    $\angle A'$ and $\angle B = \angle B'$.   (4.1)
    (a) Find a if $c = 24$.
    (b) Find b if $a = 20$.
    (c) Find c if $b = 63$.

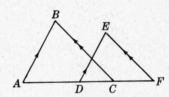

26. In each, show that the indicated triangles are similar.

(a) $\triangle ADE \sim \triangle ABC$ (b) $\triangle RQP \sim \triangle RQ'P'$ (c) $\triangle FG'H' \sim \triangle FGH$

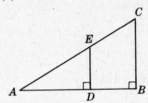 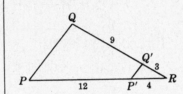

27. In each, two pairs of equal angles can be used to prove the indicated triangles similar. Determine the
    equal angles.   (4.2)

(a) $\triangle AEB \sim \triangle DEC$ (b) $\triangle BFA \sim \triangle AED$ (c) $\triangle ABC \sim \triangle DEF$

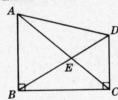 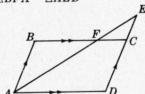 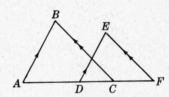

(d) $\triangle AEB \sim \triangle ADC$ (e) $\triangle ADE \sim \triangle ACB$ (f) $\triangle ABC \sim \triangle ABD$

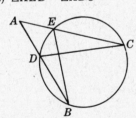 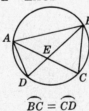 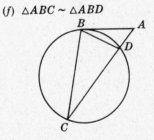

$\stackrel{\frown}{BC} = \stackrel{\frown}{CD}$

28. In each, determine the angles that can be used to prove the indicated triangles similar.   (4.3)

(a) $\triangle AED \sim \triangle FGB$ (b) $\triangle ABD \sim \triangle ABC$ (c) $\triangle AEB \sim \triangle AED$

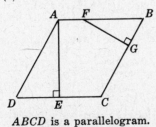 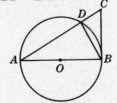 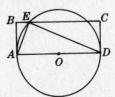

ABCD is a parallelogram.   AB is a diameter.   AD is a diameter.
                                  BC is a tangent.   ABCD is a rectangle.

29. In each, determine the pair of equal angles and the proportion needed to prove the indicated triangles similar.

(4.4)

(a) $\triangle ABC \sim \triangle DEF$

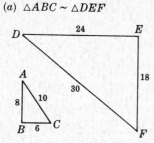

(b) $\triangle ADE \sim \triangle ABC$

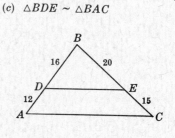

(c) $\triangle BDE \sim \triangle BAC$

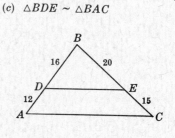

30. In each, indicate the proportion needed to prove the indicated triangles similar.

(4.5)

(a) $\triangle ABC \sim \triangle DEF$

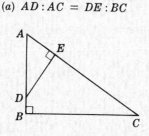

(b) $\triangle ADE \sim \triangle ABC$

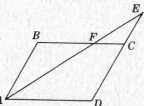

(c) $\triangle DEF \sim \triangle ABC$

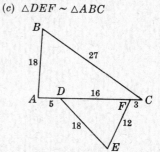

31. In each, prove the indicated proportion.

(4.9)

(a) $AD:AC = DE:BC$

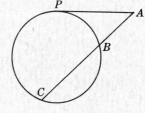

(b) $AB:EC = BF:FC$

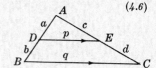

ABCD is a parallelogram.

(c) $AC:AP = AP:AB$

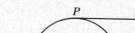

AP is a tangent.

32. In $\triangle ABC$, $DE \parallel BC$.

(4.6)

(a) $a = 4$, $AB = 8$, $p = 10$. Find q.
(b) $c = 5$, $AC = 15$, $q = 24$. Find p.
(c) $a = 7$, $p = 11$, $q = 22$. Find b.
(d) $b = 9$, $p = 20$, $q = 35$. Find a.
(e) $a = 10$, $p = 24$, $q = 84$. Find AB.
(f) $c = 3$, $p = 4$, $q = 7$. Find d.

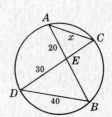

33. Find x in each.

(4.6)

(a)

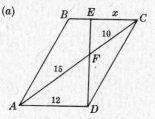

ABCD is a parallelogram.

(b)

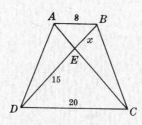

ABCD is a trapezoid.

(c)

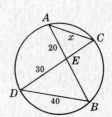

34. A 7 ft. upright pole near a vertical tree casts a 6 ft. shadow. At that time, (a) find the height of the tree if its shadow is 36 ft, (b) find the shadow of the tree if its height is 77 ft.

(4.7)

35. Prove each of the following.   (4.9)

    (a) In $\triangle ABC$, if AD and CE are altitudes, then $AD:CE = AB:BC$.

    (b) In circle O, diameter AB and tangent BC are sides of $\triangle ABC$. If AC intersects the circle in D, then $AD:AB = AB:AC$.

    (c) The diagonals of a trapezoid divide each other into proportional segments.

    (d) In right $\triangle ABC$, CD is the altitude to the hypotenuse AB; then $AC:CD = AB:BC$.

36. Prove each of the following.   (4.8)

    (a) A line parallel to one side of a triangle cuts off a triangle similar to the given triangle.

    (b) Isosceles right triangles are similar to each other.

    (c) Equilateral triangles are similar to each other.

    (d) The bases of a trapezoid form similar triangles with the segments of the diagonals.

37. Complete each statement.   (5.1)

    (a) In similar triangles, if corresponding sides are in the ratio of $8:5$, then corresponding altitudes are in the ratio of ( ).

    (b) In similar triangles, if corresponding angle bisectors are in the ratio of $3:5$, then their perimeters are in the ratio of ( ).

    (c) If the sides of a triangle are halved, then the perimeter is ( ), the angle bisectors are ( ), the medians are ( ) and the radii of the circumscribed circle are ( ).

38. (a) Corresponding sides of two similar triangles are 18 and 12. If an altitude of the smaller is 10, find the corresponding altitude of the larger.   (5.1)

    (b) Corresponding medians of two similar triangles are 25 and 15. Find the perimeter of the larger if the perimeter of the smaller is 36.

    (c) The sides of a triangle are 5, 7 and 8. If the perimeter of a similar triangle is 100, find its sides.

    (d) The bases of a trapezoid are 5 and 20 and the altitude is 12. Find the altitude of the triangle formed by the shorter base and the non-parallel sides extended to meet.

    (e) The bases of a trapezoid are 11 and 22. Its altitude is 9. Find the distance from the point of intersection of the diagonals to each of the bases.

39. Complete each statement.   (5.2)

    (a) If corresponding sides of two similar polygons are in the ratio of $3:7$, then the ratio of their corresponding altitudes is ( ).

    (b) If the perimeters of two similar hexagons are in the ratio of 56 and 16, then the ratio of their corresponding diagonals is ( ).

    (c) If each side of an octagon is quadrupled and the angles remain the same, then its perimeter is ( ).

    (d) The base of a rectangle is twice that of a similar rectangle. If the radius of the circumscribed circle of the first rectangle is 14, then the radius of the circumscribed circle of the second is ( ).

40. Prove each of the following.

    (a) Corresponding angle bisectors of two similar triangles have the same ratio as a pair of corresponding sides.

    (b) Corresponding medians of similar triangles have the same ratio as a pair of corresponding sides.

41. Prove each of the following.   (6.1)

    (a)

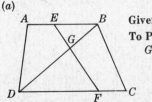

    Given: Trapezoid $ABCD$
    To Prove:
        $GB \times DF = GD \times EB$

    (b)

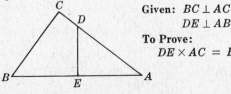

    Given: $BC \perp AC$
           $DE \perp AB$
    To Prove:
        $DE \times AC = BC \times AE$

(c)

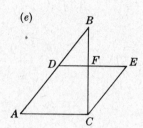

Given: Diameter BC
         $DE \perp BC$
To Prove:
         $\overline{BD}^2 = BE \times BC$

(d)

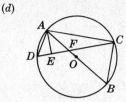

Given: Circle O
         Diameter AB
         $AE \perp CD$
To Prove:
         $AD \times BC = AB \times DE$

(e)

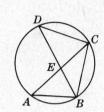

Given: $AB \parallel CE$
         $AC \perp BC$
         $DE \perp BC$
To Prove:
         $AB \times CF = BC \times EC$

(f)

Given: $\overset{\frown}{BC} = \overset{\frown}{CD}$
To Prove:
         $\overline{BC}^2 = AC \times EC$

42. Prove each of the following.   (6.1)

    (a) If two chords intersect in a circle, the product of the segments of one chord equals the product of the segments of the other.

    (b) In a right triangle, the product of the hypotenuse and the altitude upon it equals the product of the legs.

    (c) If in inscribed $\triangle ABC$ the bisector of $\angle A$ intersects BC in D and the circle in E, then $BD \times AC = AD \times EC$.

43. (a) $AE = 10$, $EB = 6$, $CE = 12$. Find ED.   (e) $OD = 10$, $OE = 8$. Find AB.   (7.1)
    (b) $AB = 15$, $EB = 8$, $ED = 4$. Find CE.   (f) $AB = 24$, $OE = 5$. Find OD.
    (c) $AE = 6$, $ED = 4$, $CD = 13$. Find EB.   (g) $OD = 25$, $EC = 18$. Find AB.
    (d) $ED = 5$, $EB = 2(AE)$, $CD = 15$. Find AE.   (h) $AB = 8$, $OD = 5$. Find EC.

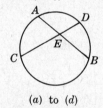

(a) to (d)

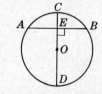

Diameter $CD \perp$ chord AB

(e) to (h)

44. A point is 12 inches from the center of a circle whose radius is 15 inches. Find the longest and shortest chords that can be drawn through this point. (Hint: The longest chord is a diameter and the shortest chord is perpendicular to this diameter.)   (7.1)

45. (a) $AC = 16$, $AD = 4$. Find AB.   (f) $AD = 6$, $OD = 9$. Find AB.   (7.2)
    (b) $CD = 5$, $AD = 4$. Find AB.   (g) $AD = 2$, $AB = 8$. Find CD.
    (c) $AB = 6$, $AD = 3$. Find AC.   (h) $AD = 5$, $AB = 10$. Find OD.
    (d) $AC = 20$, $AB = 10$. Find AD.   (i) $AB = 12$, $AC = 18$. Find OD.
    (e) $AB = 12$, $AD = 9$. Find CD.   (j) $OD = 5$, $AB = 12$. Find AD.

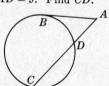

AB is a tangent.

(a) to (e)

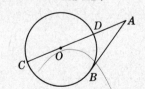

CD is a diameter, AB is a tangent.

(f) to (j)

46. (a) $AB = 14$, $AD = 4$, $AE = 7$. Find AC.

   (b) $AC = 8$, $AE = 6$, $AD = 3$. Find BD.

   (c) $BD = 5$, $AD = 7$, $AE = 4$. Find AC.

   (d) $AD = DB$, $EC = 14$, $AE = 4$. Find AD.

   (e) $OC = 3$, $AE = 6$, $AD = 8$. Find AB.   (7.3)

   (f) $BD = 7$, $AD = 5$, $AE = 2$. Find OC.

   (g) $OC = 11$, $AB = 15$, $AD = 5$. Find AE.

   (h) $OC = 5$, $AE = 6$, $BD = 4$. Find AD.

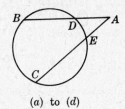

(a) to (d)

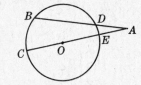

CE is a diameter.

(e) to (h)

47. CD is the altitude to the hypotenuse AB.   (8.1)

   (a) If $p = 2$ and $q = 6$, find a and h.

   (b) If $p = 4$ and $a = 6$, find c and h.

   (c) If $p = 16$ and $h = 8$, find q and b.

   (d) If $b = 12$ and $q = 6$, find p and h.

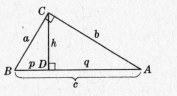

48. In a right triangle whose arms are a and b, find the hypotenuse c when: (9.1)

   (a) $a = 15$, $b = 20$ (b) $a = 15$, $b = 36$ (c) $a = 5$, $b = 4$ (d) $a = 5$, $b = 5\sqrt{3}$ (e) $a = 7$, $b = 7$.

49. In the right triangle shown, find each missing arm when: (9.1)

   (a) $a = 12$, $c = 20$ (c) $b = 15$, $c = 17$ (e) $a = 5\sqrt{2}$, $c = 10$

   (b) $b = 6$, $c = 8$ (d) $a = 2$, $c = 4$ (f) $a = \sqrt{5}$, $c = 2\sqrt{2}$

50. Find the arms of a right triangle whose hypotenuse is c if these arms have a ratio of   (9.2)

   (a) $3:4$ and $c = 15$, (b) $5:12$ and $c = 26$, (c) $8:15$ and $c = 170$, (d) $1:2$ and $c = 10$.

51. In a rectangle, find the diagonal if its sides are   (a) 9 and 40, (b) 5 and 10.   (9.1)

52. In a rectangle, find one side if the diagonal is 15 and the other side is   (a) 9, (b) 5, (c) 10.   (9.1)

53. Using the three sides given, which triangles are right triangles?

   (a) 33, 55, 44   (c) $4, 7\frac{1}{2}, 8\frac{1}{2}$ (e) 5 in., 1 ft., 1 ft. 1 in.   (g) 11 mi., 60 mi., 61 mi.

   (b) 120, 130, 50   (d) 25, 7, 24   (f) 1 yd., 1 yd. 1 ft., 1 yd. 2 ft.   (h) 5 rd., 5 rd., 7 rd.

54. Is a triangle a right triangle if its sides have the ratio of   (a) $3:4:5$, (b) $2:3:4$?

55. Find the altitude of an isosceles triangle if each of its two equal sides is 10 and its base is   (9.3)

   (a) 12, (b) 16, (c) 18, (d) 10.

56. In a rhombus, find a side if the diagonals are   (a) 18 and 24, (b) 4 and 8, (c) 6 and $6\sqrt{3}$.   (9.4)

57. In a rhombus, find a diagonal if a side and the other diagonal are respectively   (9.4)

   (a) 10 and 12, (b) 17 and 16, (c) 4 and 4, (d) 10 and $10\sqrt{3}$.

58. In isosceles trapezoid $ABCD$,   (9.5)

   (a) find a if $b = 32$, $b' = 20$ and $h = 8$.

   (b) find h if $b = 24$, $b' = 14$ and $a = 13$.

   (c) find b if $a = 15$, $b' = 10$ and $h = 12$.

   (d) find b' if $a = 6$, $b = 21$ and $h = 3\sqrt{3}$.

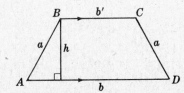

59. In trapezoid *ABCD*,   (9.5)

    (*a*) find *d* if *a* = 11, *b* = 3 and *c* = 15.

    (*b*) find *a* if *d* = 20, *b* = 12 and *c* = 36.

    (*c*) find *d* if *a* = 5, *p* = 13 and *c* = 14.

    (*d*) find *p* if *a* = 20, *c* = 28 and *d* = 17.

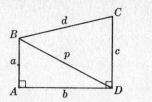

60. The radius of a circle is 15. Find   (9.6)

    (*a*) the distance from its center to a chord whose length is 18,

    (*b*) the length of a chord whose distance from its center is 9.

61. In a circle, a chord whose length is 16 is at a distance of 6 from the center. Find the length of a chord whose distance from the center is 8.   (9.6)

62. Two externally tangent circles have radii of 25 and 9. Find the length of a common external tangent.   (9.6)

63. In a 30°-60°-90° triangle, find   (10.1)

    (*a*) the legs if the hypotenuse is 20.

    (*b*) the other leg and hypotenuse if the leg opposite 30° is 7.

    (*c*) the other leg and hypotenuse if the leg opposite 60° is $5\sqrt{3}$.

64. In an equilateral triangle, find   (10.1)

    (*a*) the altitude if the hypotenuse is 22; 2*a*.

    (*b*) the side if the altitude is $24\sqrt{3}$; 24.

65. In a rhombus which has an angle of 60°, find   (10.1)

    (*a*) the diagonals if a side is 25.

    (*b*) the side and larger diagonal if the smaller diagonal is 35.

66. In an isosceles trapezoid which has base angles of 60°, find   (10.1)

    (*a*) the lower base and altitude if the upper base is 12 and the legs are 16.

    (*b*) the upper base and altitude if the lower base is 45 and the legs are 28.

67. In an isosceles right triangle, find   (10.2)

    (*a*) each leg if the hypotenuse is 34; 2*a*.

    (*b*) the hypotenuse if each leg is 34; $15\sqrt{2}$.

68. In a square, find   (10.2)

    (*a*) the side if the diagonal is 40.

    (*b*) the diagonal if the side is 40.

69. In an isosceles trapezoid which has base angles of 45°, find   (10.2)

    (*a*) the lower base and each leg if the altitude is 13 and the upper base is 19.

    (*b*) the upper base and each leg if the altitude is 27 and the lower base is 65.

    (*c*) each leg and lower base if the upper base is 25 and the altitude is 15.

70. A triangle has two angles of 30° and 45°. Find the side opposite 30° if the side opposite 45° is 8. Hint: Draw altitude to third side.   (10.1, 10.2)

71. A parallelogram has an angle of 45°. Find the distances between its pairs of opposite sides if its sides are 10 and 12.   (10.2)

Chapter 8

Trigonometry

1. Trigonometric Ratios

Trigonometry means "measurement of triangles". Consider its parts: "tri" means "three", "gon" means "angle", and "metry" means "measure". Thus, in trigonometry we study the measurement of triangles.

The following ratios relate the sides and an acute angle of a right triangle:

(1) *Tangent Ratio*: The tangent (abbreviated "tan") of an acute angle equals the leg opposite the angle divided by the leg adjacent to the angle.

(2) *Sine Ratio*: The sine (abbreviated "sin") of an acute angle equals the leg opposite the angle divided by the hypotenuse.

(3) *Cosine Ratio*: The cosine (abbreviated "cos") of an acute angle equals the leg adjacent to the angle divided by the hypotenuse.

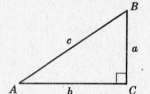

Thus in right triangle ABC,

$$\tan A = \frac{\text{leg opp. } A}{\text{leg adj. } A} = \frac{a}{b} \qquad \tan B = \frac{\text{leg opp. } B}{\text{leg adj. } B} = \frac{b}{a}$$

$$\sin A = \frac{\text{leg opp. } A}{\text{hypotenuse}} = \frac{a}{c} \qquad \sin B = \frac{\text{leg opp. } B}{\text{hypotenuse}} = \frac{b}{c}$$

$$\cos A = \frac{\text{leg adj. } A}{\text{hypotenuse}} = \frac{b}{c} \qquad \cos B = \frac{\text{leg adj. } B}{\text{hypotenuse}} = \frac{a}{c}$$

If A and B are the acute angles of a right triangle,

(1) $\sin A = \cos B$ (2) $\cos A = \sin B$ (3) $\tan A = \dfrac{1}{\tan B}$ (4) $\tan B = \dfrac{1}{\tan A}$

1.1 USING the TABLE of SINES, COSINES and TANGENTS

The following values are taken from the Table of Sines, Cosines and Tangents. State each value in proper trigonometric form and complete the last line.

	Angle	Sine	Cosine	Tangent
(a)	1°	.0175	.9998	.0175
(b)	30°	.5000	.8660	.5774
(c)	60°	.8660	.5000	1.7321
(d)	(?)	(?)	.3420	(?)

Solution:

(a) Sin 1° = .0175, cos 1° = .9998, tan 1° = .0175.

(b) Sin 30° = .5000, cos 30° = .8660, tan 30° = .5774.

(c) Sin 60° = .8660, cos 60° = .5000, tan 60° = 1.7321.

(d) Since .3420 = cos 70°, the angle is 70°. Then sin 70° = .9397 and tan 70° = 2.7475.

113

1.2 FINDING ANGLES to the NEAREST DEGREE

Find x to the nearest degree if (a) sin x = .9235, (b) cos x = 21/25 or .8400, (c) tan x = $\sqrt{5}/10$ or .2236. Use the Table of Sines, Cosines and Tangents.

Solution:

Differences

(a) sin 68° = .9272
     sin x = .9235   → .0037   Since sin x is nearer to sin 67°,
     sin 67° = .9205   → .0030 x = 67° to the nearest degree.

(b) cos 32° = .8480
     cos x = .8400   → .0080   Since cos x is nearer to cos 33°,
     cos 33° = .8387   → .0013 x = 33° to the nearest degree.

(c) tan 13° = .2309
     tan x = .2236   → .0073   Since tan x is nearer to tan 13°,
     tan 12° = .2126   → .0110 x = 13° to the nearest degree.

1.3 FINDING TRIGONOMETRIC RATIOS

In each right triangle, state trigonometric values of each acute angle.

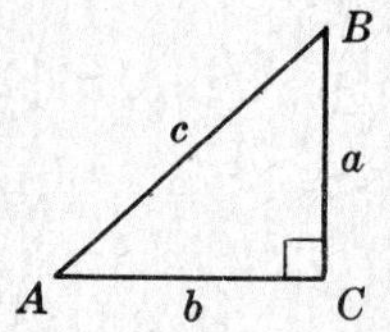

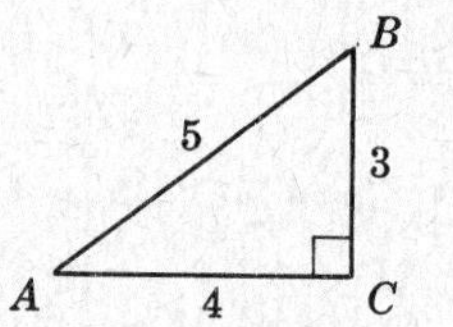

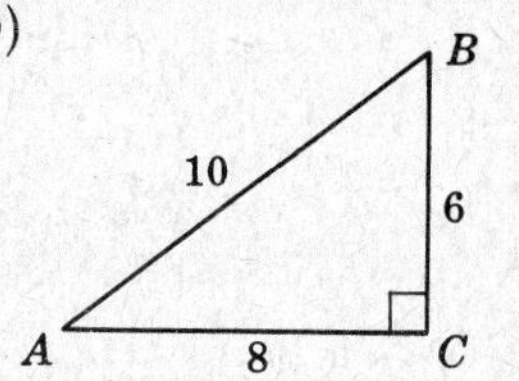

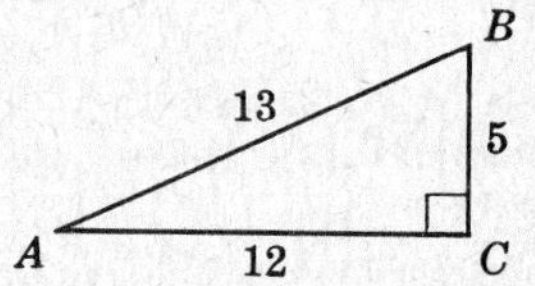

Formulas	(a) $a = 3,\ b = 4,\ c = 5$	(b) $a = 6,\ b = 8,\ c = 10$	(c) $a = 5,\ b = 12,\ c = 13$
$\tan A = \dfrac{a}{b}$	$\tan A = \dfrac{3}{4}$	$\tan A = \dfrac{6}{8} = \dfrac{3}{4}$	$\tan A = \dfrac{5}{12}$
$\tan B = \dfrac{b}{a}$	$\tan B = \dfrac{4}{3}$	$\tan B = \dfrac{8}{6} = \dfrac{4}{3}$	$\tan B = \dfrac{12}{5}$
$\sin A = \dfrac{a}{c}$	$\sin A = \dfrac{3}{5}$	$\sin A = \dfrac{6}{10} = \dfrac{3}{5}$	$\sin A = \dfrac{5}{13}$
$\sin B = \dfrac{b}{c}$	$\sin B = \dfrac{4}{5}$	$\sin B = \dfrac{8}{10} = \dfrac{4}{5}$	$\sin B = \dfrac{12}{13}$
$\cos A = \dfrac{b}{c}$	$\cos A = \dfrac{4}{5}$	$\cos A = \dfrac{8}{10} = \dfrac{4}{5}$	$\cos A = \dfrac{12}{13}$
$\cos B = \dfrac{a}{c}$	$\cos B = \dfrac{3}{5}$	$\cos B = \dfrac{6}{10} = \dfrac{3}{5}$	$\cos B = \dfrac{5}{13}$

1.4 FINDING ANGLES by TRIGONOMETRIC RATIOS

Find A, to the nearest degree, in each.

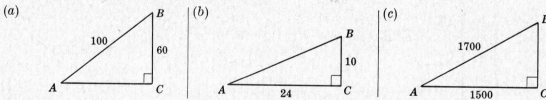

Solution:

(a) sin A = 60/100 = .6000. Since sin 37° = .6018 is the nearest sine value, A = 37°.

(b) tan A = 10/24 = .4167. Since tan 23° = .4245 is the nearest tangent value, A = 23°.

(c) cos A = 1500/1700 = .8824. Since cos 28° = .8829 is the nearest cosine value, A = 28°.

1.5 TRIGONOMETRIC RATIOS of 30° and 60°

Show that (a) tan 30° = .577　　(d) tan 60° = 1.732

　　　　　　(b) sin 30° = .500　　(e) sin 60° = .866

　　　　　　(c) cos 30° = .866　　(f) cos 60° = .500

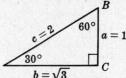

Solution:

The trigonometric ratios for 30° and 60° may be obtained by using a 30°-60°-90° triangle, in which the ratio of the sides is $a : b : c = 1 : \sqrt{3} : 2$. Thus:

(a) $\tan 30° = \dfrac{1}{\sqrt{3}} = \dfrac{1}{\sqrt{3}} \cdot \dfrac{\sqrt{3}}{\sqrt{3}} = \dfrac{\sqrt{3}}{3} = .577$　　　　(d) $\tan 60° = \dfrac{\sqrt{3}}{1} = 1.732$

(b) $\sin 30° = \dfrac{1}{2} = .500$　　　　　　　　　　　(e) $\sin 60° = \dfrac{\sqrt{3}}{2} = .866$

(c) $\cos 30° = \dfrac{\sqrt{3}}{2} = .866$　　　　　　　　　(f) $\cos 60° = \dfrac{1}{2} = .500$

1.6 FINDING SIDES by TRIGONOMETRIC RATIOS

In each triangle, solve for x and y to the nearest integer.

(a) 　　(b) 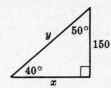　　(c)

Solution:

(a) Since $\tan 40° = \dfrac{x}{150}$, $x = 150 \tan 40° = 150(.8391) = 126.$

Since $\cos 40° = \dfrac{150}{y}$, $y = \dfrac{150}{\cos 40°} = \dfrac{150}{.766} = 196.$

(b) Since $\tan 50° = \dfrac{x}{150}$, $x = 150 \tan 50° = 150(1.1918) = 179.$

Since $\sin 40° = \dfrac{150}{y}$, $y = \dfrac{150}{\sin 40°} = \dfrac{150}{.6428} = 233.$

(c) Since $\sin 40° = \dfrac{x}{150}$, $x = 150 \sin 40° = 150(.6428) = 96.$

Since $\cos 40° = \dfrac{y}{150}$, $y = 150 \cos 40° = 150(.776) = 115.$

1.7 SOLVING TRIGONOMETRY PROBLEMS

(a) An aviator flew 70 miles east from A to C. From C, he flew 100 miles north to B. Find the angle of turn to the nearest degree that must be made at B to return to A.

(b) A road is to be constructed so that it will rise 105 feet for each 1000 feet of horizontal distance. Find the angle of rise to the nearest degree and the length of road to the nearest foot for each 1000 feet of horizontal distance.

(a) 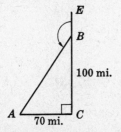　　(b)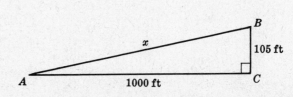

Solution:

(a) The angle of turn is $\angle EBA$. In right $\triangle ABC$, $\tan B = \dfrac{70}{100} = .7000$; hence $\angle B = 35°$ and $\angle EBA = 180° - 35° = 145°$.

(b) Since $\tan A = \dfrac{105}{1000} = .1050$, $\angle A = 6°$. Then $\cos 6° = \dfrac{1000}{x}$, $x = \dfrac{1000}{\cos 6°} = \dfrac{1000}{.9945} = 1005.5$. *Ans.* 1006 ft.

2. *Angles of Elevation and Depression*

The *line of sight* is the line from the eye of the observer to the object sighted.

A *horizontal line* is a line level with the surface of water.

An *angle of elevation (or depression)* is an angle formed by a horizontal line and a line of sight above (or below) the horizontal line and in the same vertical plane.

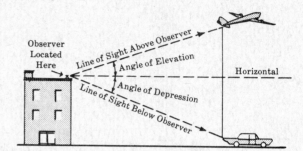

Thus in the figure, the observer is sighting an airplane above the horizontal, and the angle formed by the horizontal and the line of sight is an *angle of elevation*. In sighting the car, the angle his line of sight makes with the horizontal is an *angle of depression*.

2.1 USING an ANGLE of ELEVATION

(a) Sighting the top of a building, Henry found the angle of elevation to be 21°. The ground is level. The transit is 5 ft. above the ground and 200 ft. from the house. Find the height of the building to the nearest foot.

(b) If the angle of elevation of the sun at a certain time is 42°, find to the nearest foot the height of a tree whose shadow is 25 ft. long.

(a)   (b)

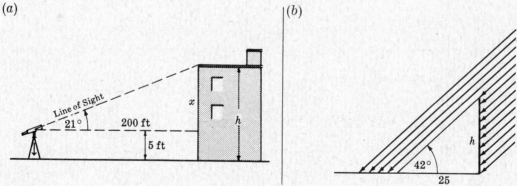

Solution:

(a) If x is the height of the part of the building above the transit,

$$\tan 21° = \frac{x}{200} \quad \text{and} \quad x = 200 \tan 21° = 200(.3839) = 77 \text{ ft.}$$

Thus the height of the building is $h = x + 5 = 77 + 5 = 82$ ft.

(b) If h is the height of the tree, $\tan 42° = \dfrac{h}{25}$ and

$$\tan 42° = \frac{h}{25}. \quad \text{and} \quad h = 25 \tan 42° = 25(.9004) = 23 \text{ ft.}$$

2.2 USING BOTH an ANGLE of ELEVATION and an ANGLE of DEPRESSION

At the top of a lighthouse 200 ft. high, a lighthouse keeper sighted an airplane, and a ship directly beneath the plane. The angle of elevation of the plane was 25°; the angle of depression of the ship was 32°. Find: (a) the distance d of the boat from the foot of the lighthouse, to the nearest 10 feet; (b) the height of the plane above the water, to the nearest 10 feet.

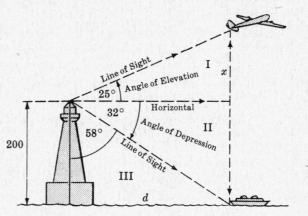

Solution:

(a) In $\triangle$III, $\tan 58° = \dfrac{d}{200}$ and $d = 200 \tan 58° = 200(1.6003) = 320$ ft.

(b) In $\triangle$I, $\tan 25° = \dfrac{x}{320}$ and $x = 320(.4663) = 150$ ft. Since the height of the tower is 200 ft., the height of the airplane is $200 + 150 = 350$ ft.

2.3 USING TWO ANGLES of DEPRESSION

An observer on the top of a hill 250 ft. above the level of a lake sighted two boats directly in line. Find to the nearest foot the distance between the boats if the angles of depression noted by the observer were 11° and 16°.

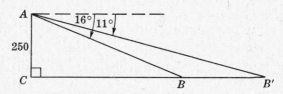

Solution:

In $\triangle AB'C$, $\angle B'AC = 90° - 11° = 79°$. Then $CB' = 250 \tan 79°$.

In $\triangle ABC$, $\angle BAC = 90° - 16° = 74°$. Then $CB = 250 \tan 74°$.

Hence $BB' = CB' - CB = 250(\tan 79° - \tan 74°) = 250(5.1446 - 3.4874) = 414$ ft.

Supplementary Problems

1. Using the table of sines, cosines and tangents, find   (1.1)

    (a) $\sin 25°$, $\sin 48°$, $\sin 59°$, $\sin 89°$.

    (b) $\cos 15°$, $\cos 52°$, $\cos 74°$, $\cos 88°$.

    (c) $\tan 4°$, $\tan 34°$, $\tan 55°$, $\tan 87°$.

    (d) which sets of values increase as the angle increases from 0° to 90°.

    (e) which set of values decreases as the angle increases from 0° to 90°.

    (f) which set has values greater than 1.

2. Using the table, find the angle in each: (1.1)

    (a) $\sin x = .3420$ (c) $\sin B = .9455$ (e) $\cos y = .7071$ (g) $\tan W = .3443$

    (b) $\sin A = .4848$ (d) $\cos A' = .9336$ (f) $\cos Q = .3584$ (h) $\tan B' = 2.3559$

3. Using the table, find x, to the nearest degree if   (1.2)

(a) $\sin x = .4400$ (e) $\cos x = .7650$ (i) $\tan x = 5.5745$ (m) $\cos x = \dfrac{3}{8}$

(b) $\sin x = .7280$ (f) $\cos x = .2675$ (j) $\sin x = \dfrac{11}{50}$ (n) $\cos x = \dfrac{\sqrt{3}}{2}$

(c) $\sin x = .9365$ (g) $\tan x = .1245$ (k) $\sin x = \dfrac{\sqrt{2}}{2}$ (o) $\tan x = \dfrac{2}{7}$

(d) $\cos x = .9900$ (h) $\tan x = .5200$ (l) $\cos x = \dfrac{13}{25}$ (p) $\tan x = \dfrac{\sqrt{3}}{10}$

4. In each right triangle, find $\sin A$, $\cos A$ and $\tan A$:   (1.3)

(a)   (b)   (c)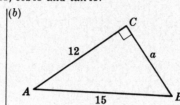

4. In each right triangle, find $\sin A$, $\cos A$ and $\tan A$:   (1.3)

(a)   (b)   (c)

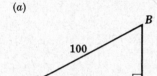
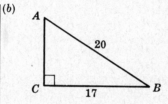
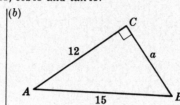

5. Find A, to the nearest degree, in each: (1.4)

(a)   (b)   (c)

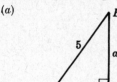
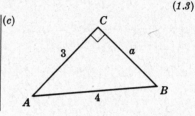
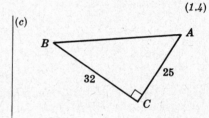

6. Find B to the nearest degree   (a) if $b = 67$ and $c = 100$, (b) if $a = 14$ and $c = 50$, (c) if $a = 22$ and $b = 55$, (d) if $a = 3$ and $b = \sqrt{3}$.   (1.4)

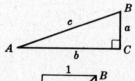

7. Using a square with a side of 1 unit, show that   (1.5)

(a) diagonal $c = \sqrt{2}$,
(b) $\tan 45° = 1$,
(c) $\sin 45° = \cos 45° = .707$.

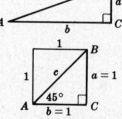

8. To the nearest degree, find each acute angle of any right triangle whose sides are in the ratio of
(a) $5:12:13$, (b) $8:15:17$, (c) $7:24:25$, (d) $11:60:61$.   (1.4)

9. In each triangle, solve for x and y to the nearest integer.   (1.6)

(a)   (b)   (c)

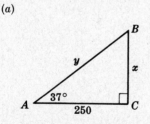
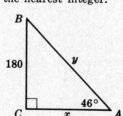
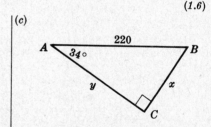

10. A ladder leans against the side of a building and makes an angle of 70° with the ground. The foot of the ladder is 30 feet from the building. Find, to the nearest foot, (a) how high up on the building the ladder reaches, (b) the length of the ladder.   (1.6)

11. To find the distance across a swamp, a surveyor took measurements as shown. AC is at right angles to BC. If $A = 24°$ and $AC = 350$ ft., find the distance BC across the swamp.   (1.6)

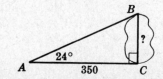

12. A plane rises from a take-off and flies at a fixed angle of 9° with the horizontal ground. When it has gained 400 ft. in altitude, find, to the nearest 10 ft., (a) the horizontal distance flown, (b) the distance the plane has actually flown. *(1.6)*

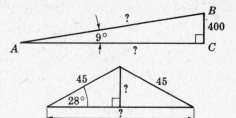

13. If the base angle of an isosceles triangle is 28° and each leg is 45 in., find, to the nearest inch, (a) the altitude drawn to the base, (b) the base. *(1.6)*

14. In a triangle, an angle of 50° is included between sides of 12 and 18. Find the altitude to the side of 12, to the nearest integer. *(1.6)*

15. Find the sides of a rectangle to the nearest inch if a diagonal of 24 inches makes an angle of 42° with a side. *(1.6)*

16. A rhombus has an angle of 76° and a long diagonal of 40 feet. Find the short diagonal to the nearest foot. *(1.6)*

17. Find the altitude to the base of an isosceles triangle to the nearest yard if its base is 40 yards and its vertex angle is 106°. *(1.6)*

18. A road is inclined uniformly at an angle of 6° with the horizontal. After driving 10,000 ft. along this road, find, to the nearest 10 ft., the (a) increase in the altitude of the driver, *(1.7)* (b) horizontal distance that has been driven.

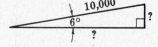

19. An airplane travels 15,000 ft. through the air at a uniform angle of climb, thereby gaining 1900 ft., in altitude. Find its angle of climb. *(1.7)*

20. Sighting the top of a monument, William found the angle of elevation to be 16°. The ground is level and the transit is 5 ft. above the ground. If the monument is 86 ft. high, find to the nearest foot the distance from William to the foot of the monument. *(2.1)*

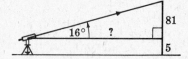

21. Find to the nearest degree the angle of elevation of the sun when a tree 60 ft. high casts a shadow of (a) 10 ft, (b) 60 ft. *(2.1)*

22. At a certain time of day, the angle of elevation of the sun is 34°. Find to the nearest foot the shadow cast by (a) a 15 ft. vertical pole, (b) a building 70 ft. high. *(2.1)*

23. A light at C is projected vertically to a cloud at B. An observer at A, 1000 ft. horizontally from C, notes the angle of elevation of B. Find the height of the cloud, to the nearest foot, if A = 37°. *(2.1)*

24. A lighthouse built at sea level is 180 ft. high. From its top, the angle of depression of a buoy is 24°. Find, to the nearest foot the distance from the buoy to the foot of the lighthouse. *(2.1)*

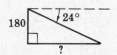

25. An observer on the top of a hill 300 ft. above the level of a lake, sighted two ships directly in line. Find, to the nearest ft., the distance between the boats if the angles of depression noted by the observer were (a) 20° and 15°, (b) 35° and 24°, (c) 9° and 6°. *(2.2)*

26. In the figure, ∠A = 43°, ∠BDC = 54°, ∠C = 90°, and DC = 170 ft. (a) Find the length of BC. (b) Using the result found in (a), find the length of AB.

27. In the figure, ∠B = 90°, ∠ACB = 58°, ∠D = 23°, and BC = 60 ft. (a) Find the length of AB. (b) Using the result of part (a), find the length of CD.

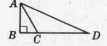

28. Tangents PA and PB are drawn to a circle from external point P. ∠APB = 40° and PA = 25. (a) Find to the nearest tenth the radius of the circle. (b) Find to the nearest integer the length of minor arc AB.

Chapter 9

Areas

1. Understanding Areas. Area of a Rectangle and a Square

A *square unit* is the surface enclosed by a square whose side is 1 unit.

The *area of a closed plane figure,* such as a polygon, is the number of square units contained in its surface. Since a rectangle 5 units long and 4 units wide can be divided into 20 unit squares, its area is 20 square units.

The *area of a rectangle* equals the product of its base and altitude.

Thus if $b = 8''$ and $h = 3''$, then $A = 24$ sq. in.

The *area of a square* equals the square of a side.

Thus if $s = 6$, then $A = s^2 = 36$.

Rectangle: $A = bh$

It follows that the area of a square also equals one-half the square of a diagonal. Since $A = s^2$ and $s = d/\sqrt{2}$, $A = \frac{1}{2}d^2$.

Square: *(1)* $A = s^2$
(2) $A = \frac{1}{2}d^2$

1.1 AREA of a RECTANGLE

(a) Find the area of a rectangle if the base is 15 and the perimeter is 50.

(b) Find the area of a rectangle if the altitude is 10 and the diagonal is 26.

(c) Find the base and altitude of a rectangle if its area is 70 and its perimeter is 34.

Solution:

(a) $p = 50$, $b = 15$. Since $p = 2b + 2h$, $50 = 2(15) + 2h$ and $h = 10$.

Hence $A = bh = 15(10) = 150$.

(b) $d = 26$, $h = 10$. In right $\triangle ACD$, $d^2 = b^2 + h^2$, $26^2 = b^2 + 10^2$ and $b = 24$.

Hence $A = bh = 24(10) = 240$.

(c) $A = 70$, $p = 34$. Since $p = 2b + 2h$, $34 = 2(b + h)$ and $h = 17 - b$.

Then $A = bh$, $70 = b(17 - b)$, $b^2 - 17b + 70 = 0$ and $b = 7$ or 10.

Since $h = 17 - b$, we obtain $h = 10$ or 7.

Ans. 10 and 7, or 7 and 10

1.2 AREA of a SQUARE

(a) Find the area of a square if the perimeter is 30.

(b) Find the area of a square if the radius of the circumscribed circle is 10.

(c) Find the side and the perimeter of a square whose area is 20.

(d) Find the number of square inches in a square foot.

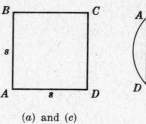

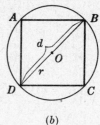

(a) and (c)   (b)

Solution:

(a) $p = 30$. Since $p = 4s$, $30 = 4s$ and $s = 7\frac{1}{2}$. Then $A = s^2 = (7\frac{1}{2})^2 = 56\frac{1}{4}$.

(b) Since $r = 10$, $d = 2r = 20$. Then $A = \frac{1}{2}d^2 = \frac{1}{2}(20)^2 = 200$.

(c) $A = 20$ and $A = s^2$; hence $s^2 = 20$ and $s = 2\sqrt{5}$. Perimeter $= 4s = 8\sqrt{5}$.

(d) $A = s^2$. Since 1 ft. = 12 in., $A = 12^2 = 144$.   *Ans.* 1 sq. ft. = 144 sq. in.

2. *Area of a Parallelogram*

(1) *The area of a parallelogram equals the product of a side and the altitude to that side.* A proof of this theorem is given in Chapter 16.

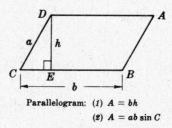

Thus in $\square ABCD$, if $b = 10$ and $h = 2.7$, then $A = 10(2.7) = 27$.

(2) *The area of a parallelogram equals the product of two adjacent sides and the sine of the included angle.*

In $\triangle DEC$, $h = a \sin C$; hence $A = bh = ab \sin C$.

Parallelogram: *(1)* $A = bh$
                (2) $A = ab \sin C$

2.1 AREA of a PARALLELOGRAM

(a) Find the area of a parallelogram if sides of 20 and 10 include an angle of 59°. (Answer to the nearest integer.)

(b) Find the area of a parallelogram if the area is represented by $x^2 - 4$, a side by $x + 4$ and the altitude to that side by $x - 3$.

(c) In a parallelogram, find the altitude if the area is 54 and the ratio of the altitude to the base is 2:3.

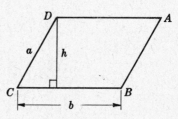

Solution:

(a) $b = 20$, $a = 10$, $\angle C = 59°$. $A = ab \sin C = (10)(20) \sin 59° = 200(.8572) = 171.44$.

Thus $A = 171$, to the nearest integer.

(b) $A = x^2 - 4$, $b = x + 4$ and $h = x - 3$.

Then $A = bh$, $x^2 - 4 = (x + 4)(x - 3)$, $x^2 - 4 = x^2 + x - 12$ and $x = 8$.

Hence $A = x^2 - 4 = 64 - 4 = 60$.

(c) Let $h = 2x$, $b = 3x$. Then $A = bh$, $54 = (3x)(2x)$, $54 = 6x^2$, $9 = x^2$ and $x = 3$.

Hence $h = 2x = 2(3) = 6$.

3. Area of a Triangle

(1) *The area of a triangle equals one-half the product of a side and the altitude to that side. A proof of this theorem is given in Chapter 16.*

(2) *The area of a triangle equals one-half the product of any two adjacent sides and the sine of their included angle.*

Thus if $a = 25$, $b = 4$ and $\sin C = .28$, then
$$A = \tfrac{1}{2}(25)(4)(.28) = 14$$

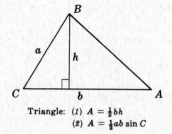

Triangle: (1) $A = \tfrac{1}{2}bh$
(2) $A = \tfrac{1}{2}ab \sin C$

3.1 AREA of a TRIANGLE

Find to the nearest integer the area of a triangle (a) if two adjacent sides of 15 and 8 include an angle of 150°, (b) if two adjacent sides of 16 and 5 include an angle of 53°.

(a)

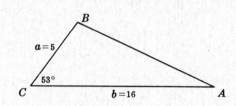

(b)

Solution:

(a) $b = 15$, $a = 8$. Since $\angle BCA = 150°$, $\angle BCD = 180° - 150° = 30°$.

In $\triangle BCD$, h is opposite $\angle BCD$; hence $h = \tfrac{1}{2}a = 4$. Then $A = \tfrac{1}{2}bh = \tfrac{1}{2}(15)(4) = 30$.

(b) $a = 5$, $b = 16$ and $\angle C = 53°$. Then $A = \tfrac{1}{2}ab \sin C = \tfrac{1}{2}(5)(16) \sin 53° = 40(.7986) = 32$.

3.2 FORMULAS for the AREA of an EQUILATERAL TRIANGLE

Derive the formula for the area of an equilateral triangle (a) of side s, (b) of altitude h.

Solution:

(a) Here $A = \tfrac{1}{2}bh$, where $b = s$ and $h^2 = s^2 - (\tfrac{1}{2}s)^2 = \tfrac{3}{4}s^2$ or $h = \tfrac{1}{2}s\sqrt{3}$.

Then $A = \tfrac{1}{2}bh = \tfrac{1}{2}s(\tfrac{1}{2}s\sqrt{3}) = \tfrac{1}{4}s^2\sqrt{3}$.

(b) Here $A = \tfrac{1}{2}bh$, where $b = s$ and $h = \tfrac{1}{2}s\sqrt{3}$ or $s = \dfrac{2h}{\sqrt{3}} = \dfrac{2h}{\sqrt{3}} \cdot \dfrac{\sqrt{3}}{\sqrt{3}} = \dfrac{2}{3}h\sqrt{3}$.

Then $A = \tfrac{1}{2}bh = \tfrac{1}{2}sh = \tfrac{1}{2}(\tfrac{2}{3}h\sqrt{3})h = \tfrac{1}{3}h^2\sqrt{3}$.

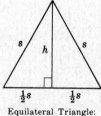

Equilateral Triangle:
(1) $A = \tfrac{1}{4}s^2\sqrt{3}$
(2) $A = \tfrac{1}{3}h^2\sqrt{3}$

3.3 AREA of an EQUILATERAL TRIANGLE

Find the area of (a) an equilateral triangle if the perimeter is 24, (b) a rhombus if the shorter diagonal is 12 and an angle is 60°, (c) a regular hexagon of side 6.

(a)

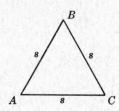

(b)

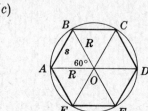

(c)

Solution:

(a) $p = 24$. Since $p = 3s$, $s = 8$. Then $A = \frac{1}{4}s^2\sqrt{3} = \frac{1}{4}(64)\sqrt{3} = 16\sqrt{3}$.

(b) Since $\angle A = 60°$, $\triangle ADB$ is equilateral and $s = d = 12$. The area of the rhombus is twice the area of $\triangle ABD$. Hence $A = 2(\frac{1}{4}s^2\sqrt{3}) = 2(\frac{1}{4})(144)\sqrt{3} = 72\sqrt{3}$.

(c) A side s of the inscribed hexagon subtends a central angle of $\frac{1}{6}(360°) = 60°$. Then, since $OA = OB =$ radius R of the circumscribed circle, $\angle OAB = \angle OBA = 60°$. Thus $\triangle AOB =$ equilateral.

   Area of hexagon $= 6(\text{area of } \triangle AOB) = 6(\frac{1}{4}s^2\sqrt{3}) = 6(\frac{1}{4})(36\sqrt{3}) = 54\sqrt{3}$.

4. Area of a Trapezoid

(1) *The area of a trapezoid equals one-half the product of its altitude and the sum of its bases.* A proof of this theorem is given in Chapter 16.

   Thus if $h = 20$, $b = 27$ and $b' = 23$, then

$$A = \frac{1}{2}(20)(27 + 23) = 500$$

(2) *The area of a trapezoid equals the product of its altitude and median.*

   Since $A = \frac{1}{2}h(b + b')$ and $m = \frac{1}{2}(b + b')$, $A = hm$.

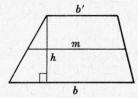

Trapezoid: (1) $A = \frac{1}{2}h(b + b')$

(2) $A = hm$

4.1   AREA of a TRAPEZOID

(a) Find the area of a trapezoid if the bases are 7.3 and 2.7 and the altitude is 3.8.

(b) Find the area of an isosceles trapezoid if the bases are 22 and 10 and the legs are each 10.

(c) Find the bases of an isosceles trapezoid if the area is $52\sqrt{3}$, the altitude is $4\sqrt{3}$ and each leg is 8.

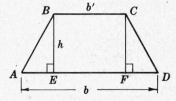

Solution:

(a) $b = 7.3$, $b' = 2.7$, $h = 3.8$. Then $A = \frac{1}{2}h(b + b') = \frac{1}{2}(3.8)(7.3 + 2.7) = 19$.

(b) $b = 22$, $b' = 10$, $AB = 10$. $EF = b' = 10$ and $AE = \frac{1}{2}(22 - 10) = 6$.

   In $\triangle BEA$, $h^2 = 10^2 - 6^2 = 64$ and $h = 8$. Then $A = \frac{1}{2}h(b + b') = \frac{1}{2}(8)(22 + 10) = 128$.

(c) $AE = \sqrt{(AB)^2 - h^2} = \sqrt{64 - 48} = 4$, $FD = AE = 4$, $b' = b - (AE + FD) = b - 8$.

   Then $A = \frac{1}{2}h(b + b') = \frac{1}{2}h(2b - 8)$ or $52\sqrt{3} = \frac{1}{2}(4\sqrt{3})(2b - 8)$, from which $26 = 2b - 8$ or $b = 17$, and $b' = b - 8 = 17 - 8 = 9$.

5. Area of a Rhombus

   The area of a rhombus equals one-half the product of its diagonals.

   Since each diagonal is the perpendicular bisector of the other, the area of triangle I is $\frac{1}{2}(\frac{1}{2}d)(\frac{1}{2}d') = \frac{1}{8}dd'$. Thus the rhombus, which consists of 4 triangles congruent to $\triangle$I, has an area of $4(\frac{1}{8}dd')$ or $\frac{1}{2}dd'$.

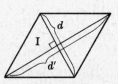

Rhombus: $A = \frac{1}{2}dd'$

5.1 AREA of a RHOMBUS

(a) Find the area of a rhombus if one diagonal is 30 and a side is 17.

(b) Find a diagonal of a rhombus if the other diagonal is 8 and the area of the rhombus is 52.

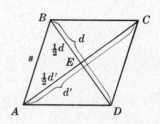

Solution:

(a) $d' = 30$, $s = 17$. In right $\triangle AEB$, $s^2 = (\tfrac{1}{2}d)^2 + (\tfrac{1}{2}d')^2$, $17^2 = (\tfrac{1}{2}d)^2 + 15^2$, $\tfrac{1}{2}d = 8$ and $d = 16$. Then $A = \tfrac{1}{2}dd' = \tfrac{1}{2}(16)(30) = 240$.

(b) $d' = 8$, $A = 52$. Then $A = \tfrac{1}{2}dd'$, $52 = \tfrac{1}{2}(d)(8)$ and $d = 13$.

6. Equal Polygons

Equal Polygons	Similar Polygons	Congruent Polygons
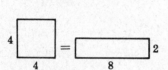	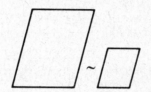	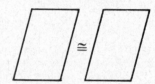
Equal polygons have the same size, that is, the same area.	Similar polygons have the same shape.	Congruent polygons have the same size and the same shape.

Pr. 1: *Parallelograms are equal if they have equal bases and equal altitudes.*

Thus the two parallelograms shown in Fig. 1 are equal.

Pr. 2: *Triangles are equal if they have equal bases and equal altitudes.*

Thus in Fig. 2, $\triangle CAB = \triangle CAD$.

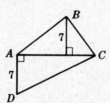

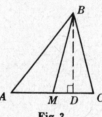

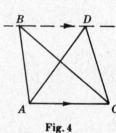

| Fig. 1 | Fig. 2 | Fig. 3 | Fig. 4 |

Pr. 3: *A median divides a triangle into two equal triangles.*

Thus in Fig. 3 where BM is a median, $\triangle AMB = \triangle BMC$ since they have equal bases ($AM = MC$) and common altitude BD.

Pr. 4: *Triangles are equal if they have a common base and their vertices lie on a line parallel to the base.*

Thus in Fig. 4, $\triangle ABC = \triangle ADC$.

6.1 PROVING an EQUAL AREAS PROBLEM

Given: Trapezoid $ABCD$ ($BC \parallel AD$)
Diagonals AC and BD

To Prove: $\triangle AEB = \triangle DEC$

Plan: Use Pr. 4 to obtain $\triangle ABD = \triangle ACD$.
Then use the subtraction axiom.

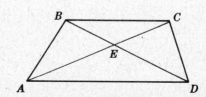

PROOF: Statements	Reasons
1. $BC \parallel AD$	1. Given
2. $\triangle ABD = \triangle ACD$	2. Triangles are equal if they have a common base and their vertices lie in a line parallel to the base.
3. $\triangle AED = \triangle AED$	3. Identity.
4. $\triangle AEB = \triangle DEC$	4. If equals are subtracted from equals, the results are equal.

6.2 PROVING an EQUAL AREAS PROBLEM STATED in WORDS

Prove: In quadrilateral $ABCD$, M is the midpoint of diagonal AC. If BM and DM are drawn, quadrilateral $ABMD$ equals quadrilateral $CBMD$.

Given: Quadrilateral $ABCD$
   M is midpoint of diagonal AC.

To Prove: Quadrilateral $ABMD$ equals quadrilateral $CBMD$.

Plan: Use Pr. 3 to obtain two pair of equal triangles.
   Then use the addition axiom.

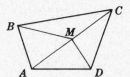

PROOF: Statements	Reasons
1. M is the midpoint of AC.	1. Given
2. BM is a median of $\triangle ACB$. DM is a median of $\triangle ACD$.	2. A line from a vertex of a triangle to the midpoint of the opposite side is a median.
3. $\triangle AMB = \triangle BMC$, $\triangle AMD = \triangle DMC$	3. A median divides a triangle into two equal triangles.
4. Quadrilateral $ABMD$ equals quadrilateral $CBMD$.	4. If equals are added to equals, the results are equal.

7. *Comparing Areas of Similar Polygons*

The areas of similar polygons are to each other as the squares of any two corresponding lines.

Thus if $\triangle ABC \sim \triangle A'B'C'$ and the area of $\triangle ABC$ is 25 times the area of $\triangle A'B'C'$, then $5:1$ is the ratio of any two corresponding sides, medians, altitudes, radii of inscribed or circumscribed circles, etc.

7.1 RATIOS of AREAS and LINES of SIMILAR TRIANGLES

Find the ratio of the areas of two similar triangles (*a*) if the ratio of two corresponding sides is $3:5$, (*b*) if their perimeters are 12 and 7. Find the ratio of a pair of (*c*) corresponding sides if the ratio of the areas is $4:9$, (*d*) corresponding medians if the areas are 250 and 10.

Solution:

(*a*) $\dfrac{A}{A'} = \left(\dfrac{s}{s'}\right)^2 = \left(\dfrac{3}{5}\right)^2 = \dfrac{9}{25}$ (*c*) $\left(\dfrac{s}{s'}\right)^2 = \dfrac{A}{A'} = \dfrac{4}{9}$ or $\dfrac{s}{s'} = \dfrac{2}{3}$

(*b*) $\dfrac{A}{A'} = \left(\dfrac{p}{p'}\right)^2 = \left(\dfrac{12}{7}\right)^2 = \dfrac{144}{49}$ (*d*) $\left(\dfrac{m}{m'}\right)^2 = \dfrac{A}{A'} = \dfrac{250}{10}$ or $\dfrac{m}{m'} = 5$

7.2 PROPORTIONS DERIVED from SIMILAR POLYGONS

(*a*) The areas of two similar polygons are 80 and 5. If a side of the smaller polygon is 2, find the corresponding side of the larger polygon.

(*b*) The corresponding diagonals of two similar polygons are 4 and 5. If the area of the larger polygon is 75, find the area of the smaller polygon.

Solution:

(a) $\left(\dfrac{s}{s'}\right)^2 = \dfrac{A}{A'}$. Then $\left(\dfrac{s}{2}\right)^2 = \dfrac{80}{5} = 16$, $\dfrac{s}{2} = 4$ and $s = 8$.

(b) $\dfrac{A}{A'} = \left(\dfrac{d}{d'}\right)^2$. Then $\dfrac{A}{75} = \left(\dfrac{4}{5}\right)^2$ and $A = 75\left(\dfrac{16}{25}\right) = 48$.

Supplementary Problems

1. Find the area of a rectangle   *(1.1)*
   (a) if the base is 11 in. and the altitude is 9 in.
   (b) if the base is 2 ft. and the altitude is 1 ft. 6 in.
   (c) if the base is 25 and the perimeter is 90.
   (d) if the base is 15 and the diagonal is 17.
   (e) if the diagonal is 12 and the angle between the diagonal and the base is 60°.
   (f) if the diagonal is 20 and the angle between the diagonal and the base is 30°.
   (g) if the diagonal is 25 and the sides are in the ratio of $3:4$.
   (h) if the perimeter is 50 and the sides are in the ratio of $2:3$.

2. Find the area of a rectangle inscribed in a circle   *(1.1)*
   (a) if the radius of the circle is 5 and the base is 6.
   (b) if the radius of the circle is 15 and the altitude is 24.
   (c) if the radius and the altitude both equal 5.
   (d) if the diameter is 26 and the base and altitude are in the ratio of $5:12$.

3. Find the base and altitude of a rectangle   *(1.1)*
   (a) if its area is 28 and the base is 3 more than the altitude.
   (b) if its area is 72 and the base is twice the altitude.
   (c) if its area is 54 and the ratio of the base to the altitude is $3:2$.
   (d) if its area is 12 and the perimeter is 16.
   (e) if its area is 70 and the base and altitude are represented by $2x$ and $x+2$ respectively.
   (f) if its area is 160 and the base and altitude are represented by $3x-4$ and x.

4. Find the area of (a) a square yard in square inches, (b) a square rod in square yards $(1\text{ rd.} = 5\frac{1}{2}\text{ yd.})$, (c) a square meter in square decimeters (1 meter $=$ 10 decimeters).   *(1.2)*

5. Find the area of a square if (a) a side is 15, (b) a side is $3\frac{1}{2}$, (c) a side is 1.8, (d) a side is $8a$, (e) the perimeter is 44, (f) the perimeter is 10, (g) the perimeter is $12b$, (h) the diagonal is 8, (i) the diagonal is 9, (j) the diagonal is $8\sqrt{2}$.   *(1.2)*

6. Find the area of a square if (a) the radius of the circumscribed circle is 8, (b) the diameter of the circumscribed circle is 12, (c) the diameter of the circumscribed circle is $10\sqrt{2}$, (d) the radius of the inscribed circle is $3\frac{1}{2}$, (e) the diameter of the inscribed circle is 20.   *(1.2)*

7. If a floor is 20 feet long and 80 feet wide, how many tiles are needed to cover it if (a) each tile is 1 sq. ft., (b) each tile is a square 2 ft. on a side, (c) each tile is a square 4 feet on a side, (d) each tile is a square 4 inches on a side.   *(1.2)*

8. If the area of a square is 81, find (a) the side, (b) the perimeter, (c) the diagonal, (d) the radius of the inscribed circle, (e) the radius of the circumscribed circle.   *(1.2)*

9. (a) Find the side of a square if the area is $6\frac{1}{4}$.   (1.2)

    (b) Find the perimeter of a square if the area is 169.

    (c) Find the diagonal of a square if the area is 50.

    (d) Find the diagonal of a square if the area is 25.

    (e) Find the radius of the inscribed circle of a square if the area is 144.

    (f) Find the radius of the circumscribed circle of a square if the area is 32.

10. Find the area of a parallelogram if the base and altitude are respectively   (2.1)

    (a) 3 ft. and $5\frac{1}{8}$ ft., (b) 4 ft. and 1 ft. 6 in., (c) 20 and 3.5, (d) 1.8 and .9.

11. Find the area of a parallelogram if the base and altitude are respectively   (2.1)

    (a) $3x$ and x, (b) $(x+3)$ and x, (c) $(x-5)$ and $(x+5)$, (d) $(4x+1)$ and $(3x+2)$.

12. Find the area of a parallelogram if two adjacent sides are respectively (a) 15 and 20 and include an angle of 30°, (b) 9 and 12 and include an angle of 45°, (c) 14 and 8 and include an angle of 60°, (d) 10 and 12 and include an angle of 150°, (e) 6 and 4 and include an angle of 28°, (f) 9 and 10 and include an angle of 38°.   (2.1)

13. Find the area of a parallelogram if   (2.1)

    (a) the area is represented by x^2, the base by $x+3$ and the altitude by $x-2$.

    (b) the area is represented by x^2-10, the base by x and the altitude by $x-2$.

    (c) the area is represented by $2x^2-34$, the base by $x+3$ and the altitude by $x-3$.

14. In a parallelogram, find   (2.1)

    (a) the base if the area is 40 and the altitude is 15.

    (b) the altitude if the area is 22 and the base is 1.1.

    (c) the base if the area is 27 and the base is three times the altitude.

    (d) the altitude if the area is 21 and the base is four more than the altitude.

    (e) the base if the area is 90 and the ratio of the base to the altitude is $5:2$.

    (f) the altitude to a side of 20 if the altitude to a side of 15 is 16.

    (g) the base if the area is 48, the base is represented by $x+3$ and the altitude by $x+1$.

    (h) the base if the area is represented by x^2+17, the base by $2x-3$ and the altitude by $x+1$.

15. Find the area of a triangle if the base and altitude are respectively (a) 6 in. and $3\frac{2}{3}$ in., (b) 1 yd. and 2 ft., (c) 8 and $(x-7)$, (d) $5x$ and $4x$, (e) $4x$ and $(x+9)$, (f) $(x+4)$ and $(x-4)$, (g) $(2x-6)$ and $(x+3)$.
                                                                                                              (3.1)

16. Find the area of a triangle if two adjacent sides are respectively (a) 8 and 5 and include an angle of 30°, (b) 5 and 4 and include an angle of 45°, (c) 8 and 12 and include an angle of 60°, (d) 25 and 10 and include an angle of 150°, (e) 20 and 5 and include an angle of 36°, (f) 16 and 9 and include an angle of 140°.   (3.1)

17. Find the area of a right triangle if (a) the legs are 5 and 6, (b) the legs are equal and the hypotenuse is 16, (c) the leg opposite a 30° angle is 6, (d) an acute angle is 60° and the hypotenuse is 8, (e) its sides are 6, 8 and 10, (f) the altitude to the hypotenuse cuts off segments of 2 and 8.   (3.1)

18. Find the area of   (3.1)

    (a) a triangle if two sides are 13 and 15 and the altitude to the third side is 12,

    (b) a triangle whose sides are 10, 10 and 16,

    (c) a triangle whose sides are 5, 12 and 13,

    (d) an isosceles triangle whose base is 30 and whose legs are each 17,

    (e) an isosceles triangle whose base is 20 and whose vertex angle is 68°,

    (f) an isosceles triangle whose base is 30 and whose base angle is 62°,

    (g) a triangle inscribed in a circle of radius 4 if one side is a diameter and another side makes an angle of 30° with the diameter,

    (h) a triangle cut off by a line parallel to the base of a triangle if the base and altitude of the larger triangle are 10 and 5, respectively, and the line parallel to the base is 6.

19. Find the altitude of a triangle if   (3.1)
   (a) its base is 10 and the triangle is equal to a parallelogram whose base and altitude are 15 and 8,
   (b) its base is 8 and the triangle is equal to a square whose diagonal is 4.
   (c) its base is 12 and the triangle is equal to another triangle whose sides are 6, 8 and 10.

20. In a triangle, find   (3.1)
   (a) a side if the area is 40 and the altitude to that side is 10,
   (b) an altitude if the area is 25 and the side to which the altitude is drawn is 5,
   (c) a side if the area is 24 and the side is 2 more than its altitude,
   (d) a side if the area is 108 and the side and its altitude are in the ratio of $3:2$,
   (e) the altitude to a side of 20, if the sides of the triangle are 12, 16 and 20,
   (f) the altitude to a side of 12 if another side and its altitude are 10 and 15,
   (g) a side represented by $4x$ if the altitude to that side is represented by $x+7$ and the area is 60,
   (h) a side if the area is represented by x^2-55, the side by $2x-2$, and its altitude by $x-5$.

21. Find the area of an equilateral triangle if (a) a side is 10, (b) the perimeter is 36, (c) an altitude is 6,
   (d) an altitude is $5\sqrt{3}$, (e) a side is $2b$, (f) the perimeter is $12x$, (g) an altitude is $3r$.   (3.3)

22. Find the area of a rhombus having an angle of $60°$ if (a) a side is 2, (b) the shorter diagonal is 7,
   (c) the longer diagonal is 12, (d) the longer diagonal is $6\sqrt{3}$.   (3.3)

23. Find the area of a regular hexagon if (a) a side is 4, (b) if the radius of the circumscribed circle is 6,
   (c) if the diameter of the circumscribed circle is 20.   (3.3)

24. Find the side of an equilateral triangle whose area equals   (3.3)
   (a) the sum of two equilateral triangles whose sides are 9 and 12,
   (b) the difference of two equilateral triangles whose sides are 17 and 15,
   (c) the area of a trapezoid whose bases are 6 and 2 and whose altitude is $9\sqrt{3}$,
   (d) twice the area of a right triangle having a hypotenuse of 5 and an acute angle of $30°$.

25. Find the area of trapezoid $ABCD$ (4.1)
   (a) if $b=25$, $b'=15$, and $h=7$.
   (b) if $m=10$ and $h=6.9$.
   (c) if $AB=30$, $\angle A=30°$, $b=24$, and $b'=6$.
   (d) if $AB=12$, $\angle A=45°$, $b=13$, and $b'=7$.
   (e) if $AB=10$, $\angle A=70°$, and $b+b'=20$.

26. Find the area of isosceles trapezoid $ABCD$ (4.1)
   (a) if $b'=17$, $l=10$, and $h=6$.
   (b) if $b=22$, $b'=12$, and $l=13$.
   (c) if $b=16$, $b'=10$, and $\angle A=45°$.
   (d) if $b=20$, $l=8$, and $\angle A=60°$.
   (e) if $b=40$, $b'=20$, and $\angle A=28°$.

27. (a) Find the altitude of a trapezoid if the bases are 13 and 7 and the area is 40.   (4.1)
   (b) Find the altitude of a trapezoid if the sum of the bases is twice the altitude and the area is 49.
   (c) Find the sum of the bases and the median of a trapezoid if the area is 63 and the altitude is 7.
   (d) Find the bases of a trapezoid if the upper base is 3 less than the lower base, the altitude is 4, and
       the area is 30.
   (e) Find the bases of a trapezoid if the lower base is twice the upper base, the altitude is 6, and the
       area is 45.

28. In an isosceles trapezoid,   *(4.1)*

    (*a*) find the bases if each leg is 5, the altitude is 3, and the area is 39.

    (*b*) find the bases if the altitude is 5, each base angle is 45°, and the area is 90.

    (*c*) find the bases if the area is $42\sqrt{3}$, the altitude is $3\sqrt{3}$, and each base angle is 60°.

    (*d*) find each leg if the bases are 24 and 32 and the area is 84.

    (*e*) find each leg if the area is 300, the median is 25 and the lower base is 30.

29. Find the area of a rhombus if   *(5.1)*

    (*a*) the diagonals are 8 and 9,   (*f*) the perimeter is 40 and a diagonal is 12,

    (*b*) the diagonals are 11 and 7,   (*g*) the side is 6 and an angle is 30°,

    (*c*) the diagonals are 4 and $6\sqrt{3}$,   (*h*) the perimeter is 28 and an angle is 45°,

    (*d*) the diagonals are $3x$ and $8x$,   (*i*) the perimeter is 32 and the short diagonal equals

    (*e*) one diagonal is 10 and a side is 13,   a side,

                                (*j*) a side is 14 and an angle is 120°.

30. Find the area of a rhombus to the nearest integer if (*a*) the side is 30 and an angle is 55°, (*b*) the perimeter is 20 and an angle is 33°, (*c*) the side is 10 and an angle is 130°.   *(5.1)*

31. In a rhombus, find   *(5.1)*

    (*a*) a diagonal if the other diagonal is 7 and the area is 35,

    (*b*) the diagonals if their ratio is 4 : 3 and the area is 54,

    (*c*) the diagonals if the longer is twice the shorter and the area is 100,

    (*d*) the side if the area is 24 and one diagonal is 6,

    (*e*) the side if the area is 6 and one diagonal is four more than the other.

32. A rhombus is equivalent to a trapezoid whose lower base is 26 and whose other three sides are 10. Find the altitude of the rhombus if its perimeter is 36.   *(5.1)*

33. Prove each of the following: *(6.1)*

    (*a*)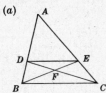

        Given: $\triangle ABC$
             $DB = \tfrac{1}{3}AB$
             $EC = \tfrac{1}{3}AC$
        To Prove: $\triangle DFB = \triangle FEC$

    (*b*)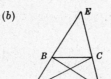

        Given: Trapezoid $ABCD$
             AB and CD extended
             meet at E.
        To Prove: $\triangle ECA = \triangle EBD$

34. Prove each of the following: *(6.1)*

    (*a*)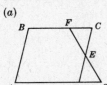

        Given: $\square ABCD$
             E is midpoint of CD.
        To Prove:
             $\square ABCD$ = trapezoid $BFGA$

    (*b*)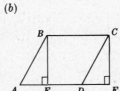

        Given: $\square ABCD$
             BE and $CF \perp AF$.
        To Prove:
        $BCFE$ is a rectangle and
        equal to $\square ABCD$.

35. Prove each of the following: *(6.2)*

    (*a*) A median divides a triangle into two equal triangles.

    (*b*) Triangles are equal if they have a common base and their vertices lie in a line parallel to the base.

    (*c*) In a triangle, if lines are drawn from a vertex to the trisection points of the opposite sides, the triangle is trisected.

    (*d*) In trapezoid $ABCD$, base AD is twice base BC. If M is the midpoint of AD, then $ABCM$ and $BCDM$ are equal parallelograms.

36. Prove each of the following: *(6.2)*

    (a) In $\triangle ABC$, E is a point on BM, the median to AC. Prove that $\triangle BEA = \triangle BEC$.

    (b) In $\triangle ABC$, Q is a point on BC, M is the midpoint of AB and P is the midpoint of AC. Prove that $\triangle BQM + \triangle PQC = $ quadrilateral $APQM$.

    (c) In quadrilateral $ABCD$, diagonal AC bisects diagonal BD. Prove that $\triangle ABC = \triangle ACD$.

    (d) The diagonals of a parallelogram divide the parallelogram into four equal triangles.

37. Find the ratio of the areas of two similar triangles if the ratio of two corresponding sides is   *(7.1)*
    (a) $1:7$,   (b) $7:2$,   (c) $1:\sqrt{3}$,   (d) $a:5a$,   (e) $9:x$,   (f) $3:\sqrt{x}$,   (g) $s:s\sqrt{2}$.

38. Find the ratio of the areas of two similar triangles   *(7.1)*

    (a) if the ratio of two corresponding medians is $7:10$.

    (b) if an altitude of the first is $\frac{2}{3}$ of a corresponding altitude of the second.

    (c) if two corresponding angle bisectors are 10 and 12.

    (d) if each side of the first is one-third of each corresponding side of the second.

    (e) if the radii of their circumscribed circles are $7\frac{1}{2}$ and 5.

    (f) if their perimeters are 30 and $30\sqrt{2}$.

39. Find the ratio of any two corresponding sides of two similar triangles if the ratio of the areas is
    (a) $100:1$,   (b) $1:49$,   (c) $400:81$,   (d) $25:121$,   (e) $4:y^2$,   (f) $9x^2:1$,   (g) $3:4$,   (h) $1:2$,   (i) $x^2:5$,   (j) $x:16$.
                                                 (7.1)

40. In similar triangles, find the ratio of   *(7.1)*

    (a) corresponding sides if the areas are 72 and 50.

    (b) corresponding medians if the ratio of the areas is $9:49$.

    (c) corresponding altitudes if the areas are 18 and 6.

    (d) the perimeters if the areas are 50 and 40.

    (e) radii of the inscribed circles if the ratio of the areas is $1:3$.

41. The areas of two similar triangles are in the ratio of $25:16$. Find   *(7.1)*

    (a) a side of the larger if the corresponding side of the smaller is 80,

    (b) a median of the larger if the corresponding median of the smaller is 10,

    (c) an angle bisector of the smaller if the corresponding angle bisector of the larger is 15,

    (d) the perimeter of the smaller if the perimeter of the larger is 125,

    (e) the circumference of the inscribed circle of the larger if the circumference of the inscribed circle of the smaller is 84,

    (f) the diameter of the circumscribed circle of the smaller if the diameter of the circumscribed circle of the larger is 22.5,

    (g) an altitude of the larger if the corresponding altitude of the smaller is $16\sqrt{3}$.

42. (a) The areas of two similar triangles are 36 and 25. If a median of the smaller triangle is 10, find the corresponding median of the larger.   *(7.2)*

    (b) Two corresponding altitudes of similar triangles are 3 and 4. If the area of the larger triangle is 112, find the area of the smaller.

    (c) Two similar polygons have perimeters of 32 and 24. If the area of the smaller is 27, find the area of the larger.

    (d) The areas of two similar pentagons are 88 and 22. If a diagonal of the larger is 5, find the corresponding diagonal of the smaller.

    (e) In two similar polygons, the ratio of two corresponding sides is $\sqrt{3}:1$. If the area of the smaller is 15, find the area of the larger.

Chapter 10

Regular Polygons and the Circle

1. Regular Polygons

A *regular polygon* is an equilateral and equiangular polygon.

The *center of a regular polygon* is the common center of its inscribed and circumscribed circles.

A *radius of a regular polygon* is a line joining its center to any vertex. A radius of a regular polygon is also a radius of the circumscribed circle.

A *central angle of a regular polygon* is an angle included between two radii drawn to successive vertices.

An *apothem of a regular polygon* is a line from its center perpendicular to one of its sides. An apothem is also a radius of the inscribed circle.

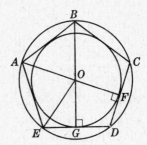

Fig. 1

Thus for the regular pentagon shown in Fig. 1, $AB = BC = CD = DE = EA$, $\angle A = \angle B = \angle C = \angle D = \angle E$. Also, its center is O, OA and OB are its radii, $\angle AOB$ is a central angle, and OG and OF are apothems.

A. Regular Polygon Principles

Pr. 1: *If a regular polygon of n sides has a side s, the perimeter is $p = ns$.*

Pr. 2: *A circle may be circumscribed about any regular polygon.*

Pr. 3: *A circle may be inscribed in any regular polygon.*

Pr. 4: *The center of the circumscribed circle of a regular polygon is also the center of its inscribed circle.*

Pr. 5: *An equilateral polygon inscribed in a circle is a regular polygon.*

Pr. 6: *Radii of a regular polygon are equal.*

Pr. 7: *A radius of a regular polygon bisects the angle to which it is drawn.*
    Thus in Fig. 1, OB bisects $\angle ABC$.

Pr. 8: *Apothems of a regular polygon are equal.*

Pr. 9: *An apothem of a regular polygon bisects the side to which it is drawn.*
    Thus in Fig. 1, OF bisects CD and OG bisects ED.

Pr. 10: *For a regular polygon of n sides:*

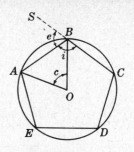

(a) *each central angle c equals* $\dfrac{360°}{n}$.

(b) *each interior angle i equals* $\dfrac{(n-2)180°}{n}$.

(c) *each exterior angle e equals* $\dfrac{360°}{n}$.

Fig. 2

Thus for the regular pentagon *ABCDE* of Fig. 2,

$$\angle AOB = \angle ABS = \frac{360°}{n} = \frac{360°}{5} = 72°, \quad \angle ABC = \frac{(n-2)180°}{n} = \frac{(5-2)180°}{5} = 108°,$$

$$\angle ABC + \angle ABS = 180°.$$

1.1 FINDING LINES and ANGLES in a REGULAR POLYGON

(a) Find a side *s* of a regular pentagon if the perimeter *p* is 35.

(b) Find the apothem *r* of a regular pentagon if the radius of the inscribed circle is 21.

(c) In a regular polygon of 5 sides, find the central angle *c*, the exterior angle *e* and the interior angle *i*.

(d) If an interior angle of a regular polygon is 108°, find the exterior angle, the central angle and the number of sides.

Solution:

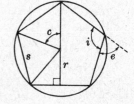

(a) $p = 35$. Since $p = 5s$, $35 = 5s$ and $s = 7$.

(b) Since an apothem *r* is a radius of the inscribed circle, it equals 21.

(c) $n = 5$. Then $c = \dfrac{360°}{n} = \dfrac{360°}{5} = 72°$, $e = \dfrac{360°}{n} = 72°$, $i = 180° - e = 108°$.

(d) $i = 108°$. Then $c = e = 180° - i = 72°$. Since $c = \dfrac{360°}{n}$, $n = 5$.

1.2 PROVING a REGULAR POLYGON PROBLEM STATED in WORDS

Prove: A vertex angle of a regular pentagon is trisected by diagonals drawn from that vertex.

Given: Regular pentagon *ABCDE*
Diagonals *AC* and *AD*

To Prove: *AC* and *AD* trisect $\angle A$.

Plan: Circumscribe a circle and show that angles *BAC*, *CAD* and *DAE* are equal.

PROOF: Statements	Reasons
1. *ABCDE* is a regular pentagon.	1. Given
2. Circumscribe a circle about *ABCDE*.	2. A circle may be circumscribed about any regular polygon.
3. $BC = CD = DE$	3. A regular polygon is equilateral.
4. $\overarc{BC} = \overarc{CD} = \overarc{DE}$	4. In a circle, equal chords have equal arcs.
5. $\angle BAC = \angle CAD = \angle DAE$	5. In a circle, inscribed angles having equal arcs are equal.
6. $\angle A$ is trisected.	6. To divide into three equal parts is to trisect.

2. *Line Relationships in Regular Polygons of 3, 4 and 6 Sides*

In the case of the regular hexagon, square, and equilateral triangle, special right triangles are formed when the apothem r and a radius R terminating in the same side are drawn. In the case of the square we obtain a 45°-45°-90° triangle while in the other two cases we obtain the 30°-60°-90° triangle. The following formulas relate the sides and radii of these regular polygons.

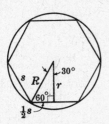
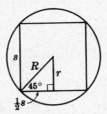
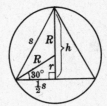

Regular Hexagon	Square	Equilateral Triangle
$s = R$	$s = R\sqrt{2}$	$s = R\sqrt{3}, \ h = r + R$
$r = \frac{1}{2}R\sqrt{3}$	$r = \frac{1}{2}s = \frac{1}{2}R\sqrt{2}$	$r = \frac{1}{3}h, \ R = \frac{2}{3}h, \ r = \frac{1}{2}R$

2.1 APPLYING LINE RELATIONSHIPS in a REGULAR HEXAGON

In a regular hexagon, (*a*) find the side and apothem if its radius is 12, (*b*) find its radius and apothem if the side is 8.

Solution:

(*a*) Since $R = 12$, $s = R = 12$ and $r = \frac{1}{2}R\sqrt{3} = 6\sqrt{3}$.

(*b*) Since $s = 8$, $R = s = 8$ and $r = \frac{1}{2}R\sqrt{3} = 4\sqrt{3}$.

2.2 APPLYING LINE RELATIONSHIPS in a SQUARE

In a square, (*a*) find the side and apothem if its radius is 16, (*b*) find its radius and apothem if its side is 10.

Solution:

(*a*) Since $R = 16$, $s = R\sqrt{2} = 16\sqrt{2}$ and $r = \frac{1}{2}s = 8\sqrt{2}$.

(*b*) Since $s = 10$, $r = \frac{1}{2}s = 5$ and $R = s/\sqrt{2} = \frac{1}{2}s\sqrt{2} = 5\sqrt{2}$.

2.3 APPLYING LINE RELATIONSHIPS in an EQUILATERAL TRIANGLE

In an equilateral triangle, (*a*) find its radius, apothem and side if the altitude is 6, (*b*) find the side, apothem and altitude if its radius is 9.

Solution:

(*a*) Since $h = 6$, then $r = \frac{1}{3}h = 2$, $R = \frac{2}{3}h = 4$ and $s = R\sqrt{3} = 4\sqrt{3}$.

(*b*) Since $R = 9$, then $s = R\sqrt{3} = 9\sqrt{3}$, $r = \frac{1}{2}R = 4\frac{1}{2}$ and $h = \frac{3}{2}R = 13\frac{1}{2}$.

3. *Area of a Regular Polygon*

The area of a regular polygon equals one-half the product of its perimeter and apothem.

As shown, by drawing radii a regular polygon of n sides and perimeter $p = ns$ can be divided into n triangles, each of area $\frac{1}{2}rs$. Hence the area of the regular polygon $= n(\frac{1}{2}rs) = \frac{1}{2}nsr = \frac{1}{2}pr$.

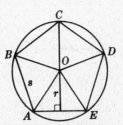

Regular Polygon
$A = \frac{1}{2}nsr = \frac{1}{2}pr$

3.1 FINDING the AREA of a REGULAR POLYGON

(a) Find the area of a regular hexagon if the apothem is $5\sqrt{3}$.

(b) Find the area of a regular pentagon to the nearest integer if the apothem is 20.

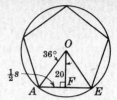

Solution:

(a) In a regular hexagon, $r = \frac{1}{2}s\sqrt{3}$; since $r = 5\sqrt{3}$, then $s = 10$ and $p = 6(10) = 60$. Then $A = \frac{1}{2}pr = \frac{1}{2}(60)(5\sqrt{3}) = 150\sqrt{3}$.

(b) In the figure, $\angle AOE = 360°/5 = 72°$, $\angle AOF = \frac{1}{2}\angle AOE = 36°$; then $\tan 36° = \frac{1}{2}s \div 20 = s/40$ or $s = 40 \tan 36°$.

$\qquad$ Now $A = \frac{1}{2}pr = \frac{1}{2}nsr = \frac{1}{2}(5)(40\tan 36°)(20) = 1453$.

4. Ratios of Lines and Areas of Regular Polygons

Pr. 1: *Regular polygons having the same number of sides are similar.*

Pr. 2: *Corresponding lines of regular polygons having the same number of sides are in proportion. "Lines" includes sides, perimeters, radii or circumferences of circumscribed or inscribed circles, etc.*

Pr. 3: *Areas of regular polygons having the same number of sides are to each other as the squares of any two corresponding lines.*

4.1 RATIOS of LINES and AREAS of REGULAR POLYGONS

(a) In two regular polygons having the same number of sides, find the ratio of the apothems if the perimeters are in the ratio of 5:3.

(b) In two regular polygons having the same number of sides, find a side of the smaller if the apothems are 20 and 50 and a side of the larger is 32.5.

(c) In two regular polygons having the same number of sides, find the ratio of the areas if the sides are in the ratio of 1:5.

(d) In two regular polygons having the same number of sides, find the area of the smaller if the sides are 4 and 12 and the area of the larger is 10,260.

Solution:

(a) By Pr. 2: $r:r' = p:p' = 5:3$.

(b) By Pr. 2: $s:s' = r:r'$, $s:32.5 = 20:50$ and $s = 13$.

(c) By Pr. 3: $\dfrac{A}{A'} = \left(\dfrac{s}{s'}\right)^2 = \left(\dfrac{1}{5}\right)^2 = \dfrac{1}{25}$.

(d) By Pr. 3: $\dfrac{A}{A'} = \left(\dfrac{s}{s'}\right)^2$, $\dfrac{A}{10,260} = \left(\dfrac{4}{12}\right)^2$ and $A = 1140$.

5. Circumference and Area of a Circle

$\qquad \pi$ (pi) is the ratio of the circumference C of any circle to its diameter d; that is, $\pi = C/d$. Hence

$$C = \pi d \quad \text{or} \quad C = 2\pi r$$

Approximate values for π are 3.1416, 3.14 or 22/7. Unless otherwise stated, use 3.14 for π in solving problems.

Circle: *(1)* $C = 2\pi r$
$\qquad\qquad$ *(2)* $A = \pi r^2$

A circle may be regarded as a regular polygon having an infinite number of sides. If a square is inscribed in a circle, and the number of sides continually doubled (to form an octagon, a 16-gon, etc.), the perimeters of the resulting polygons will very closely approximate the circumference of the circle.

Thus to find the area of a circle, the formula $A = \frac{1}{2}pr$ can be used with C substituted for p; hence

$$A = \tfrac{1}{2}Cr = \tfrac{1}{2}(2\pi r)(r) \quad \text{and} \quad A = \pi r^2$$

Circles are similar figures since they have the same shape. As similar figures, (1) corresponding lines of circles are in proportion and (2) the areas of two circles are to each other as the squares of their radii or circumferences.

5.1 FINDING the CIRCUMFERENCE and AREA of a CIRCLE

In a circle, (a) find the circumference and area if the radius is 6, (b) find the radius and area if the circumference is 18π, (c) find the radius and circumference if the area is 144π. (Answer in terms of π and also to the nearest integer.)

Solution:

(a) $r = 6$.   Then $C = 2\pi r = 12\pi$ and $A = \pi r^2 = 36\pi = 36(3.14) = 113$.

(b) $C = 18\pi$.   Since $C = 2\pi r$, $18\pi = 2\pi r$ and $r = 9$; then $A = \pi r^2 = 81\pi = 254$.

(c) $A = 144\pi$.   Since $A = \pi r^2$, $144\pi = \pi r^2$ and $r = 12$; then $C = 2\pi r = 24\pi = 75$.

5.2 CIRCUMFERENCE and AREA of CIRCUMSCRIBED and INSCRIBED CIRCLES

Find the circumference and area of the circumscribed circle and inscribed circle (a) of a regular hexagon if the side is 8, (b) of an equilateral triangle if the altitude is $9\sqrt{3}$. (Answer may be given in terms of π.)

(a)

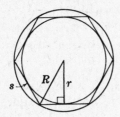

(b)

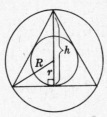

Solution:

(a) $R = s = 8$.   Then $C = 2\pi R = 16\pi$ and $A = \pi R^2 = 64\pi$.

   $r = \frac{1}{2}R\sqrt{3} = 4\sqrt{3}$.   Then $C = 2\pi r = 8\pi\sqrt{3}$ and $A = \pi r^2 = 48\pi$.

(b) $R = \frac{2}{3}h = 6\sqrt{3}$.   Then $C = 2\pi R = 12\pi\sqrt{3}$ and $A = \pi R^2 = 108\pi$.

   $r = \frac{1}{3}h = 3\sqrt{3}$.   Then $C = 2\pi r = 6\pi\sqrt{3}$ and $A = \pi r^2 = 27\pi$.

5.3 RATIOS of LINES and AREAS in CIRCLES

(a) If the circumferences of two circles are in the ratio of $2:3$, find the ratio of the diameters and the ratio of the areas.

(b) If the areas of two circles are in the ratio of $1:25$, find the ratio of the diameters and the ratio of the circumferences.

Solution:

(a) $\dfrac{d}{d'} = \dfrac{C}{C'} = \dfrac{2}{3}$ and $\dfrac{A}{A'} = \left(\dfrac{C}{C'}\right)^2 = \left(\dfrac{2}{3}\right)^2 = \dfrac{4}{9}$.

(b) Since $\dfrac{A}{A'} = \left(\dfrac{d}{d'}\right)^2$, $\dfrac{1}{25} = \left(\dfrac{d}{d'}\right)^2$ and $\dfrac{d}{d'} = \dfrac{1}{5}$.   Also, $\dfrac{C}{C'} = \dfrac{d}{d'} = \dfrac{1}{5}$.

6. *Length of an Arc. Area of a Sector and a Segment*

A *sector of a circle* is a part of a circle bounded by two radii and their intercepted arc.

Thus in Fig. 1, the shaded section of circle O is sector OAB.

                                              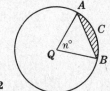

Fig. 1   **Fig. 2**

A *segment of a circle* is a part of a circle bounded by a chord and its arc.

Thus in Fig. 2, the shaded section of circle Q is segment ACB.

Pr. 1:   *In a circle of radius r, the length l of an arc of n° equals $\frac{n}{360}$ of the circumference of the circle, or $l = \frac{n}{360}(2\pi r) = \frac{\pi n r}{180}$.*

Pr. 2:   *In a circle of radius r, the area K of a sector of n° equals $\frac{n}{360}$ of the area of the circle, or $K = \frac{n}{360}(\pi r^2)$.*

Pr. 3: $\dfrac{\text{Area of a sector of } n°}{\text{Area of the circle}} = \dfrac{\text{Length of an arc of } n°}{\text{Circumference of the circle}} = \dfrac{n}{360}$

Pr. 4:   *The area of a minor segment of a circle equals the area of its sector less the area of the triangle formed by its radii and chord.*

Pr. 5:   *If a regular polygon is inscribed in a circle, each segment equals the difference between the area of the circle and the area of the polygon divided by the number of sides.*

6.1 LENGTH of an ARC

(a) Find the length of a 36° arc in a circle whose circumference is 45π.

(b) Find the radius of a circle if a 40° arc has a length of 4π.

Solution:

(a) $n° = 36°$, $C = 2\pi r = 45\pi$. Then $l = \dfrac{n}{360}(2\pi r) = \dfrac{36}{360}(45\pi) = \dfrac{9}{2}\pi$.

(b) $l = 4\pi$, $n° = 40°$. Then $l = \dfrac{n}{360}(2\pi r)$, $4\pi = \dfrac{40}{360}(2\pi r)$ and $r = 18$.

6.2 AREA of a SECTOR

(a) Find the area K of a 300° sector of a circle whose radius is 12.

(b) Find the central angle of a sector whose area is 6π if the area of the circle is 9π.

(c) Find the radius of a circle if an arc of length 2π has a sector of area 10π.

Solution:

(a) $n° = 300°$, $r = 12$. Then $K = \dfrac{n}{360}(\pi r^2) = \dfrac{300}{360}(144\pi) = 120\pi$.

(b) $\dfrac{\text{Area of sector}}{\text{Area of circle}} = \dfrac{n}{360}$, $\dfrac{6\pi}{9\pi} = \dfrac{n}{360}$, and $n = 240$; thus the central angle is 240°.

(c) $\dfrac{\text{Length of arc}}{\text{Circumference}} = \dfrac{\text{Area of sector}}{\text{Area of circle}}$, $\dfrac{2\pi}{2\pi r} = \dfrac{10\pi}{\pi r^2}$ and $r = 10$.

6.3   AREA of a SEGMENT

(a) Find the area of a segment if its central angle is 60° and the radius of the circle is 12.

(b) Find the area of a segment if its central angle is 90° and the radius of the circle is 8.

(c) Find each segment formed by an inscribed equilateral triangle if the radius of the circle is 8.

(a)

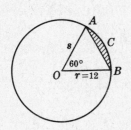

(b)

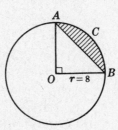

(c)

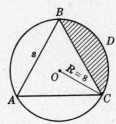

Solution:

(a) $n° = 60°$, $r = 12$.   Area of sector $OAB = \dfrac{n}{360}(\pi r^2) = \dfrac{60}{360}(144\pi) = 24\pi$.

Area of equilateral $\triangle OAB = \frac{1}{4}s^2\sqrt{3} = \frac{1}{4}(144)\sqrt{3} = 36\sqrt{3}$.

Hence the area of segment $ACB = 24\pi - 36\sqrt{3}$.

(b) $n° = 90°$, $r = 8$.   Area of sector $OAB = \dfrac{n}{360}(\pi r^2) = \dfrac{90}{360}(64\pi) = 16\pi$.

Area of right $\triangle OAB = \frac{1}{2}bh = \frac{1}{2}(8)(8) = 32$.

Hence the area of segment $ACB = 16\pi - 32$.

(c) $R = 8$.   Since $s = R\sqrt{3} = 8\sqrt{3}$, the area of $\triangle ABC = \frac{1}{4}s^2\sqrt{3} = 48\sqrt{3}$.

Area of circle $O = \pi R^2 = 64\pi$.   Hence the area of segment $BDC = \frac{1}{3}(64\pi - 48\sqrt{3})$.

6.4   AREA of a SEGMENT FORMED by an INSCRIBED REGULAR POLYGON

Find each segment formed by an inscribed regular polygon of 12 sides (dodecagon) if the radius of the circle is 12.

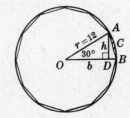

Solution:

Area of sector $OAB = \dfrac{n}{360}(\pi r^2) = \dfrac{30}{360}(144\pi) = 12\pi$.

To find the area of $\triangle OAB$, draw altitude AD to base OB. Since $\angle AOB = 30°$, $h = AD = \frac{1}{2}r = 6$. Then the area of $\triangle OAB = \frac{1}{2}bh = \frac{1}{2}(12)(6) = 36$.

Hence the area of segment $ACB = 12\pi - 36$.

7. *Areas of Combination Figures*

The areas of combination figures may be found by determining the separate areas in the figure and then performing whatever operations are required to find the total area.

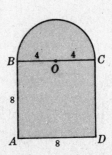

Thus the shaded area equals the sum of the areas of the square and the semicircle:

$$A = 8^2 + \frac{1}{2}(16\pi) = 64 + 8\pi$$

7.1 FINDING AREAS of COMBINATION FIGURES

Find the shaded area in each. In (a) the circles A, B and C are tangent externally and each has a radius 3. In (b) each arc is part of a circle of radius 9.

(a)

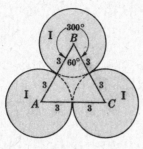

(b)

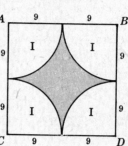

Solution:

(a) Area of $\triangle ABC = \frac{1}{4}s^2\sqrt{3} = \frac{1}{4}(6^2)\sqrt{3} = 9\sqrt{3}$.   Area of sector I $= \dfrac{n^\circ}{360^\circ}(\pi r^2) = \dfrac{300}{360}(9\pi) = \dfrac{15}{2}\pi$.

     Shaded area $= 9\sqrt{3} + 3(\frac{15}{2}\pi) = 9\sqrt{3} + \frac{45}{2}\pi$.

(b) Area of square $= 18^2 = 324$.   Area of sector I $= \dfrac{n^\circ}{360^\circ}(\pi r^2) = \dfrac{90}{360}(81\pi) = \dfrac{81}{4}\pi$.

     Shaded area $= 324 - 4(\frac{81}{4}\pi) = 324 - 81\pi$.

Supplementary Problems

1. In a regular polygon, find   *(1.1)*
   (a) the perimeter if a side is 8 and the number of sides is 25.
   (b) the perimeter if a side is 2.45 and the number of sides is 10.
   (c) the perimeter if a side is $4\frac{2}{3}$ and the number of sides is 24.
   (d) the number of sides if the perimeter is 325 and a side is 25.
   (e) the number of sides if the perimeter is $27\sqrt{3}$ and a side is $3\sqrt{3}$.
   (f) the side if the number of sides is 30 and the perimeter is 100.
   (g) the side if the perimeter is 67.5 and the number of sides is 15.

2. In a regular polygon, find   *(1.1)*
   (a) the apothem if the diameter of an inscribed circle is 25.
   (b) the apothem if the radius of the inscribed circle is 23.47.
   (c) the radius of the inscribed circle if the apothem is $7\sqrt{3}$.
   (d) the radius of the regular polygon if the diameter of the circumscribed circle is 37.
   (e) the radius of the circumscribed circle if the radius of the regular polygon is $3\sqrt{2}$.

3. In a regular polygon of 15 sides, find   *(1.1)*
   (a) the central angle, (b) the exterior angle, (c) the interior angle.

4. If an exterior angle of a regular polygon is 40°, find   *(1.1)*
   (a) the central angle, (b) the number of sides, (c) the interior angle.

5. If an interior angle of a regular polygon is 165°, find   (1.1)
   (a) the exterior angle, (b) the central angle, (c) the number of sides.

6. If a central angle of a regular polygon is 5°, find   (1.1)
   (a) the exterior angle, (b) the number of sides, (c) the interior angle.

7. Name the regular polygon if   (1.1)
   (a) its central angle is 45°,   (d) its exterior angle is 36°,
   (b) its central angle is 60°,   (e) its interior angle equals its central angle,
   (c) its exterior angle is 120°,   (f) its interior angle is 150°.

8. Prove each of the following: (1.2)
   (a) The diagonals of a regular pentagon are equal.
   (b) A diagonal of a regular pentagon forms an isosceles trapezoid with three of its sides.
   (c) If two diagonals of a regular pentagon intersect, the longer segment of each diagonal equals a side
       of the regular pentagon.

9. In a regular hexagon, find   (2.1)
   (a) the side if its radius is 9,   (e) the apothem if the side is 26,
   (b) the perimeter if its radius is 5,   (f) its radius if the apothem is $3\sqrt{3}$,
   (c) the apothem if its radius is 12,   (g) the side if the apothem is 30,
   (d) its radius if the side is 6,   (h) the perimeter if the apothem is $5\sqrt{3}$.

10. In a square, find   (2.2)
    (a) the side if its radius is 18,   (e) the side if the apothem is 1.7,
    (b) the apothem if its radius is 14,   (f) the perimeter if the apothem is $3\frac{1}{2}$,
    (c) the perimeter if its radius is $5\sqrt{2}$, (g) its radius if the perimeter is 40,
    (d) its radius if the side is 16,   (h) the apothem if the perimeter is $16\sqrt{2}$.

11. In an equilateral triangle, find   (2.3)
    (a) the side if its radius is 30,   (g) its altitude if the side is 96,
    (b) the apothem if its radius is 28,   (h) its radius if the apothem is 21,
    (c) the altitude if its radius is 18,   (i) its side if the apothem is $\sqrt{3}$,
    (d) the perimeter if its radius is $2\sqrt{3}$, (j) its altitude if the apothem is $3\frac{1}{3}$,
    (e) its radius if the side is 24,   (k) the altitude if the perimeter is 15,
    (f) its apothem if the side is 24,   (l) the apothem if the perimeter is 54.

12. (a) Find the area of a regular pentagon to the nearest integer if the apothem is 15.   (3.1)
    (b) Find the area of a regular decagon to the nearest integer if the side is 20.

13. Find the area of a regular hexagon, in radical form, if   (3.1)
    (a) the side is 6, (b) its radius is 8, (c) the apothem is $10\sqrt{3}$.

14. Find the area of a square if   (3.1)
    (a) the apothem is 12, (b) its radius is $9\sqrt{2}$, (c) its perimeter is 40.

15. Find the area of an equilateral triangle, in radical form, if   (3.1)
    (a) the apothem is $2\sqrt{3}$,   (c) the altitude is 4,   (e) the perimeter is $6\sqrt{3}$,
    (b) its radius is 6,   (d) the altitude is $12\sqrt{3}$, (f) the apothem is 4.

16. If the area of a regular hexagon is $150\sqrt{3}$, find (a) the side, (b) its radius, (c) the apothem.   (3.1)

17. If the area of an equilateral triangle is $81\sqrt{3}$, find   (3.1)
    (a) the side, (b) the altitude, (c) its radius, (d) the apothem.

18. Find the ratio of the perimeters of two regular polygons having the same number of sides if   (4.1)
    (a) the ratio of the sides is 1 : 8,   (e) the side of the larger is triple the side of the smaller,
    (b) the ratio of their radii is 4 : 9,   (f) the smaller apothem is two-fifths of the larger,
    (c) their radii are 18 and 20,   (g) the apothems are $20\sqrt{2}$ and 15,
    (d) their apothems are 16 and 22,   (h) the circumference of the larger circumscribed circle
                                                      is $2\frac{1}{2}$ times that of the smaller.

19. Find the ratio of the perimeters of two equilateral triangles if (a) the sides are 20 and 8, (b) their radii are 12 and 60, (c) their apothems are $2\sqrt{3}$ and $6\sqrt{3}$, (d) the circumferences of their inscribed circles are 120 and 160, (e) their altitudes are $5x$ and x.   (4.1)

20. Find the ratio of the sides of two regular polygons having the same number of sides if the ratio of the areas is (a) 25 : 1, (b) 16 : 49, (c) $x^2 : 4$, (d) 2 : 1, (e) $3 : y^2$, (f) $x : 18$.   (4.1)

21. Find the ratio of the areas of two regular hexagons if (a) the sides are 14 and 28, (b) the apothems are 3 and 15, (c) the radii are $6\sqrt{3}$ and $\sqrt{3}$, (d) the perimeters are 75 and 250, (e) the circumferences of the circumscribed circles are 28 and 20.   (4.1)

22. Find the circumference of a circle in terms of π if (a) the radius is 6, (b) the diameter is 14, (c) the area is 25π, (d) the area is 3π.   (5.1)

23. Find the area of a circle in terms of π if (a) the radius is 3, (b) the diameter is 10, (c) the circumference is 16π, (d) the circumference is π, (e) the circumference is $6\pi\sqrt{2}$.   (5.1)

24. In a circle (a) find the circumference and area if the radius is 5, (b) find the radius and area if the circumference is 16π, (c) find the radius and circumference if the area is 16π.   (5.1)

25. In a regular hexagon, find the circumference of the circumscribed circle if (a) the apothem is $3\sqrt{3}$, (b) the perimeter is 12, (c) the side is $3\frac{1}{2}$. Also find the circumference of its inscribed circle if (d) the apothem is 13, (e) the side is 8, (f) the perimeter is $6\sqrt{3}$.   (5.2)

26. In a square, find the area in terms of π of the   (5.2)
    (a) circumscribed circle if the apothem is 7,   (d) inscribed circle if the apothem is 5,
    (b) circumscribed circle if the perimeter is 24,   (e) inscribed circle if the side is $12\sqrt{2}$,
    (c) circumscribed circle if the side is 8,   (f) inscribed circle if the perimeter is 80.

27. Find the circumference and area of the (1) circumscribed circle and (2) inscribed circle of a   (5.2)
    (a) regular hexagon if the side is 4,   (d) equilateral triangle if the apothem is 4,
    (b) regular hexagon if the apothem is $4\sqrt{3}$,   (e) square if the side is 20,
    (c) equilateral triangle if the altitude is 9,   (f) square if the apothem is 3.

28. Find the radius of a pipe having the same capacity as two pipes whose radii are (a) 6 ft. and 8 ft., (b) 8 ft. and 15 ft., (c) 3 ft. and 6 ft. *Hint:* Find the areas of their circular cross-sections.

29. In a circle, find the length of a 90° arc if   (6.1)
    (a) the radius is 4,   (d) the circumference is 44π,
    (b) the diameter is 40,   (e) an inscribed hexagon has a side of 12,
    (c) the circumference is 32,   (f) an inscribed equilateral triangle has an altitude of 30.

30. Find the length of   (6.1)
    (a) a 90° arc if the radius of the circle is 6,   (c) a 30° arc if the circumference is 60π,
    (b) a 180° arc if the circumference is 25,   (d) a 40° arc if the diameter is 18,
    (e) an arc intercepted by the side of a regular hexagon inscribed in a circle of radius 3,
    (f) an arc intercepted by a chord of 12 in a circle of radius 12.

31. In a circle, find the area of a 60° sector if   (6.2)
    (a) the radius is 6,   (e) the area of the circle is 27,
    (b) the diameter is 2,   (f) the area of a 240° sector is 52,
    (c) the circumference is 10π,   (g) an inscribed hexagon has a side of 12,
    (d) the area of the circle is 150π,   (h) an inscribed hexagon has an area of $24\sqrt{3}$.

32. Find the area of a   (6.2)
    (a) 60° sector if the radius of the circle is 6,   (c) 15° sector if the area of the circle is 72π,
    (b) 240° sector if the area of the circle is 30,   (d) 90° sector if its arc is 4π.

33. Find the central angle of an arc whose length is   (6.1)
    (a) 3 in. if the circumference is 9 in.,   (d) 6π if the circumference is 12π,
    (b) 2 ft. if the circumference is 1 yd.,   (e) three-eighths of the circumference,
    (c) 25 if the circumference is 250,   (f) equal to the radius.

34. Find the central angle of a sector whose area is   (6.2)
    (a) 10, if the area of the circle is 50,
    (b) 15 sq. in. if the area of the circle is 20 sq. in.,
    (c) 1 sq. ft. if the area of the circle is 1 sq. yd.,
    (d) 5π if the area of the circle is 12π,
    (e) eight-ninths of the area of the circle.

35. Find the central angle of   (6.1, 6.2)
    (a) an arc whose length is 5π if the area of its sector is 25π,
    (b) an arc whose length is 12π if the area of its sector is 48π,
    (c) a sector whose area is 2π if the length of its arc is π,
    (d) a sector whose area is 10π if the length of its arc is 2π.

36. Find the radius of a circle if a   (6.1, 6.2)
    (a) 120° arc has a length of 8π,   (d) 30° sector has an area of 3π,
    (b) 40° arc has a length of 2π,   (e) 36° sector has an area of $2\frac{1}{2}π$,
    (c) 270° arc has a length of 15π,   (f) 120° sector has an area of 6π.

37. Find the radius of a circle if a   (6.2)
    (a) sector of area 12π has an arc of length 6π,   (c) sector of 25 sq. in. has an arc of 5 in.,
    (b) sector of area 10π has an arc of length 2π,   (d) sector of area 162 has an arc of length 36.

38. Find the area of a segment if its central angle is 60° and the radius of the circle is (a) 6, (b) 12,
    (c) 3, (d) r, (e) 2r.   (6.3)

39. Find the area of a segment of a circle if   (6.3)
    (a) the radius of the circle is 4 and the central angle is 90°,
    (b) the radius of the circle is 30 and the central angle is 60°,
    (c) the radius of the circle and the chord of the segment are each equal to 12,
    (d) the central angle is 90° and the length of the arc is 4π,
    (e) its chord of 20 units is 10 units from the center of the circle.

40. Find the area of a segment if the radius of the circle is 8 and the central angle is (a) 120°, (b) 135°,
    (c) 150°.   (6.3)

41. If the radius of a circle is 4, find each segment formed by an inscribed (a) equilateral triangle,
    (b) regular hexagon, (c) square.   (6.3)

42. Find each segment formed by an inscribed   (6.3)

  (a) equilateral triangle if the radius of the circle is 6,

  (b) regular hexagon if the radius of the circle is 3,

  (c) square if the radius of the circle is 6.

43. Find the shaded area in each.   (7.1)

Each heavy dot represents the center of an arc or a circle.

(a)

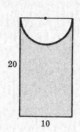

20

10

(b)

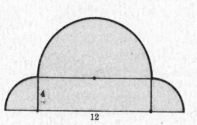

4

12

(c)

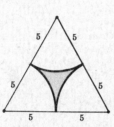

5   5

5   5

5   5

(d)

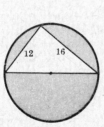

12   16

(e)

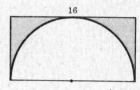

16

(f)

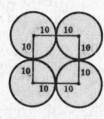

10 10

10   10

10   10

10 10

(g)

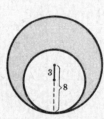

3

8

(h)

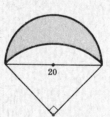

20

44. Find the shaded area in each. Each heavy dot represents the center of an arc or circle.   (7.1)

(a)

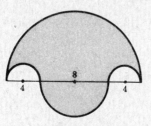

8

4   4

(b)

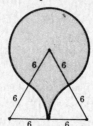

6   6

6   6

6   6

(c)

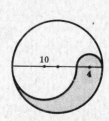

10

4

Chapter 11

Locus

1. Determining a Locus

Locus, in Latin, means location. The plural is loci.

The *locus of a point* is the set of points, and only those points, that satisfy given conditions.

Thus the locus of a point that is 1 inch from a given point P is the set of points 1 inch from P. These points lie on a circle with its center at P and a radius of 1 inch, and hence this circle is the required locus.

To determine a locus, (1) state the given and the condition to be satisfied, (2) find **several** points satisfying the condition which indicate the shape of the locus, (3) connect the points and describe the locus fully.

All geometric constructions **require** the use of straightedge and compasses. Hence if a locus is to be constructed, such drawing instruments must be used.

A. Fundamental Locus Theorems

In each of the following, the required locus is shown as a dashed line.

Pr. 1: *The locus of a point equidistant from two given points is the perpendicular bisector of the line joining the two points.*

Pr. 2: *The locus of a point equidistant from two given parallel lines is a line parallel to the two lines and midway between them.*

Pr. 3: *The locus of a point equidistant from the sides of a given angle is the bisector of the angle.*

Pr. 4: *The locus of a point equidistant from two given intersecting lines is the bisectors of the angles formed by the lines.*

Pr. 5: *The locus of a point equidistant from two concentric circles is the circle concentric with the given circles and midway between them.*

Pr. 6: *The locus of a point at a given distance from a given point is a circle whose center is the given point and whose radius is the given distance.*

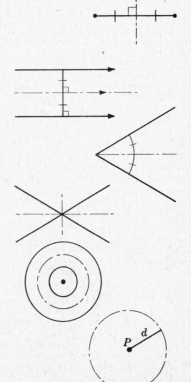

143

Pr. 7: *The locus of a point at a given distance from a given line is a pair of lines, parallel to the given line and at the given distance from the given line.*

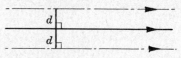

Pr. 8: *The locus of a point at a given distance from a given circle whose radius is greater than that distance is a pair of concentric circles, one on either side of the given circle and at the given distance from it.*

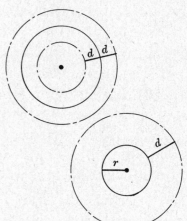

Pr. 9: *The locus of a point at a given distance from a given circle whose radius is less than the distance, is a circle outside the given circle and concentric to it.*

　Note: If $r = d$, the locus also includes the center of the given circle.

1.1 DETERMINING LOCI

Determine the locus of (a) a runner moving equidistant from the sides of a straight track, (b) a plane flying equidistant from two separated aircraft batteries, (c) a satellite 100 miles above the earth, (d) the furthermost point reached by a gun with a range of 10 miles.

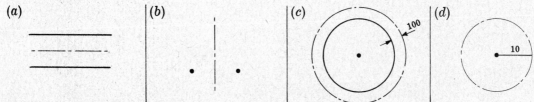

Solution:

(a) See Fig. (a). The locus is a line parallel to the two given lines and midway between them.

(b) See Fig. (b). The locus is the perpendicular bisector of the line joining the two points.

(c) See Fig. (c). The locus is a circle concentric with the earth and of radius 100 miles greater than that of the earth.

(d) See Fig. (d). The locus is a circle of radius 10 miles with its center at the gun.

1.2 DETERMINING the LOCUS of the CENTER of a CIRCLE

Determine the locus of the center of a circular disk (a) moving so that it touches each of two parallel lines, (b) moving tangentially to two concentric circles, (c) moving so that its rim passes through a fixed point, (d) rolling along a large fixed circular hoop.

(a)

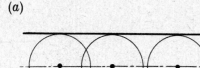

(b)

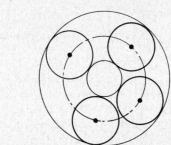

(c) 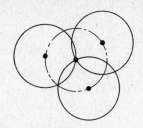   (d)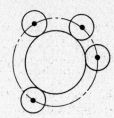

Solution:

(a) See Fig. (a). The locus is a line parallel to the two given lines and midway between them.

(b) See Fig. (b). The locus is a circle concentric to the given circles and midway between them.

(c) See Fig. (c). The locus is a circle whose center is the given point and whose radius is the radius of the circular disk.

(d) See Fig. (d). The locus is a circle outside the given circle and concentric to it.

1.3 CONSTRUCTING LOCI

Construct each of the following loci:

(a) The locus of a point equidistant from two given points.

(b) The locus of a point equidistant from two given parallel lines.

(c) The locus of a point at a given distance from a given circle whose radius is less than that distance.

Solution:

(a) 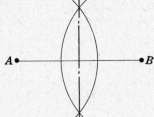   (b) 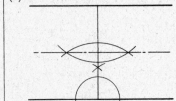   (c)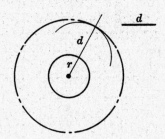

2. Locating Points by Means of Intersecting Loci

A point or points which satisfy two conditions may be found by drawing the locus for each condition. The required points are the points of intersection of the two loci.

2.1 LOCATING POINTS that SATISFY TWO CONDITIONS

On a map locate buried treasure that is 3 feet from a tree (T) and equidistant from two points (A and B).

Solution:

The required loci are (1) the perpendicular bisector of AB and (2) a circle with its center at T and of radius 3 feet.

As shown, these meet in P_1 and P_2, which are the locations of the treasure.

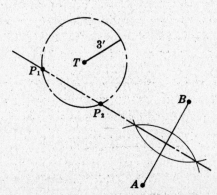

Note: The diagram shows the two loci intersecting at P_1 and P_2. However, there are three possible answers, depending on the location of T with respect to A and B:

(a) There are two points if the loci intersect.

(b) There is one point if the perpendicular bisector is tangent to the circle.

(c) There is no point if the perpendicular bisector does not meet the circle.

3. *Proving a Locus*

To prove that a locus satisfies a given condition, it **is necessary** to prove the locus theorem *and* its converse or inverse.

Thus to prove that a circle *A* of radius 2 inches is the **locus of a** point two inches from *A*, it is necessary to prove either (*a*) or (*b*) as follows:

a. (1) A point on circle *A* is 2 inches from *A*.
  (2) A point 2 inches from *A* is on circle *A*. (Converse of 1)

b. (1) A point on circle *A* is 2 inches from *A*.
  (2) A point not on circle *A* is not 2 inches from *A*. (Inverse of 1)

These statements are easily proved using the principle that **a point is outside, on** or inside a circle according as its distance from the center is greater than, equal to or **less than the** radius of the circle.

3.1 PROVING a LOCUS THEOREM

Prove: The locus of a point equidistant from two **given points** is the perpendicular bisector of the line joining the two **points.**

(*a*) First prove that any point on the locus satisfies the **condition.**

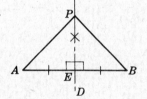

Given: Points *A* and *B*. *CD* is the perpendicular bisector of *AB*.

To Prove: Any point *P* on *CD* is equidistant from *A* and *B*; that is, $PA = PB$.

Plan: Prove $\triangle PEA \cong \triangle PEB$ to obtain $PA = PB$.

PROOF: Statements	Reasons
1. *CD* is the perpendicular bisector of *AB*.	1. Given
2. $\angle PEA = \angle PEB$	2. Perpendiculars **form right angles; all right angles are** equal.
3. $AE = EB$	3. To bisect is **to divide into two equal parts.**
4. $PE = PE$	4. Identity
5. $\triangle PEA \cong \triangle PEB$	5. s.a.s. = s.a.s.
6. $PA = PB$	6. Corresponding **parts of congruent triangles are equal.**

(*b*) Then prove that any point satisfying the condition **is on the locus.**

Given: Any point *Q* which is equidistant from points *A* and *B* ($QA = QB$).

To Prove: *Q* is on the perpendicular bisector of *AB*.

Plan: Draw *QG* perpendicular to *AB* and prove by congruent triangles that *QG* bisects *AB*.

PROOF: Statements	Reasons
1. Draw $QG \perp AB$	1. Through an **external point, a line can be drawn perpendicular to a given line.**
2. $QA = QB$	2. Given
3. $\angle QGA = \angle QGB$	3. Perpendiculars **form right angles; all right angles are** equal.
4. $QG = QG$	4. Identity
5. $\triangle QGA \cong \triangle QGB$	5. hy. leg = hy. leg
6. $AG = GB$	6. Corresponding **parts of congruent triangles are equal.**
7. *QG* bisects *AB*	7. To bisect is **to divide into two equal parts.**
8. *QG* is $\perp$ bisector of *AB*.	8. A line perpendicular **to another line and bisecting it is** its perpendicular **bisector.**

Supplementary Problems

1. Determine the locus of (*a*) the midpoint of a radius of a given circle, (*b*) the midpoint of a chord of a given circle parallel to a given line, (*c*) the midpoint of a chord of fixed length in a given circle, (*d*) the vertex of the right angle of a triangle having a given hypotenuse, (*e*) the vertex of an isosceles triangle having a given base, (*f*) the center of a circle which passes through two given points, (*g*) the center of a circle tangent to a given line at a given point on that line, (*h*) the center of a circle tangent to the sides of a given angle.   (*1.1*)

2. Determine the locus of (*a*) a boat moving so that it is equidistant from the parallel banks of a stream, (*b*) a swimmer maintaining the same distance from two floats, (*c*) a police helicopter in pursuit of a car which has just passed the junction of two straight roads and which may be on either one of them, (*d*) a treasure buried at the same distance from two intersecting straight roads.   (*1.1*)

3. Determine the locus of (*a*) a planet moving at a fixed distance from its sun, (*b*) a boat moving at a fixed distance from the coast of a circular island, (*c*) plants being laid at a distance of 20 ft. from a straight row of other plants, (*d*) the outer extremity of a clock hand.   (*1.1*)

4. Excluding points lying outside of rectangle *ABCD*, as shown in Fig. (*a*), state the locus of a point which is   (*1.1*)

(*a*) equidistant from *AD* and *BC*,   (*e*) 5 units from *BC*,

(*b*) equidistant from *AB* and *CD*,   (*f*) 10 units from *AB*,

(*c*) equidistant from *A* and *B*,   (*g*) 20 units from *CD*,

(*d*) equidistant from *B* and *C*,   (*h*) 10 units from *B*.

5. State the locus of a point in rhombus *ABCD* shown in Fig. (*b*) which is equidistant from   (*1.1*)

(*a*) *AB* and *AD*,   (*c*) *A* and *C*,   (*e*) each of the four sides.

(*b*) *AB* and *BC*,   (*d*) *B* and *D*,

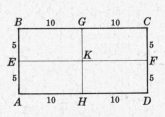

Fig. (*a*)

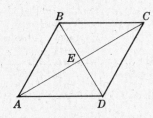

Fig. (*b*)

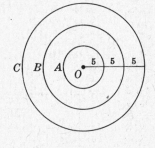

Fig. (*c*)

6. Refer to Fig. (*c*). State the locus of a point which is on or inside circle *C* and is   (*1.1, 1.2*)

(*a*) 5 units from *O*,   (*e*) 10 units from circle *A*,

(*b*) 15 units from *O*,   (*f*) 5 units from circle *B*,

(*c*) equidistant from circles *A* and *C*,   (*g*) the center of a circle tangent to circles *A* and *C*.

(*d*) 10 units from circle *C*,

7. Determine the locus of the center of (*a*) a coin rolling around and touching a smaller coin, (*b*) a coin rolling around and touching a larger coin, (*c*) a wheel moving between two parallel bars and touching both of them, (*d*) a wheel moving along a straight metal bar and touching it.   (*1.2*)

8. State the locus of a point that is in rectangle *ABCD* and is the center of a circle   (*1.2*)

(*a*) tangent to *AD* and *BC*,   (*d*) of radius 10, tangent to *BC*,

(*b*) tangent to *AB* and *CD*,   (*e*) of radius 20, tangent to *AD*,

(*c*) tangent to *AD* and *EF*,   (*f*) tangent to *BC* at *G*.

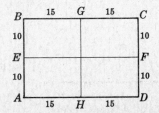

9. Locate each of the following. (2.1)

(a) Treasure that is buried 5 ft. from a straight fence and equidistant from two given points where the fence meets the ground.

(b) Points that are 3 ft. from a circle whose radius is 2 ft. and also equidistant from two lines which are parallel to each other and tangent to the circle.

(c) A point equidistant from the three vertices of a given triangle.

(d) A point equidistant from two given points and also equidistant from two given parallels.

(e) Points equidistant from two given intersecting lines and also 5 ft. from their intersection.

(f) A point that is equidistant from the sides of an angle and $\frac{1}{2}$ inch from their intersection.

10. Locate the point or points which satisfy the following conditions with respect to △ABC: (2.1)

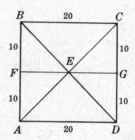

(a) Equidistant from its sides.

(b) Equidistant from its vertices.

(c) Equidistant from A and B and also from AB and BC.

(d) Equidistant from BC and AC and 5 units from C.

(e) 5 units from B and 10 units from A.

11. Excluding points lying outside of square $ABCD$, how many points are there that are (2.1)

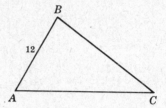

(a) equidistant from its vertices,

(b) equidistant from its sides,

(c) 5 units from E and on one of the diagonals,

(d) 5 units from E and equidistant from AD and BC,

(e) 5 units from FG and equidistant from AB and CD,

(f) 20 units from A and 10 units from B.

12. Prove that the locus of a point equidistant from the sides of the angle is the bisector of the angle.

(3.1)

Chapter 12

Coordinate Geometry

1. Graphs

A *number scale* is a line on which distances from a point are numbered in equal units, positively in one direction and negatively in the other.

The *origin* is the zero point from which distances are numbered.

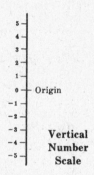

Vertical Number Scale

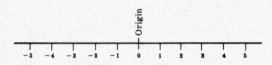

Horizontal Number Scale

The *graph* shown is formed by combining two number scales at right angles to each other so that their zero points coincide. The horizontal number scale is called the *x*-axis and the vertical number scale is the *y*-axis. The origin is the point where the two scales cross each other.

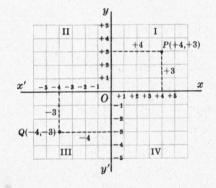

A *point* is located on a graph by its *coordinates*, which are its distances from the axes. The *abscissa* or *x-coordinate* of a point is its distance from the *y*-axis. The *ordinate* or *y-coordinate* of a point is its distance from the *x*-axis.

In stating the coordinates of a point, the *x*-coordinate precedes the *y*-coordinate. Thus the coordinates for P are written $(4, 3)$, those for Q, $(-4, -3)$.

The quadrants of a graph are the four parts cut off by the axes. These are numbered I, II, III and IV in a counterclockwise direction, as shown.

1.1 LOCATING POINTS on a GRAPH

Locate each point on the graph shown.

(a) B (c) O (e) N (g) P (i) R

(b) M (d) L (f) G (h) Q (j) S

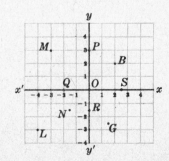

Ans. (a) $(2, 2)$ (c) $(0, 0)$ (e) $(-1\frac{1}{2}, -1\frac{1}{2})$ (g) $(0, 3)$ (i) $(0, -1\frac{1}{2})$

     (b) $(-3, 3)$ (d) $(-4, -3)$ (f) $(1\frac{1}{2}, -2\frac{1}{2})$ (h) $(-2, 0)$ (j) $(2\frac{1}{2}, 0)$

1.2 COORDINATES of POINTS in the FOUR QUADRANTS

State the signs of the coordinates of (a) a point in quadrant I, (b) a point in quadrant II, (c) a point in quadrant III, (d) a point in quadrant IV. State the sign or zero value of the coordinates of a point between quadrants (e) IV and I, (f) I and II, (g) II and III, (h) III and IV.

Ans.   (a) $(+,+)$ (b) $(-,+)$ (c) $(-,-)$ (d) $(+,-)$ (e) $(+,0)$ (f) $(0,+)$ (g) $(-,0)$ (h) $(0,-)$

1.3 GRAPHING a QUADRILATERAL

If the vertices of a rectangle have the coordinates $A(3,1)$, $B(-5,1)$, $C(-5,-3)$ and $D(3,-3)$, find its perimeter and area.

Solution:

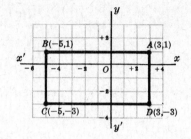

The base and height of the rectangle are 8 and 4. Hence the perimeter $= 2b + 2h = 2(8) + 2(4) = 24$, and the area $= bh = (8)(4) = 32$.

1.4 GRAPHING a TRIANGLE

If the vertices of a triangle have the coordinates $A(4\frac{1}{2}, -2)$, $B(-2\frac{1}{2}, -2)$ and $C(1,5)$, find its area.

Solution:

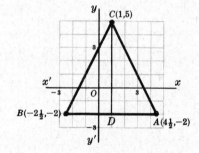

The base, $BA = 7$. The height, $CD = 7$.

Then $A = \frac{1}{2}bh = \frac{1}{2}(7)(7) = 24\frac{1}{2}$.

2. *Midpoint of a Line Segment*

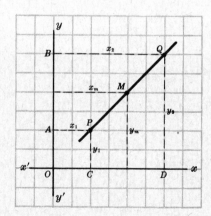

The coordinates (x_m, y_m) of the midpoint M of the line segment joining $P(x_1, y_1)$ to $Q(x_2, y_2)$ are

$$x_m = \tfrac{1}{2}(x_1 + x_2) \quad \text{and} \quad y_m = \tfrac{1}{2}(y_1 + y_2)$$

In the diagram shown, y_m is the median of trapezoid $CPQD$ whose bases are y_1 and y_2. Since a median is one-half the sum of the bases, $y_m = \frac{1}{2}(y_1 + y_2)$. Similarly, x_m is the median of trapezoid $ABQP$ whose bases are x_1 and x_2; hence $x_m = \frac{1}{2}(x_1 + x_2)$.

2.1 APPLYING the MIDPOINT FORMULA

If M is the midpoint of PQ, find the coordinates of (a) M if the coordinates of P and Q are $P(3,4)$ and $Q(5,8)$, (b) Q if the coordinates of P and M are $P(1,5)$ and $M(3,4)$.

Solution:

(a) $x_m = \frac{1}{2}(x_1 + x_2) = \frac{1}{2}(3 + 5) = 4$; $y_m = \frac{1}{2}(y_1 + y_2) = \frac{1}{2}(4 + 8) = 6$.

(b) $x_m = \frac{1}{2}(x_1 + x_2)$, $3 = \frac{1}{2}(1 + x_2)$, $x_2 = 5$; $y_m = \frac{1}{2}(y_1 + y_2)$, $4 = \frac{1}{2}(5 + y_2)$, $y_2 = 3$.

2.2 DETERMINING if LINES BISECT EACH OTHER

The vertices of a quadrilateral are $A(0,0)$, $B(0,3)$, $C(4,3)$ and $D(4,0)$.

(a) Show that $ABCD$ is a rectangle.

(b) Show that the midpoint of AC is also the midpoint of BD.

(c) Do the diagonals bisect each other? Why?

Solution:

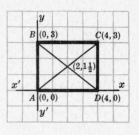

(a) $AB = CD = 3$, $BC = AD = 4$; hence $ABCD$ is a parallelogram.

Since $\angle BAD$ is a right angle, $ABCD$ is a rectangle.

(b) Coordinates of midpoint of AC: $x = \frac{1}{2}(0+4) = 2$, $y = \frac{1}{2}(0+3) = 1\frac{1}{2}$.

Coordinates of midpoint of BD: $x = \frac{1}{2}(0+4) = 2$, $y = \frac{1}{2}(3+0) = 1\frac{1}{2}$.

Hence $(2, 1\frac{1}{2})$ is the midpoint of AC and BD.

(c) Yes, since the midpoint of each diagonal is their common point.

3. *Distance Between Two Points*

Pr. 1: *The distance between two points having the same ordinate (or y-value) is the absolute value of the difference of their abscissas.* Hence, the distance between two points must be *positive*.

Thus the distance between the points $P(6,1)$ and $Q(9,1)$ is $9 - 6 = 3$.

Pr. 2: *The distance between two points having the same abscissa (or x-value) is the absolute value of the difference of their ordinates.*

Thus the distance between the points $P(2,1)$ and $Q(2,4)$ is $4 - 1 = 3$.

Pr. 3: *The distance d between the points $P_1(x_1, y_1)$ and $P_2(x_2, y_2)$ is*

$$d = \sqrt{(x_2 - x_1)^2 + (y_2 - y_1)^2} \quad or \quad d = \sqrt{(\Delta x)^2 + (\Delta y)^2}$$

The difference $x_2 - x_1$ is denoted by the symbol Δx; the difference $y_2 - y_1$ is denoted by Δy. Delta is the fourth letter of the Greek alphabet, corresponding to our d. These differences, Δx or Δy, may be positive or negative.

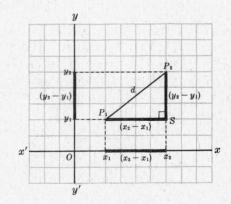

By Pr. 1, $P_1 S = x_2 - x_1 = \Delta x$.

By Pr. 2, $P_2 S = y_2 - y_1 = \Delta y$.

In right triangle $P_1 S P_2$,

$$(P_1 P_2)^2 = (P_1 S)^2 + (P_2 S)^2$$

or $\qquad d^2 = (x_2 - x_1)^2 + (y_2 - y_1)^2$

and $\qquad d = \sqrt{(x_2 - x_1)^2 + (y_2 - y_1)^2}$

Thus the distance from point $(2,5)$ to point $(6,8)$ equals 5, as follows:

$$(x, y)$$

$P_2 \quad (6,8) \rightarrow x_2 = 6, \quad y_2 = 8$

$P_1 \quad (2,5) \rightarrow x_1 = 2, \quad y_1 = 5$

$$d^2 = (x_2 - x_1)^2 + (y_2 - y_1)^2$$
$$d^2 = (6 - 2)^2 + (8 - 5)^2$$
$$d^2 = 4^2 + 3^2 = 25, \quad d = 5$$

3.1 FINDING the DISTANCE BETWEEN TWO POINTS

Find the distance between the points (a) $(-3, 5)$ and $(1, 5)$, (b) $(3, -2)$ and $(3, 4)$, (c) $(3, 4)$ and $(6, 8)$, (d) $(-3, 2)$ and $(9, -3)$.

Solution:

(a) Since both points have the same ordinate (or y-value), $d = x_2 - x_1 = 1 - (-3) = 4.$

(b) Since both points have the same abscissa (or x-value), $d = y_2 - y_1 = 4 - (-2) = 6.$

(c) $d = \sqrt{(x_2 - x_1)^2 + (y_2 - y_1)^2} = \sqrt{(6-3)^2 + (8-4)^2} = \sqrt{3^2 + 4^2} = 5$

(d) $d = \sqrt{(x_2 - x_1)^2 + (y_2 - y_1)^2} = \sqrt{[9 - (-3)]^2 + (-3 - 2)^2} = \sqrt{12^2 + (-5)^2} = 13$

3.2 APPLYING the DISTANCE FORMULA to a TRIANGLE

(a) Find the lengths of the sides of a triangle whose vertices are $A(1, 1)$, $B(1, 4)$ and $C(5, 1)$.

(b) Show that the triangle whose vertices are $G(2, 10)$, $H(3, 2)$ and $J(6, 4)$ is a right triangle.

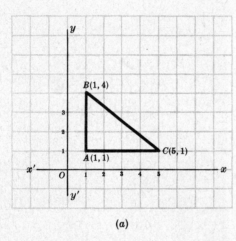

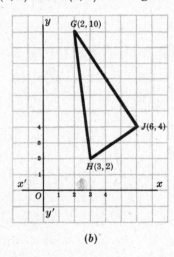

(a)   (b)

Solution:

(a) $AC = 5 - 1 = 4,$ $AB = 4 - 1 = 3,$ $BC = \sqrt{(5-1)^2 + (1-4)^2} = \sqrt{4^2 + (-3)^2} = 5.$

(b) $(GJ)^2 = (6-2)^2 + (4-10)^2 = 52,$ $(HJ)^2 = (6-3)^2 + (4-2)^2 = 13,$ $(GH)^2 = (2-3)^2 + (10-2)^2 = 65.$

Thus since $\overline{GJ}^2 + \overline{HJ}^2 = \overline{GH}^2,$ $\triangle GHJ$ is a right triangle.

3.3 APPLYING the DISTANCE FORMULA to a PARALLELOGRAM

The coordinates of the vertices of a quadrilateral are $A(2, 2)$, $B(3, 5)$, $C(6, 7)$ and $D(5, 4)$. Show that $ABCD$ is a parallelogram.

Solution:

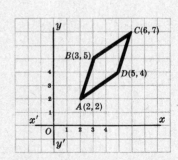

$$AB = \sqrt{(3-2)^2 + (5-2)^2} = \sqrt{1^2 + 3^2} = \sqrt{10}.$$

$$CD = \sqrt{(6-5)^2 + (7-4)^2} = \sqrt{1^2 + 3^2} = \sqrt{10}.$$

$$BC = \sqrt{(6-3)^2 + (7-5)^2} = \sqrt{3^2 + 2^2} = \sqrt{13}.$$

$$AD = \sqrt{(5-2)^2 + (4-2)^2} = \sqrt{3^2 + 2^2} = \sqrt{13}.$$

Hence $AB = CD$ and $BC = AD$. Since the opposite sides are equal, $ABCD$ is a parallelogram.

3.4 APPLYING the DISTANCE FORMULA to a CIRCLE

A circle is tangent to the x-axis and has its center at $(6, 4)$. Where is the point $(9, 7)$ with respect to the circle?

Solution:

Since the circle is tangent to the x-axis, AQ is a radius. By Pr. 2, $AQ = 4$.

By Pr. 3, $BQ = \sqrt{(9-6)^2 + (7-4)^2} = \sqrt{3^2 + 3^2} = \sqrt{18}$.

Since $\sqrt{18}$ is greater than 4, BQ is greater than a radius and B is outside the circle.

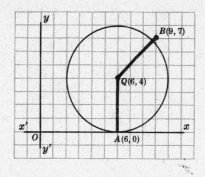

4. Slope of a Line

Pr. 1: *If a line passes through the points $P_1(x_1, y_1)$ and $P_2(x_2, y_2)$, then*

$$slope\ of\ P_1P_2 \;=\; \frac{y_2 - y_1}{x_2 - x_1} \;=\; \frac{\Delta y}{\Delta x}$$

Pr. 2: *The slope of the line whose equation is $y = mx + b$ is m.*

Pr. 3: *The slope of a line equals the tangent of its inclination, where the inclination i is the angle above the x-axis and included between the line and the positive direction of the x-axis. Then*

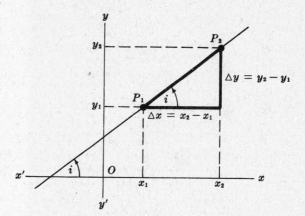

$$slope\ of\ P_1P_2 \;=\; \frac{y_2 - y_1}{x_2 - x_1} \;=\; \frac{\Delta y}{\Delta x} \;=\; m \;=\; \tan i$$

The slope is independent of the order in which the endpoints are selected. Thus

$$slope\ of\ P_1P_2 \;=\; \frac{y_2 - y_1}{x_2 - x_1} \;=\; \frac{y_1 - y_2}{x_1 - x_2} \;=\; slope\ of\ P_2P_1$$

A. Positive and Negative Slope

Pr. 4: *If a line slants upward from left to right, its inclination i is an acute angle and its slope is positive. (Fig. 1)*

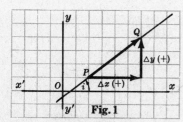

Slope of $PQ = \dfrac{\Delta y}{\Delta x} = \dfrac{+}{+} = +$

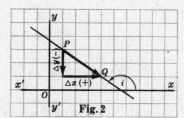

Slope of $PQ = \dfrac{\Delta y}{\Delta x} = \dfrac{-}{+} = -$

Pr. 5: *If a line slants downward from left to right, its inclination i is an obtuse angle and its slope is negative. (Fig. 2)*

Pr. 6: *If a line is parallel to the x-axis, its inclination is 0° and its slope is 0.* (Fig. 3)

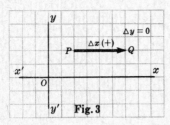

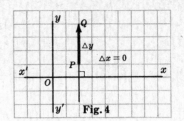

Slope of $PQ = \dfrac{\Delta y}{\Delta x} = \dfrac{0}{+} = 0$ Slope of $PQ = \dfrac{\Delta y}{\Delta x} = \dfrac{+}{0}$ (meaningless)

Pr. 7: *If a line is perpendicular to the x-axis, its inclination is 90° and it has no slope.* (Fig. 4)

B. Slopes of Parallel and Perpendicular Lines

Pr. 8: *Parallel lines have the same slope.*

    In Fig. 5, $l \parallel l'$; hence corresponding angles i and i' are equal, and $\tan i = \tan i'$ or $m = m'$, where m and m' are the slopes of l and l'.

Pr. 9: *Lines having the same slope are parallel to each other.* (Converse of Pr. 8.)

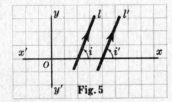

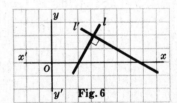

Pr. 10: *Perpendicular lines have slopes that are negative reciprocals of each other.* Negative reciprocals are numbers, such as 2/5 and −5/2, whose product is −1.

    Thus in Fig. 6, if $l \perp l'$, then $m = -1/m'$ or $mm' = -1$, where m and m' are the slopes of l and l'.

Pr. 11: *Lines whose slopes are negative reciprocals of each other are perpendicular.* (Converse of Pr. 10.)

C. Collinear Points

    Collinear points are points which lie on the same straight line.

    Thus A, B and C on PQ are collinear points.

Pr. 12: *The slope of a straight line is a constant between any two points of the line.*

    Thus if PQ is a straight line, the slope of the segment from A to B equals the slope of the segment from B to C.

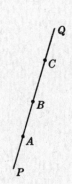

Pr. 13: *If the slope of a segment between a first point and a second equals the slope of the segment between either point and a third, then the points are collinear.*

4.1 SLOPE and INCLINATION of a LINE

(a) Find the slope of the line through $(-2, -1)$ and $(4, 3)$.

(b) Find the slope of the line whose equation is $3y - 4x = 15$.

(c) Find the inclination of the line whose equation is $y = x + 4$.

Solution:

(a) $m = \dfrac{y_2 - y_1}{x_2 - x_1} = \dfrac{3 - (-1)}{4 - (-2)} = \dfrac{4}{6} = \dfrac{2}{3}$ (b) Since $3y - 4x = 15$, $y = \frac{4}{3}x + 5$ and $m = \frac{4}{3}$.

(c) Since $y = x + 4$, we have $m = 1$, $\tan i = 1$ and $i = 45°$.

4.2 SLOPES of PARALLEL or PERPENDICULAR LINES

Find the slope of CD if (a) $AB \parallel CD$ and the slope of AB is 2/3, (b) $AB \perp CD$ and the slope of AB is 3/4.

Solution:

(a) By Pr. 8:   slope of CD = slope of AB = 2/3.

(b) By Pr. 10:   slope of CD = $-\dfrac{1}{\text{slope of } AB}$ = $-1 \div 3/4$ = $-\dfrac{4}{3}$.

4.3 APPLYING PRINCIPLES 9 and 11 to TRIANGLES and QUADRILATERALS

Complete each of the following statements.

(a) In quadrilateral $ABCD$, if the slopes of AB, BC, CD and DA are $\frac{1}{2}$, -2, $\frac{1}{2}$ and -2 respectively, the quadrilateral is a ( ).

(b) In triangle LMP, if the slopes of LM and MP are 5 and $-1/5$, then LMP is a ( ) triangle.

Solution:

(a) Since the slopes of the opposite sides are equal, $ABCD$ is a parallelogram.

The slopes of the adjacent sides are negative reciprocals; hence $ABCD$ is a rectangle.

(b) Since the slopes of LM and MP are negative reciprocals, $LM \perp MP$ and the triangle is a right triangle.

4.4 APPLYING PRINCIPLE 12

(a) Determine the line segments that have a slope of 2 if AB has a slope of 2 and the points A, B and C are collinear.

(b) Find y if $G(1, 4)$, $H(3, 2)$ and $J(9, y)$ are collinear.

Solution:

(a) By Pr. 12: AC and BC have a slope of 2.

(b) By Pr. 12:   slope of GJ = slope of GH.   Hence $\dfrac{y - 4}{9 - 1} = \dfrac{2 - 4}{3 - 1}$, $\dfrac{y - 4}{8} = \dfrac{-2}{2} = -1$ and $y = -4$.

5. *Locus in Coordinate Geometry*

The locus of a point is the set of points, and only the points, satisfying a given condition. In coordinate geometry, a line or curve (or set of lines or curves) on a graph is the locus of a point that satisfies the equation of the line or curve.

Think also of the locus as the path of a point moving according to a given condition or as the set of points satisfying a given condition.

Pr. 1: *The locus of a point whose abscissa is a constant k is a line parallel to the y-axis; its equation is x = k.*

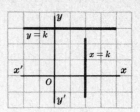

Pr. 2: *The locus of a point whose ordinate is a constant k is a line parallel to the x-axis; its equation is y = k.*

Pr. 3: *The locus of a point whose ordinate equals the product of a constant m and its abscissa is a straight line passing through the origin; its equation is y = mx.* The constant *m* is the slope of the line.

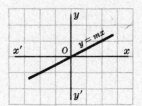

Pr. 4: *The locus of a point whose ordinate and abscissa are related by either of the equations*

$$y = mx + b, \quad \frac{y - y_1}{x - x_1} = m$$

where m and b are constants, is a straight line.

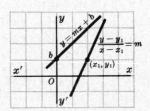

    In the equation $y = mx + b$, *m* is the slope and *b* is the *y*-intercept. In the equation $\frac{y - y_1}{x - x_1} = m$, the line passes through a fixed point (x_1, y_1) and has a slope of *m*.

Pr. 5: *The locus of a point such that the sum of the squares of its coordinates is a constant is a circle whose center is the origin.* The constant is the square of the radius, and the equation of the circle is

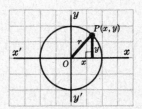

$$x^2 + y^2 = r^2$$

    Note that for any point $P(x, y)$ on the circle, $x^2 + y^2 = r^2$.

5.1 APPLYING PRINCIPLES 1 and 2

    Graph and state the equation of the locus of a point   (*a*) whose ordinate is −2, (*b*) 3 units from the *y*-axis, (*c*) equidistant from the points (3, 0) and (5, 0).

(*a*)   (*b*)   (*c*)

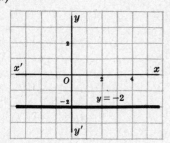   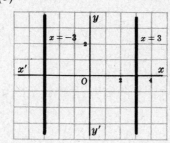   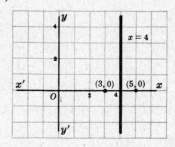

Ans.   (*a*) $y = -2$ (*b*) $x = 3$ and $x = -3$ (*c*) $x = 4$

5.2   APPLYING PRINCIPLES 3 and 4

Graph and describe the locus whose equation is

(a) $y = \frac{1}{3}x + 1$,   (b) $y = \frac{3}{2}x$,   (c) $\dfrac{y-1}{x-1} = \dfrac{3}{4}$.

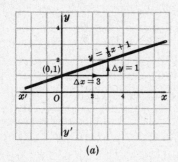   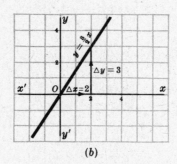   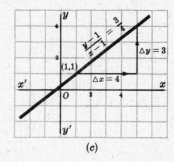

| (a) | (b) | (c) |

Solution:

(a)   The locus is a line whose y-intercept is 1 and whose slope equals 1/3.

(b)   The locus is a line which passes through the origin and has a slope of 3/2.

(c)   The locus is a line which passes through the point $(1, 1)$ and has a slope of 3/4.

5.3   APPLYING PRINCIPLE 5

Graph and state the equation of the locus of a point   (a) 2 units from the origin, (b) 2 units from the locus of $x^2 + y^2 = 9$.

(a)   (b)

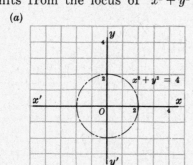   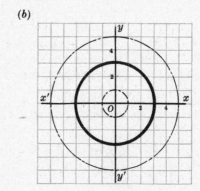

Solution:

(a)   The locus is a circle whose equation is $x^2 + y^2 = 4$.

(b)   The locus is a pair of circles whose equations are $x^2 + y^2 = 25$ and $x^2 + y^2 = 1$.

6.   *Areas in Coordinate Geometry*

A.   *Area of a Triangle*

(1)   If one side of a triangle is parallel to either axis, the length of that side and the length of the altitude to that side can be found readily, and the formula $A = \frac{1}{2}bh$ used.

(2)   If no side of a triangle is parallel to either axis,

    (a)   the triangle can be enclosed in a rectangle whose sides are parallel to the axes,

    (b)   trapezoids whose bases are parallel to the y-axis can be formed by dropping perpendiculars to the x-axis.

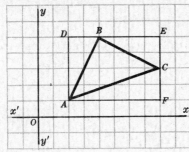
(a)

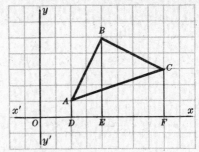
(b)

In Fig. (a), $\triangle ABC$ = rectangle $ADEF - (\triangle ABD + \triangle BCE + \triangle ACF)$.

In Fig. (b), $\triangle ABC$ = trapezoid $ABED$ + trapezoid $BEFC$ − trapezoid $DFCA$.

B. Area of a Quadrilateral

The trapezoid method shown above can be extended to find the area of a quadrilateral if its vertices are given.

6.1   AREA of a TRIANGLE HAVING a SIDE PARALLEL to EITHER AXIS

Find the area of the triangle whose vertices are $A(1,2)$, $B(7,2)$ and $C(5,4)$.

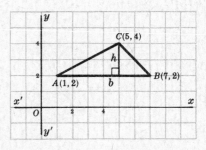

Solution:

$b = 7 - 1 = 6, \quad h = 4 - 2 = 2.$

Then $A = \frac{1}{2}bh = \frac{1}{2}(6)(2) = 6.$

6.2   AREA of a TRIANGLE HAVING NO SIDE PARALLEL to EITHER AXIS

Find the area of $\triangle ABC$ whose vertices are $A(2,4)$, $B(5,8)$ and $C(8,2)$ (a) using the rectangle method, (b) using the trapezoid method.

(a)

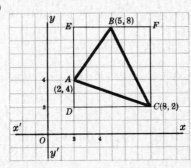

(b)

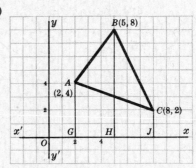

Solution:

(a) Area of rectangle $DEFC = bh = 6(6) = 36$. Area of $\triangle DAC = \frac{1}{2}bh = \frac{1}{2}(2)(6) = 6$.

Area of $\triangle ABE = \frac{1}{2}bh = \frac{1}{2}(3)(4) = 6$. Area of $\triangle BCF = \frac{1}{2}bh = \frac{1}{2}(3)(6) = 9$.

Area of $\triangle ABC = DEFC - (\triangle DAC + \triangle ABE + \triangle BCF) = 36 - (6 + 6 + 9) = 15$.

(b) Area of trapezoid $ABHG = \frac{1}{2}h(b + b') = \frac{1}{2}(3)(4 + 8) = 18$.

Area of trapezoid $BCJH = \frac{1}{2}(3)(2 + 8) = 15$.

Area of trapezoid $ACJG = \frac{1}{2}(6)(2 + 4) = 18$.

Area of $\triangle ABC = ABHG + BCJH - ACJG = 18 + 15 - 18 = 15$.

7. *Proving Theorems by Coordinate Geometry*

Many theorems of plane geometry can be proved by coordinate geometry. The procedure of proving a theorem has two major steps, as follows.

(1) *Place each figure in a convenient graphic position.*

Thus for the triangle, rectangle and parallelogram, place one vertex at the origin and one side of the figure on the x-axis. Indicate the coordinates of each vertex.

Triangle

Rectangle

Parallelogram

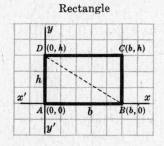 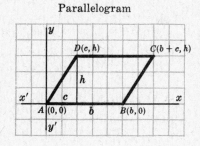

(2) *Apply the principles of coordinate geometry*

Thus: (*a*) Prove that lines are parallel by showing that their slopes are equal.

(*b*) Prove that lines are perpendicular by showing that their slopes are negative reciprocals of each other; that is, their product equals -1.

(*c*) Use the midpoint formula when the midpoint of the line is involved.

(*d*) Use the distance formula to obtain lengths of lines.

7.1 PROVING a THEOREM by COORDINATE GEOMETRY

Using coordinate geometry, prove that the diagonals of a parallelogram bisect each other.

Given: $\square ABCD$, diagonals AC and BD.

To Prove: AC and BD bisect each other.

Plan: Use midpoint formula to obtain the coordinates of midpoints of the diagonals.

Solution:

Placing $\square ABCD$ with vertex A at the origin and side AD along the x-axis, the vertices have the coordinates $A(0,0)$, $B(a,b)$, $C(a+c,b)$ and $D(c,0)$.

Using the midpoint formula, the midpoint of AC has the coordinates $\left(\dfrac{a+c}{2}, \dfrac{b}{2}\right)$ and the midpoint of BD has the coordinates $\left(\dfrac{a+c}{2}, \dfrac{b}{2}\right)$. Hence the diagonals bisect each other since the midpoint of each diagonal is their common point.

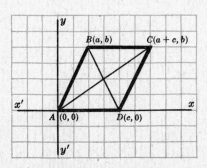

Supplementary Problems

1. State the coordinates of each lettered point on the graph shown in Fig. 1.   *(1.1)*

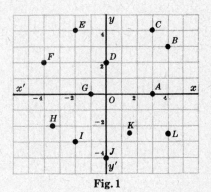

Fig. 1

2. Plot each point: *(1.2)*

   $A(-2, -3)$ $C(0, -1)$ $E(3, -4)$ $G(0, 3)$

   $B(-3, 2)$ $D(-3, 0)$ $F(1\frac{1}{2}, 2\frac{1}{2})$ $H(3\frac{1}{2}, 0)$

3. Plot each point: $A(2, 3)$, $B(-3, 3)$, $C(-3, -2)$, $D(2, -2)$. Find the perimeter and area of the square $ABCD$.   *(1.3)*

4. Plot each point: $A(4, 3)$, $B(-1, 3)$, $C(-3, -3)$, $D(2, -3)$. Find the area of parallelogram $ABCD$ and triangle BCD.

                                           (1.3, 1.4)

5. Find the midpoint of the line segment joining   *(2.1)*

   (*a*) $(0, 0)$ and $(8, 6)$ (*e*) $(-20, -5)$ and $(0, 0)$ (*i*) $(3, 4)$ and $(7, 6)$

   (*b*) $(0, 0)$ and $(5, 7)$ (*f*) $(0, 4)$ and $(0, 16)$ (*j*) $(-2, -8)$ and $(-4, -12)$

   (*c*) $(0, 0)$ and $(-8, 12)$ (*g*) $(8, 0)$ and $(0, -2)$ (*k*) $(7, 9)$ and $(3, 3)$

   (*d*) $(14, -10)$ and $(0, 0)$ (*h*) $(-10, 0)$ and $(0, -5)$ (*l*) $(2, -1)$ and $(-2, -5)$.

6. Find the midpoints of the sides of a triangle whose vertices are   *(2.1)*

   (*a*) $(0, 0)$, $(8, 0)$, $(0, 6)$ (*c*) $(12, 0)$, $(0, -4)$, $(0, 0)$ (*e*) $(4, 0)$, $(0, -6)$, $(-4, 10)$

   (*b*) $(-6, 0)$, $(0, 0)$, $(0, 10)$ (*d*) $(3, 5)$, $(5, 7)$, $(3, 11)$ (*f*) $(-1, -2)$, $(0, 2)$, $(1, -1)$.

7. Find the midpoints of the sides of the quadrilateral whose successive vertices are   *(2.1)*

   (*a*) $(0, 0)$, $(0, 4)$, $(2, 10)$, $(6, 0)$ (*c*) $(-2, 0)$, $(0, 4)$, $(6, 2)$, $(0, -10)$

   (*b*) $(-3, 5)$, $(-1, 9)$, $(7, 3)$, $(5, -1)$ (*d*) $(-3, -7)$, $(-1, 5)$, $(9, 0)$, $(5, -8)$

8. Find the midpoints of the diagonals of the quadrilateral whose successive vertices are   *(2.1)*

   (*a*) $(0, 0)$, $(0, 5)$, $(4, 12)$, $(8, 1)$,   (*b*) $(-4, -1)$, $(-2, 3)$, $(6, 1)$, $(2, -8)$ (*c*) $(0, -5)$, $(0, 1)$, $(4, 9)$, $(4, 3)$

9. Find the center of a circle if the endpoints of a diameter are   *(2.1)*

   (*a*) $(0, 0)$ and $(-4, 6)$ (*c*) $(-3, 1)$ and $(0, -5)$ (*e*) (a, b) and $(3a, 5b)$

   (*b*) $(-1, 0)$ and $(-5, -12)$ (*d*) $(0, 0)$ and $(2a, 2b)$ (*f*) $(a, 2b)$ and $(a, 2c)$.

10. If M is the midpoint of AB, find the coordinates of   *(2.1)*

   (*a*) M if the coordinates of A and B are $A(2, 5)$ and $B(6, 11)$,

   (*b*) A if the coordinates of M and B are $M(1, 3)$ and $B(3, 6)$,

   (*c*) B if the coordinates of A and M are $A(-2, 1)$ and $M(2, -1)$.

11. The trisection points of AD are B and C. Find the coordinates of   *(2.1)*

   (*a*) B if the coordinates of A and C are $A(1, 2)$ and $C(3, 5)$,

   (*b*) D if the coordinates of B and C are $B(0, 5)$ and $C(1\frac{1}{2}, 4)$,

   (*c*) A if the coordinates of B and C are $B(0, 6)$ and $C(2, 3)$.

12. $A(0, 0)$, $B(0, 5)$, $C(6, 5)$ and $D(6, 0)$ are the vertices of quadrilateral $ABCD$.   *(2.2)*

   (*a*) Prove that $ABCD$ is a rectangle.

   (*b*) Show that the midpoints of AC and BD have the same coordinates.

   (*c*) Do the diagonals bisect each other? Why?

13. The vertices of $\triangle ABC$ are $A(0, 0)$, $B(0, 4)$ and $C(6, 0)$.   *(2.2)*

   (*a*) If AD is the median to BC, find the coordinates of D and the midpoint of AD.

   (*b*) If CE is the median to AB, find the coordinates of E and the midpoint of CE.

   (*c*) Do the medians, AD and CE, bisect each other? Why?

14. Find the distance between each of the following pairs of points: $\qquad$ *(3.1)*
    (a) $(0,0)$ and $(0,5)$ (c) $(0,-3)$ and $(0,7)$ (e) $(5,3)$ and $(5,8.4)$ (g) $(-3,-4\frac{1}{2})$ and $(-3,4\frac{1}{2})$
    (b) $(4,0)$ and $(-2,0)$ (d) $(-6,-1)$ and $(-6,11)$ (f) $(-1.5,7)$ and $(6,7)$ (h) (a,b) and $(2a,b)$.

15. Find the distances between the following collinear points: $\qquad$ *(3.1)*
    (a) $(5,-2)$, $(5,1)$, $(5,4)$;   (b) $(0,-6)$, $(0,-2)$, $(0,12)$;   (c) $(-4,2)$, $(-3,2)$, $(0,2)$;   (d) $(0,b)$, (a,b), $(3a,b)$.

16. Find the distance between each of the following pairs of points: $\qquad$ *(3.1)*
    (a) $(0,0)$ and $(5,12)$ (d) $(4,1)$ and $(7,5)$ (g) $(2,2)$ and $(5,5)$ (j) $(-1,-1)$ and $(1,3)$
    (b) $(-3,-4)$ and $(0,0)$ (e) $(-3,-6)$ and $(3,2)$ (h) $(0,5)$ and $(-5,0)$ (k) $(-3,0)$ and $(0,\sqrt{7})$
    (c) $(0,-6)$ and $(9,6)$ (f) $(2,3)$ and $(-10,12)$ (i) $(3,4)$ and $(4,7)$ (l) $(a,0)$ and $(0,a)$

17. Show that the triangles having the following vertices are isosceles triangles: $\qquad$ *(3.2)*
    (a) $A(3,5)$, $B(6,9)$ and $C(2,6)$ (c) $G(5,-5)$, $H(-2,-2)$ and $J(8,2)$
    (b) $D(2,0)$, $E(6,0)$ and $F(4,4)$ (d) $K(7,0)$, $L(3,4)$ and $M(2,-1)$

18. Which of the triangles having the following vertices are right triangles? $\qquad$ *(3.2)*
    (a) $A(7,0)$, $B(6,3)$ and $C(12,5)$ (c) $G(1,-1)$, $H(5,0)$ and $J(3,8)$
    (b) $D(2,0)$, $E(5,2)$ and $F(1,8)$ (d) $K(-4,0)$, $L(-2,4)$ and $M(4,-1)$

19. The vertices of $\triangle ABC$ are $A(-2,2)$, $B(4,4)$ and $C(8,2)$. Find the length of the median to $\qquad$ *(3.2)*
    (a) AB, (b) AC, (c) BC.

20. (a) The vertices of quadrilateral $ABCD$ are $A(0,0)$, $B(3,2)$, $C(7,7)$, $D(4,5)$. Show that $ABCD$ is a parallelogram. $\qquad$ *(3.3)*
    (b) The vertices of quadrilateral $DEFG$ are $D(3,5)$, $E(1,1)$, $F(5,3)$, $G(7,7)$. Show that $DEFG$ is a rhombus.
    (c) The vertices of quadrilateral $HJKL$ are $H(0,0)$, $J(4,4)$, $K(0,8)$, $L(-4,4)$. Show that $HJKL$ is a square.

21. Find the radius of a circle that has its center at $\qquad$ *(3.4)*
    (a) $(0,0)$ and passes through $(-6,8)$,   (d) $(2,0)$ and passes through $(7,-12)$,
    (b) $(0,0)$ and passes through $(3,-4)$,   (e) $(4,3)$ and is tangent to the y-axis,
    (c) $(0,0)$ and passes through $(-5,5)$,   (f) $(-1,7)$ and is tangent to the line, $x=-4$.

22. A circle has its center at the origin and a radius of 10. State whether each of the following points is on, inside or outside of this circle: $\qquad$ *(3.4)*
    (a) $(6,8)$, (b) $(-6,8)$, (c) $(0,11)$, (d) $(-10,0)$, (e) $(7,7)$, (f) $(-9,4)$, (g) $(9,\sqrt{19})$.

23. Find the slope of the line through each pair of points: $\qquad$ *(4.1)*
    (a) $(0,0)$ and $(5,9)$ (d) $(2,3)$ and $(6,15)$ (g) $(3,-4)$ and $(5,6)$ (j) $(0,-2)$ and $(8,10)$
    (b) $(0,0)$ and $(9,5)$ (e) $(-2,-3)$ and $(7,15)$ (h) $(0,0)$ and $(-4,8)$ (k) $(-1,-5)$ and $(1,-7)$
    (c) $(0,0)$ and $(6,15)$ (f) $(-2,-3)$ and $(2,1)$ (i) $(3,-9)$ and $(0,0)$ (l) $(-3,-4)$ and $(-1,-2)$.

24. Find the slope of the line whose equation is $\qquad$ *(4.1)*
    (a) $y=3x-4$ (e) $y=5x$ (i) $3y=-12x+6$ (m) $\frac{1}{5}y=x-3$
    (b) $y=4x-3$ (f) $y=5$ (j) $3y=12-2x$ (n) $\frac{1}{3}y=2x-6$
    (c) $y=-\frac{1}{2}x+5$ (g) $2y=6x-10$ (k) $y+x=21$ (o) $\frac{1}{4}y=7-x$
    (d) $y=8-7x$ (h) $2y=10x-6$ (l) $2x=12-y$ (p) $\frac{1}{4}y+2x=1$.

25. Find the inclination, to the nearest degree, of each line: $\qquad$ *(4.1)*
    (a) $y=3x-1$ (c) $2y=5x+10$ (e) $5y=5x-3$
    (b) $y=\frac{1}{3}x-1$ (d) $y=\frac{2}{5}x+5$ (f) $y=-3$

26. Find the slope of a line whose inclination is $\qquad$ *(4.1)*
    (a) $5°$, (b) $17°$, (c) $20°$, (d) $35°$, (e) $45°$, (f) $73°$, (g) $85°$.

27. Find the inclination, to the nearest degree, of a line whose slope is $\qquad$ *(4.1)*
    (a) 0, (b) .4663, (c) 1, (d) 1.4281, (e) $\frac{1}{8}$, (f) $\frac{1}{2}$, (g) $\frac{3}{4}$, (h) $1\frac{1}{3}$, (i) $2\frac{1}{5}$.

28. In the hexagon $ABCDEF$, $CD \parallel AF$. Which sides or diagonals have
      a) positive slope, (b) negative slope, (c) zero slope, (d) no slope?

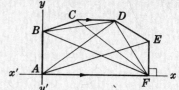

29. Find the slope of a line parallel to a line whose slope is   (4.2)
      (a) 0,   (b) no slope,   (c) 5,   (d) −5,   (e) .5,   (f) −.0005.

30. Find the slope of a line parallel to line whose equation is   (4.2)
      (a) $y = 0$,   (b) $x = 0$,   (c) $x = 7$,   (d) $y = 7$,   (e) $y = 5x - 2$,   (f) $x + y = 5$,   (g) $3y - 6x = 12$.

31. Find the slope of a line parallel to a line which passes through each pair of points: (4.2)
      (a) $(0,0)$ and $(2,3)$;   (b) $(2,-1)$ and $(5,6)$;   (c) $(3,4)$ and $(5,2)$;   (d) $(1,2)$ and $(0,-4)$.

32. Find the slope of a line perpendicular to a line whose slope is   (4.2)
      (a) $\frac{1}{2}$,   (b) 1,   (c) 3,   (d) $2\frac{1}{2}$,   (e) .1,   (f) −1,   (g) $-\frac{4}{5}$,   (h) $-3\frac{1}{4}$,   (i) 0,   (j) no slope.

33. Find the slope of a line perpendicular to a line which passes through each pair of points: (4.2)
      (a) $(0,0)$ and $(0,5)$;   (b) $(0,0)$ and $(2,1)$;   (c) $(0,0)$ and $(3,-1)$;   (d) $(1,1)$ and $(3,3)$.

34. In rectangle $DEFG$, the slope of $DE = \frac{2}{3}$. State the slope of   (a) EF,   (b) FG,   (c) DG.   (4 3)

35. In $\square ABCD$, the slope of $AB = 1$ and the slope of $BC = -\frac{1}{2}$. State the slope of   (4.3)
      (a) AD,   (b) CD,   (c) altitude to AD,   (d) altitude to CD.

36. The vertices of $\triangle ABC$ are $A(0,5)$, $B(3,7)$ and $C(5,-1)$. State the slope of the altitude to   (4 3)
      (a) AB,   (b) BC,   (c) AC.

37. Which of the following sets of points are collinear?   (4.4)
      (a) $(2,1)$, $(4,4)$, $(8,10)$;   (b) $(-1,1)$, $(2,4)$, $(6,8)$;   (c) $(1,-1)$, $(3,4)$, $(5,8)$.

38. State the value of k which will make each set of three points collinear: (4.4)
      (a) $A(0,1)$, $B(2,7)$, $C(6,k)$;   (b) $D(-1,5)$, $E(3,k)$, $F(5,11)$;   (c) $G(0,k)$, $H(1,1)$, $I(3,-1)$.

39. State the equation of the line or pair of lines which is the locus of a point   (5.1)
      (a) whose abscissa is −5,   (d) 5 units below x-axis,   (g) 6 units above line $y = -2$,
      (b) whose ordinate is $3\frac{1}{2}$,   (e) 4 units from y-axis,   (h) 1 unit to the right of y-axis,
      (c) 3 units from x-axis,   (f) 3 units from line $x = 2$,   (i) equidistant from lines $x = 5$ and $x = 13$.

40. State the equation of the locus of the center of a circle which   (5.3)
      (a) is tangent to the x-axis at $(6,0)$,   (d) passes through the origin and $(10,0)$,
      (b) is tangent to the y-axis at $(0,5)$,   (e) passes through $(3,7)$ and $(9,7)$,
      (c) is tangent to the lines $x = 4$ and $x = 8$,   (f) passes through $(3,-2)$ and $(3,8)$.

41. State the equation of the line or pair of lines which is the locus of a point   (5.1)
      (a) whose coordinates are equal,   (e) the sum of whose coordinates is 12,
      (b) whose ordinate is 5 more than its abscissa,   (f) the difference of whose coordinates is 2,
      (c) whose abscissa is 4 less than its ordinate,   (g) equidistant from the x-axis and y-axis,
      (d) whose ordinate exceeds its abscissa by 10,   (h) equidistant from $x + y = 3$ and $x + y = 7$.

42. Describe the locus of each of the following equations: (5.2)
      (a) $y = 2x + 5$,   (b) $\dfrac{y-3}{x-2} = 4$,   (c) $\dfrac{y+3}{x+2} = \dfrac{5}{4}$,   (d) $y = \frac{1}{2}x$,   (e) $x + y = 7$,   (f) $3y = x$

43. State the equation of a line which passes through the origin and has a slope of   (5.2)
      (a) 4,   (b) −2,   (c) 3/2,   (d) −2/5,   (e) 0.

44. State the equation of a line which has a y-intercept of   (5.2)
      (a) 5 and a slope of 4,   (d) 8 and is parallel to $y = 3x - 2$,
      (b) 2 and a slope of −3,   (e) −3 and is parallel to $y = 7 - 4x$,
      (c) −1 and a slope of $\frac{1}{3}$,   (f) 0 and is parallel to $y - 2x = 8$.

45. State the equation of a line which has a slope of 2 and passes through   (5.2)
    (a) $(1, 4)$, (b) $(-2, 3)$, (c) $(-4, 0)$, (d) $(0, -7)$.

46. State the equation of a line   (5.2)
    (a) which passes through the origin and has a slope of 4,
    (b) which passes through $(0, 3)$ and has a slope of $\frac{1}{2}$,
    (c) which passes through $(1, 2)$ and has a slope of 3,
    (d) which passes through $(-1, -2)$ and has a slope of $\frac{1}{3}$,
    (e) which passes through the origin and is parallel to a line which has a slope of 2.

47. (a) Describe the locus of the equation $x^2 + y^2 = 49$.   (5.3)
    (b) State the equation of the locus of a point 4 units from the origin.
    (c) State the equations of the locus of a point 3 units from the locus of $x^2 + y^2 = 25$.

48. State the equation of the locus of a point 5 units from   (5.3)
    (a) the origin, (b) the circle $x^2 + y^2 = 16$, (c) the circle $x^2 + y^2 = 49$.

49. State the radius of the circle whose equation is   (5.3)
    (a) $x^2 + y^2 = 9$, (b) $x^2 + y^2 = 16/9$, (c) $9x^2 + 9y^2 = 36$, (d) $x^2 + y^2 = 3$.

50. State the equation of a circle whose center is the origin and whose radius is   (5.3)
    (a) 4, (b) 11, (c) $\frac{2}{3}$, (d) $1\frac{1}{2}$, (e) $\sqrt{5}$, (f) $\frac{1}{2}\sqrt{3}$.

51. Find the area of $\triangle ABC$ whose vertices are $A(0, 0)$ and   (6.1)
    (a) $B(0, 5)$ and $C(4, 5)$ (c) $B(0, 8)$ and $C(-5, 8)$ (e) $B(6, 2)$ and $C(7, 0)$
    (b) $B(0, 5)$ and $C(4, 2)$ (d) $B(0, 8)$ and $C(-5, 12)$ (f) $B(6, -5)$ and $C(10, 0)$

52. Find the area of   (6.1)
    (a) a triangle whose vertices are $A(0, 0)$, $B(3, 4)$ and $C(8, 0)$,
    (b) a triangle whose vertices are $D(1, 1)$, $E(5, 6)$ and $F(1, 7)$,
    (c) a rectangle three of whose vertices are $H(2, 2)$, $J(2, 6)$ and $K(7, 2)$,
    (d) a parallelogram three of whose vertices are $L(3, 1)$, $M(9, 1)$ and $P(5, 5)$.

53. Find the area of $\triangle DEF$ whose vertices are $D(0, 0)$ and   (6.2)
    (a) $E(6, 4)$ and $F(8, 2)$ (b) $E(3, 2)$ and $F(6, -4)$ (c) $E(-2, 3)$ and $F(10, 7)$.

54. Find the area of a triangle whose vertices are: (6.2)
    (a) $(0, 0)$, $(2, 3)$ and $(4, 1)$;   (b) $(1, 1)$, $(7, 3)$ and $(3, 6)$;   (c) $(-1, 2)$, $(0, -2)$ and $(3, 1)$.

55. The vertices of $\triangle ABC$ are $A(2, 1)$, $B(8, 9)$ and $C(5, 7)$.   (6.2)
    (a) Find the area of $\triangle ABC$. (b) Find the length of AB. (c) Find the length of the altitude to AB.

56. Find the area of a quadrilateral whose vertices are: (6.2)
    (a) $(3, 3)$, $(10, 4)$, $(8, 7)$, $(5, 6)$;   (c) $(0, 1)$, $(2, 4)$, $(8, 10)$ and $(12, 2)$.
    (b) $(0, 4)$, $(5, 8)$, $(10, 6)$, $(14, 0)$;

57. Find the area of the quadrilateral formed by the lines   (6.1)
    (a) $x = 0$, $x = 5$, $y = 0$ and $y = 6$,   (d) $x = 0$, $x = 6$, $y = 0$ and $y = x + 1$,
    (b) $x = 0$, $x = 7$, $y = -2$ and $y = 5$,   (e) $y = 0$, $y = 4$, $y = x$ and $y = x + 4$,
    (c) $x = -3$, $x = 5$, $y = 3$ and $y = -8$,   (f) $y = 0$, $y = 6$, $y = 2x$ and $y = 2x + 6$.

58. Prove by coordinate geometry: (7.1)
    (a) A line joining the midpoints of two sides of a triangle is parallel to the third side.
    (b) The diagonals of a rhombus are perpendicular to each other.
    (c) The median to the base of an isosceles triangle is perpendicular to the base.
    (d) A line joining the midpoints of two sides of a triangle equals one-half of the third side.
    (e) If the midpoints of the sides of a rectangle taken in succession are joined, the quadrilateral formed
        is a rhombus.
    (f) In a right triangle, the median to the hypotenuse is one-half of the hypotenuse.

Chapter 13

Inequalities and Indirect Reasoning

1. Inequalities

An inequality is a statement that quantities are not equal. If two quantities are unequal, the first is either greater than or less than the second. The inequality symbols are: (a) $\neq$, meaning unequal to; (b) $>$, meaning greater than; (c) $<$, meaning less than. Thus $4 \neq 3$ is read "four is unequal to three", $7 > 2$ is read "seven is greater than two" and $1 < 5$ is read "one is less than five".

Two inequalities may be of the same order or of opposite order. In inequalities of the same order, the same inequality symbol is used; in inequalities of the opposite order, opposite inequality symbols are used. Thus $5 > 3$ and $10 > 7$ are inequalities of the same order, and $5 > 3$ and $7 < 10$ are inequalities of opposite order.

Inequalities of the same order may be combined, as follows. $x < y$ and $y < z$ may be combined into $x < y < z$; that is, y is greater than x and less than z. $a > b$ and $b > c$ may be combined into $a > b > c$; that is, b is less than a and greater than c.

A. Inequality Axioms

Ax. 1: *A quantity may be substituted for its equal in any inequality.*

Thus if $x > y$ and $y = 10$, then $x > 10$.

Ax. 2: *If the first of three quantities is greater than the second, and the second is greater than the third, then the first is greater than the third.*

Thus if $x > y$ and $y > z$, then $x > z$.

Ax. 3: *The whole is greater than any of its parts.*

Thus $AB > AM$ and $\angle BAD > \angle BAC$.

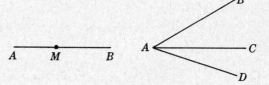

B. Inequality Axioms of Operation

Ax. 4: *If equals are added to unequals, the sums are unequal in the same order.*

Thus since $5 > 4$ and $4 = 4$, then $5 + 4 > 4 + 4$ or $9 > 8$. If $x - 4 < 5$, then $x - 4 + 4 < 5 + 4$ or $x < 9$.

Ax. 5: *If unequals are added to unequals of the same order, the sums are unequal in the same order.*

Thus since $5 > 3$ and $4 > 1$, then $5 + 4 > 3 + 1$ or $9 > 4$. If $2x - 4 < 5$ and $x + 4 < 8$, then $2x - 4 + x + 4 < 5 + 8$ or $3x < 13$.

Ax. 6: *If equals are subtracted from unequals, the differences are unequal in the same order.*

Thus since $10 > 5$ and $3 = 3$, then $10 - 3 > 5 - 3$ or $7 > 2$. If $x + 6 < 9$ and $6 = 6$, then $x + 6 - 6 < 9 - 6$ or $x < 3$.

Ax. 7: *If unequals are subtracted from equals, the differences are unequal in the opposite order.*

Thus since $10 = 10$ and $5 > 3$, then $10 - 5 < 10 - 3$ or $5 < 7$. If $x + y = 12$ and $y > 5$, then $x + y - y < 12 - 5$ or $x < 7$.

Ax. 8: *If unequals are multiplied by the same positive number, the products are unequal in the same order.*

Thus if $\frac{1}{4}x < 5$, then $4(\frac{1}{4}x) < 4(5)$ or $x < 20$.

Ax. 9: *If unequals are multiplied by the same negative number, the results are unequal in the opposite order.*

Thus if $\frac{1}{2}x < 5$, then $(-2)(\frac{1}{2}x) > (-2)(5)$ or $-x > -10$ or $x < 10$.

Ax. 10: *If unequals are divided by the same positive number, the results are unequal in the same order.*

Thus if $4x > 20$, then $\frac{4x}{4} > \frac{20}{4}$ or $x > 5$.

Ax. 11: *If unequals are divided by the same negative number, the results are unequal in the opposite order.*

Thus if $-7x < 42$, then $\frac{-7x}{-7} > \frac{42}{-7}$ or $x > -6$.

C. Inequality Postulate

Post. 1: *A straight line is the shortest distance between two points.*

D. Triangle Inequality Theorems

Pr. 1: *The sum of two sides of a triangle is greater than the third side.* (Corollary: The longest side of a triangle is less than the sum of the other two sides and greater than their difference.)

Thus in Fig. 1, $BC + CA > AB$.

Pr. 2: *In a triangle, an exterior angle is larger than either nonadjacent interior angle.*

Thus in Fig. 1, $\angle BCD > \angle BAC$ and $\angle BCD > \angle ABC$.

Pr. 3: *If two sides of a triangle are unequal, the angles opposite these sides are unequal, the larger angle being opposite the longer side.* (Corollary: The largest angle of a triangle is opposite the longest side.)

Thus in Fig. 1, if $BC > AC$, then $\angle A > \angle B$.

Pr. 4: *If two angles of a triangle are unequal, the sides opposite these angles are unequal, the longer side being opposite the larger angle.* (Corollary: The longest side of a triangle is opposite the largest angle.)

Thus in Fig. 1, if $\angle A > \angle B$, then $BC > AC$.

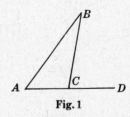

Fig. 1

Pr. 5: *The perpendicular from a point to a line is the shortest line from the point to the line.*

Thus in Fig. 2, if $PC \perp AB$ and PD is any other line from P to AB, then $PC < PD$.

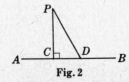

Fig. 2

Pr. 6: *If two sides of a triangle equal two sides of another triangle, the triangle having the greater included angle has the greater third side.*

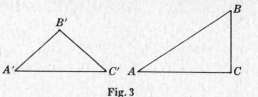
Fig. 3

Thus in Fig. 3, if $BC = B'C'$, $AC = A'C'$ and $\angle C > \angle C'$, then $AB > A'B'$.

Pr. 7: *If two sides of a triangle equal two sides of another triangle, the triangle having the greater third side has the greater angle opposite this side.*

Thus in Fig. 3, if $BC = B'C'$, $AC = A'C'$ and $AB > A'B'$, then $\angle C > \angle C'$.

E. Circle Inequality Theorems

Pr. 8: *In the same or equal circles, the greater central angle has the greater arc.*

Thus in Fig. 4, if $\angle AOB > \angle COD$, then $\overset{\frown}{AB} > \overset{\frown}{CD}$.

Pr. 9: *In the same or equal circles, the greater arc has the greater central angle.* (Converse of **Pr. 8.**)

Thus in Fig. 4, if $\overset{\frown}{AB} > \overset{\frown}{CD}$, then $\angle AOB > \angle COD$.

Pr. 10: *In the same or equal circles, the greater chord has the greater minor arc.*

Thus in Fig. 5, if $AB > CD$, then $\overset{\frown}{AB} > \overset{\frown}{CD}$.

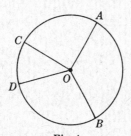
Fig. 4

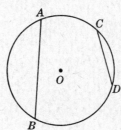
Fig. 5

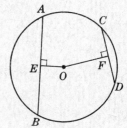
Fig. 6

Pr. 11: *In the same or equal circles, the greater minor arc has the greater chord.* (Converse of **Pr. 10.**)

Thus in Fig. 5, if $\overset{\frown}{AB} > \overset{\frown}{CD}$, then $AB > CD$.

Pr. 12: *In the same or equal circles, the greater chord is at a smaller distance from the center.*

Thus in Fig. 6, if $AB > CD$, then $OE < OF$.

Pr. 13: *In the same or equal circles, the chord at the smaller distance from the center is the greater chord.* (Converse of **Pr. 12.**)

Thus in Fig. 6, if $OE < OF$, then $AB > CD$.

1.1 SELECTING INEQUALITY SYMBOLS

Determine the inequality symbol needed in each, $>$ or $<$.

(a) $5\ (\ )\ 3$ (c) $-5\ (\ )\ 3$ (e) If $x = 3$, then $x^2\ (\ )\ x$.

(b) $6\ (\ )\ 9$ (d) $-5\ (\ )\ -3$ (f) If $x > 10$, then $10\ (\ )\ x$.

Ans.   (a) $>$ (b) $<$ (c) $<$ (d) $<$ (e) $>$ (f) $<$

1.2 APPLYING INEQUALITY AXIOMS

Complete each of the following statements.

(a) If $a > b$ and $b > 8$, then a ( ) 8.

(b) If $x > y$ and $y = 15$, then x ( ) 15.

(c) If $c < 20$ and $d < 5$, then $c + d$ ( ) 25.

(d) If $x > y$ and $y > 6$, then x ( ) y ( ) 6.

(e) If $x > y$, then $\frac{1}{2}x$ ( ) $\frac{1}{2}y$.

(f) If $e < \frac{1}{4}f$, then $4e$ ( ) f.

(g) If $-y < z$ then y ( ) $-z$.

(h) If $-4x > p$, then x ( ) $-\frac{1}{4}p$.

(i) If Paul and Jack have equal amounts of money and Paul spends more than Jack, then Paul will have ( ) than Jack.

(j) If Anne is now older than Helen, then 10 years ago, Anne was ( ) than Helen.

Ans. (a) >   (c) <   (e) >   (g) >   (i) less

     (b) >   (d) >,>   (f) <   (h) <   (j) older

1.3 APPLYING TRIANGLE INEQUALITY THEOREMS

(a) Determine the integral values that a side of a triangle can have if the other two sides are 3 and 7.

(b) Determine the longest side of a triangle if two angles are 59° and 60°.

(c) Determine the longest side of parallelogram $ABCD$ if E is the midpoint of the diagonals and $\angle AEB > \angle AED$.

(a)

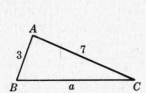

(b)

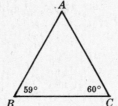

(c)

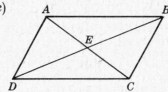

Solution:

(a) Since a must be less than $3 + 7$ or 10 and greater than $7 - 3$ or 4, a can have the integral values of 5, 6, 7, 8, 9.

(b) Since $\angle B = 59°$ and $\angle C = 60°$, $\angle A = 180° - (59° + 60°) = 61°$. Then the longest side is opposite the largest angle, $\angle A$, or the longest side is BC.

(c) In $\square ABCD$, $AE = AE$ and $DE = EB$. Since $\angle AEB > \angle AED$, then $AB > AD$ or AB ($= DC$) is the longest side. (Pr. 6)

1.4 APPLYING CIRCLE INEQUALITY THEOREMS

Compare

(a) OD and OF if $\angle C$ is the largest angle of $\triangle ABC$,

(b) AC and BC if $\overset{\frown}{AC} > \overset{\frown}{BC}$,

(c) $\overset{\frown}{BC}$ and $\overset{\frown}{AC}$ if $OF > OE$,

(d) $\angle AOB$ and $\angle BOC$ if $\overset{\frown}{AB} > \overset{\frown}{BC}$.

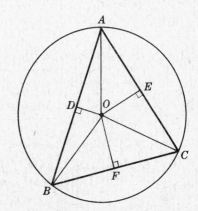

Solution:

(a) Since $\angle C$ is the largest angle of the triangle, AB is the longest side, or $AB > BC$; hence $OD < OF$ by Pr. 12.

(b) Since $\overset{\frown}{AC} > \overset{\frown}{BC}$, $AC > BC$, the greater arc having the greater chord.

(c) Since $OF > OE$, $BC < AC$ by Pr. 13; hence $\overset{\frown}{BC} < \overset{\frown}{AC}$ by Pr. 10.

(d) Since $\overset{\frown}{AB} > \overset{\frown}{BC}$, $\angle AOB > \angle BOC$, the greater arc having the greater central angle.

1.5 PROVING an INEQUALITY PROBLEM

In △*ABC*, if *M* is the midpoint of
AC and *BM* > *AM*, then ∠*A* + ∠*C* > ∠*B*.

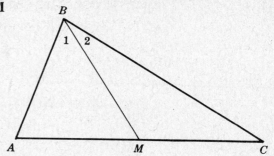

Given: △*ABC*, *M* is midpoint of *AC*.
       BM > *AM*

To Prove: ∠*A* + ∠*C* > ∠*B*

Plan: Prove ∠*A* > ∠1 and ∠*C* > ∠2 and
      then add unequals.

PROOF: Statements	Reasons
1. *M* is the midpoint of *AC*.	1. Given
2. *AM* = *MC*	2. A midpoint divides a line into two equal parts.
3. *BM* > *AM*	3. Given
4. *BM* > *MC*	4. A quantity may be substituted for its equal in any inequality.
5. In △*AMB*, ∠*A* > ∠1 In △*BMC*, ∠*C* > ∠2	5. In a triangle, the larger angle lies opposite the longer side.
6. ∠*A* + ∠*C* > ∠*B*	6. If unequals are added to unequals, the sums are unequal in the same order.

2. *Indirect Reasoning*

We often arrive at a correct conclusion by a method of indirect reasoning. In this form of reasoning, the correct conclusion is reached by eliminating all the possible conclusions except one. The remaining possibility must be the correct one. Suppose we are given the years 1492, 1809 and 1960 and are assured that one of these years is the year in which a President of the United States was born. By eliminating 1492 and 1960 as impossibilities, we know by indirect reasoning that 1809 is the correct answer. (Had we known that 1809 was the year in which Lincoln was born, the reasoning would have been direct.)

In proving a theorem by indirect reasoning, a possible conclusion may be eliminated by assuming it is true and arriving at a contradiction with a given or a known fact.

2.1 APPLYING INDIRECT REASONING in LIFE SITUATIONS

Explain how indirect reasoning is used in each of the following situations.

(*a*) A detective determines the murderer of a slain person.

(*b*) A librarian determines which volume of a set of books is in use.

Solution:

(*a*) The detective, using a list of all those who could have been a murderer in the case, eliminates all except one. He concludes that the remaining one is the murderer.

(*b*) The librarian finds all the books of the set except one by looking on the shelf and checking her records. She concludes the missing one is the one in use.

2.2 PROVING an INEQUALITY THEOREM by the INDIRECT METHOD

Prove: In the same or equal circles, unequal chords are unequally distant from the center.

Given: Circle O, $AB \neq CD$
$OE \perp AB$, $OF \perp CD$

To Prove: $OE \neq OF$

Plan: Assume the other possible conclusion $OE = OF$ and arrive at a contradiction.

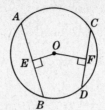

PROOF: Statements	Reasons
1. Either $OE = OF$ or $OE \neq OF$.	1. Two quantities are either equal or unequal.
2. Assume $OE = OF$.	2. This is one of the possible conclusions.
3. If $OE = OF$, then $AB = CD$.	3. In the same or equal circles, chords equally distant from the center are equal.
4. But $AB \neq CD$.	4. Given
5. The assumption $OE = OF$ is not valid.	5. It leads to a contradiction.
6. Hence $OE \neq OF$.	6. This is the only remaining possibility.

Supplementary Problems

1. Select the inequality symbol needed in each, $>$ or $<$. *(1.1)*
   (a) If $y > 15$, then 15 ( ) y.
   (b) If $x = 2$, then $3x - 1$ ( ) 4.
   (c) If $x = 2$ and $y = 3$, then xy ( ) 5.
   (d) If $a = 4$ and $b = \frac{1}{4}$, then a/b ( ) 15.
   (e) If $a = 5$, then a^2 ( ) $4a$.
   (f) If $b = \frac{1}{2}$, then b^2 ( ) b.

2. Complete each of the following statements. *(1.2)*
   (a) If $y > x$ and $x = z$, then y ( ) z.
   (b) If $a + b > c$ and $b = d$, then $a + d$ ( ) c.
   (c) If $a < b$ and $b < 15$, then a ( ) 15.
   (d) If $z > y$, $y > x$ and $x = 10$, then z ( ) 10.

3. Complete each of the following statements, using the diagram at the right. *(1.2)*
   (a) BC ( ) BD
   (b) $\angle BAD$ ( ) $\angle BAC$
   (c) $\triangle ADC$ ( ) $\triangle ABC$
   (d) If $\angle A = \angle C$, then AB ( ) BD.

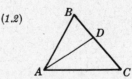

4. Complete each of the following statements. *(1.2)*
   (a) If Mary and Ann earn the same weekly wage and Mary is to receive a larger increase than Ann, then Mary will earn ( ) than Ann.
   (b) If Bernice, who is the same weight as Helen, reduces more than Helen, then Bernice will weigh ( ) than Helen.

5. Complete each of the following statements. *(1.2)*
    (a) If $a > 3$, then $4a$ ( ) 12.
    (b) If $x - 3 > 15$, then x ( ) 18.
    (c) If $3x < 18$, then x ( ) 6.
    (d) If $f > 8$, then $f + 7$ ( ) 15.
    (e) If $x = y$, then $x + 5$ ( ) $y + 6$.
    (f) If $g = h$, then $g - 10$ ( ) $h - 9$.

6. Which of the following sets can be the lengths of the sides of a triangle? *(1.3)*
    (a) 3, 4, 8   (b) 5, 7, 12   (c) 3, 4, 6   (d) 2, 7, 8   (e) 50, 50, 5

7. What integral values can the third side of a triangle have if the two sides are *(1.3)*
    (a) 2 and 6, (b) 3 and 8, (c) 4 and 7, (d) 4 and 6, (e) 4 and 5, (f) 7 and 7?

8. Arrange the indicated angles or sides in descending order of size. *(1.3)*
    (a) Angles of $\triangle ABC$

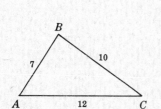

    (b) Sides of $\triangle DEF$

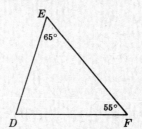

    (c) Angles 1, 2 and 3

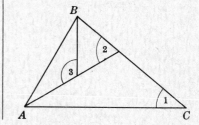

9. (a) In quadrilateral $ABCD$, compare $\angle BAC$ and $\angle ACD$ if $AB = CD$ and $BC > AD$. *(1.3)*
    (b) In $\triangle ABC$, compare AB and BC if BM is the median to AC and $\angle AMB > \angle BMC$.
    (a)

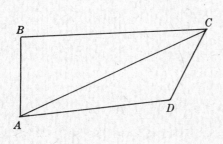

    (b)

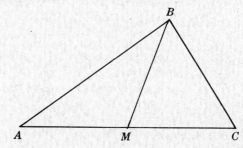

10. Arrange the indicated angles or sides in descending order of magnitude. *(1.4)*
    (a) Sides of $\triangle ABC$.
    (b) Central angles AOB, BOC and AOC.

    (c) Sides of trapezoid $ABCD$.

    (d) Distances of sides of $\triangle DEF$ from center.

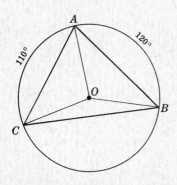

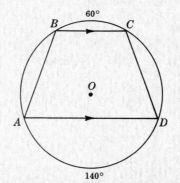

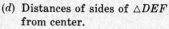

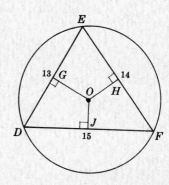

11. Prove each of the following: *(1.5)*

(a) **Given:** Parallelogram $ABCD$
　　　$AC > BD$
　To Prove: $\angle BDA > \angle CAD$

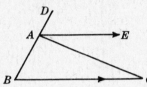

(b) **Given:** Rhombus $FGHJ$
　　　$\angle G > \angle F$
　To Prove: $FL > GL$

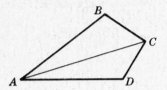

(c) **Given:** AD bisects $\angle A$
　　　CD bisects $\angle C$, $AB > BC$
　To Prove: $AD > CD$

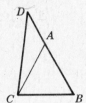

(d) **Given:** $AE \parallel BC$
　　　$\angle DAE > \angle EAC$
　To Prove: $AC > AB$

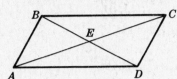

(e) **Given:** Quad. $ABCD$
　　　$AB > BC$, $AD > CD$
　To Prove: $\angle C > \angle A$

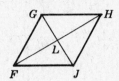

(f) **Given:** $AB = AC$
　To Prove: $BD > CD$

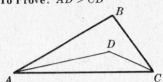

12. Explain how indirect reasoning is used in each of the following situations. *(2.1)*

　(a) A person determines which of his ties has been borrowed by his roommate.

　(b) A boy determines that the electric motor in his train set is not defective even though his toy trains do not run.

　(c) A teacher finds which of her students did not do their assigned, homework.

　(d) A mechanic finds the reason why the battery in a car does not work.

　(e) A person accused of a crime proves his innocence by means of an alibi.

13. Prove each of the following. *(2.2)*

　(a) The base angles of an isosceles triangle cannot be right angles.

　(b) A scalene triangle cannot have two equal angles.

　(c) The median to the base of a scalene triangle cannot be perpendicular to the base.

　(d) If the diagonals of a parallelogram are not equal, then it is not a rectangle.

　(e) If a diagonal of a parallelogram does not bisect a vertex angle, then the parallelogram is not a rhombus.

　(f) If two angles of a triangle are unequal, the sides opposite are unequal, the longer side being opposite the larger angle.

Chapter 14

Improvement of Reasoning

1. Definitions

"Was Lincoln an educated man?" is a question that cannot be properly answered unless we agree upon the meaning of "an educated man." In any discussion or problem, understanding cannot exist and progress cannot be made unless the terms involved are properly defined or, by agreement, are to be undefined.

A. Requirements of a Good Definition

Pr. 1: *All terms in a definition must have been previously defined.*

Thus to define a regular polygon as an equilateral and equiangular polygon, it is necessary that equilateral, equiangular and polygon be previously defined.

Pr. 2: *The term being defined should be placed in the next larger set or class to which it belongs.*

Thus the terms polygon, quadrilateral, parallelogram and rectangle should be defined in that order. Once the term polygon has been defined, the term quadrilateral is then defined as a kind of polygon. Then the term parallelogram is defined as a kind of quadrilateral and lastly, the term rectangle is defined as a kind of parallelogram.

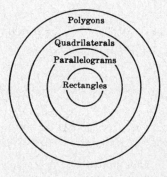

Proper sequence in definition can be understood by using a circle to represent a set of objects. In the figure at the right, the set of rectangles is in the next larger set of parallelograms. In turn, the set of parallelograms is in the next larger set of quadrilaterals and finally, the set of quadrilaterals is in the next larger set of polygons.

Pr. 3: *The term being defined should be distinguished from all other members of its class.*

Thus the definition of a triangle as a polygon of three sides is a good one since it shows how the triangle differs from all other polygons.

Pr. 4: *The distinguishing characteristics of a defined term should be reduced to a minimum.*

Thus a right triangle should be defined as a triangle having a right angle and not as a triangle having a right angle and two acute angles.

1.1 TERMS REQUIRING DEFINITION

State the terms to be defined in order to answer each question properly.

(*a*) Is a worker on strike entitled to unemployment insurance?

(*b*) Is a motorist speeding when he travels 50 miles per hour on a road that does not have a speed limit posted?

(*c*) Is Yugoslavia a communist country?

Solution:

(*a*) "Unemployment" needs to be defined to see if it covers a worker who is not working because he is on strike.

(*b*) "Speeding" must be defined.

(*c*) "Communist country" needs definition to see if Yugoslavia can be classified as one.

1.2 OBSERVING PROPER SEQUENCE in DEFINITION

State the order in which the terms in each of the following sets should be defined.

(*a*) Englishman European, Londoner.   (*b*) Quadrilateral, square, rectangle, parallelogram.

Ans. (*a*) European, Englishman, Londoner.   (*b*) Quadrilateral, parallelogram, rectangle, square.

1.3 CORRECTING FAULTY DEFINITIONS

Correct each of the following definitions.

(*a*) An uncle of a person is a relative of his father or mother.

(*b*) A trapezoid is a quadrilateral having two parallel sides.

Solution:

(*a*) "Relative" is too large a class. The correct definition is that an uncle of a person is a brother of his or her father or mother.

(*b*) The given definition is incomplete. The correct definition is that a trapezoid is a quadrilateral having only two parallel sides. The correct definition distinguishes a trapezoid from a parallelogram.

2. *Deductive Reasoning in Geometry*

The following kinds of statements comprise the deductive structure of geometry.

A. *Undefined and Defined Terms*

Point, line and surface are the terms in geometry which are not defined. These undefined terms begin the process of definition in geometry and underlie the definitions of all other geometric terms.

Thus we can define a triangle in terms of a polygon, a polygon in terms of a geometric figure, and a geometric figure as a figure composed of lines. However, the process of definition cannot be defined further because the term "line" is undefined.

B. *Assumptions*

Axioms and postulates are the statements in geometry which are not proved. They are called assumptions since we willingly accept them as true. These assumptions enable us to begin the process of proof in the same way that undefined terms enable us to begin the process of definition.

Thus when we draw a line between two points, we justify this by using as a reason the postulate "One and only one straight line can be drawn between two points." This reason is an assumption since we assume it to be true without requiring further justification.

C. *Theorems*

Theorems are the statements in geometry which are proved. By using definitions and assumptions as reasons, we deduce or prove the basic theorems. Using each new theorem to prove still more theorems, the process of deduction grows. However, if a new theorem is used to prove a previous one, the logical sequence is violated.

For example, the theorem "the sum of the angles of a triangle equals 180°" is used to prove that "the sum of the angles of a pentagon is 540°". This, in turn, enables us to prove that "each angle of a regular pentagon is 108°". However, it would be violating logical sequence if we tried to use the last theorem to prove either of the first two.

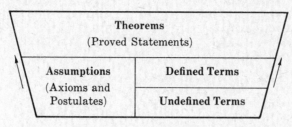

DEDUCTIVE STRUCTURE OF GEOMETRY

2.1 OBSERVING LOGICAL SEQUENCE

State the nature and logical order of the statements in the following set.

(1) Parallel lines are lines in the same plane that do not meet no matter how far extended.

(2) Through a given point not on a given line, one and only one line can be constructed parallel to the given line.

(3) The sum of the angles of a polygon of n sides equals $(n-2)180°$.

(4) The sum of the angles of a triangle is 180°.

Solution:

Statement (1) is a definition, (2) is a postulate, (3) and (4) are theorems. Their logical order or sequence is (1), then (2), then (4) and lastly (3).

3. *Converse, Inverse and Contrapositive of a Statement*

A. *Converse.* The converse of a statement is formed by interchanging the hypothesis and conclusion.

Thus the converse of the statement "Lions are wild animals" is "Wild animals are lions". Note that the converse is not necessarily true.

B. *Negative and Inverse of a Statement.*

(1) The negative of a statement is the denial of the statement.

Thus the negative of the statement "A burglar is a criminal" is "A burglar is not a criminal".

(2) The inverse of a statement is formed by denying both the hypothesis and the conclusion.

Thus the inverse of the statement "A burglar is a criminal" is "A person who is not a burglar is not a criminal". Note that the inverse is not necessarily true.

C. *Contrapositive.* The contrapositive of a statement is formed by interchanging the negative of the hypothesis with the negative of the conclusion. Hence the contrapositive is the converse of the inverse and the inverse of the converse.

Thus the contrapositive of the statement "If you live in New York City, then you live in New York State" is "If you do not live in New York State, then you do not live in New York City". Note that both statements are true.

A. Converse, Inverse and Contrapositive Principles

Pr. 1:   *A statement is considered false if one false instance of the statement exists.*

Pr. 2:   *The converse of a definition is true.*

> Thus the definition "A quadrilateral is a four-sided polygon" and its converse "A four-sided polygon is a quadrilateral" are both true.

Pr. 3:   *The converse of a true statement other than a definition is not necessarily true.*

> **Example.**   Statement: Vertical angles are equal angles.   (True)
> Converse: Equal angles are vertical angles.   (Not necessarily **true**)

Pr. 4:   *The inverse of a true statement is not necessarily true.*

> **Example.**   Statement: A square is a quadrilateral.   (True)
> Inverse: A non-square is not a quadrilateral.   (Not necessarily true)

Pr. 5:   *The contrapositive of a true statement is true and the contrapositive of a false statement is false.*

> **Example.**   Statement: A triangle is a square.   (False)
> Contrapositive: A non-square is not a triangle.   (False)

> **Example.**   Statement: Right angles are equal angles.   (True)
> Contrapositive: Angles that are not equal are not right angles. (True)

D. Logically Equivalent Statements.

Logically equivalent statements are kinds of statements that are either both true or all false.   Thus according to Principle 5, a statement and its contrapositive are logically equivalent statements.   Also, the converse and inverse of a statement are logically equivalent since each is the contrapositive of the other.

The relationships among the statement and its inverse, converse and contrapositive are summed up in the following rectangle of logical equivalency.

(1) Logically equivalent statements are at diagonally opposite vertices.

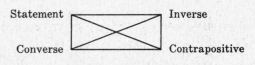

Rectangle of Logical Equivalency

> Thus the logically equivalent pairs of statements are either (*a*) a statement and its contrapositive, and (*b*) the inverse and converse of the same statement.

(2) Statements that are not logically equivalent are at adjacent vertices.

> Thus pairs of statements that are not logically equivalent are (*a*) a statement and its **inverse**, (*b*) a statement and its converse, (*c*) the converse and contrapositive of the same statement, **and** (*d*) the inverse and contrapositive of the same statement.

3.1 CONVERSE of a STATEMENT

State the converse of each statement and indicate whether or not it is true.

(*a*) Supplementary angles are two angles whose sum is 180°.

(*b*) A square is a parallelogram with a right angle.

(*c*) A regular polygon is an equilateral and equiangular polygon.

Solution:

(*a*) Two angles whose sum is 180° are supplementary. (True)

(*b*) A parallelogram with a right angle is a square. (False)

(*c*) An equilateral and equiangular polygon is a regular polygon. (True)

3.2 NEGATIVE of a STATEMENT

State the negative of each: (*a*) *a = b*, (*b*) ∠*B* ≠ ∠*C*, (*c*) ∠*C* is the complement of ∠*D*, (*d*) The point does not lie on the line.

Solution:

(*a*) *a* ≠ *b*, (*b*) ∠*B* = ∠*C*, (*c*) ∠*C* is not the complement of ∠*D*, (*d*) The point lies on the line.

3.3 INVERSE of a STATEMENT

State the inverse of each statement and indicate whether or not it is true.

(*a*) A person born in the United States is a citizen of the United States.

(*b*) A sculptor is a talented person.   (*c*) A triangle is a polygon.

Solution:

(*a*) A person who is not born in the United States is not a citizen of the United States. (False, since there are naturalized citizens.)

(*b*) One who is not a sculptor is not a talented person. (False, since one may be a fine musician, etc.)

(*c*) A figure that is not a triangle is not a polygon. (False, since the figure may be a quadrilateral, etc.)

3.4 FORMING the CONVERSE, INVERSE and CONTRAPOSITIVE of a STATEMENT

State the converse, inverse and contrapositive of the statement "a square is a rectangle." Determine the truth or falsity of each form, and check the logical equivalence of the statement and its contrapositive and also of the converse and inverse.

Solution:

Statement: A square is a rectangle. (True)

Converse: A rectangle is a square. (False)

Inverse: A figure that is not a square is not a rectangle. (False)

Contrapositive: A figure that is not a rectangle is not a square. (True)

Thus the statement and its contrapositive are true, and the converse and inverse are false.

4. Partial Converse and Partial Inverse of a Theorem

A *partial converse of a theorem* is formed by interchanging any one condition in the hypothesis with one consequence in the conclusion.

A *partial inverse of a theorem* is formed by denying one condition in the hypothesis and one consequence in the conclusion.

Thus from the theorem "If a line bisects the vertex angle of an isosceles triangle, then it is an altitude to the base", we can form a partial inverse or partial converse as follows:

(*a*) Theorem | (*b*) Partial Converse | (*c*) Partial Inverse

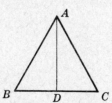

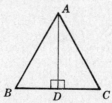

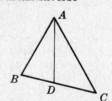

Given: △*ABC*
    (*1*) *AB = AC*
    (*2*) *AD* bisects ∠*A*.
To Prove: (*3*) *AD* is the altitude to *BC*.

Given: △*ABC*
    (*2*) *AD* bisects ∠*A*.
    (*3*) *AD* is the altitude to *BC*.
To Prove: (*1*) *AB = AC*

Given: △*ABC*
    (*1'*) *AB ≠ AC*
    (*2*) *AD* bisects ∠*A*.
To Prove: (*3'*) *AD* is not the altitude to *BC*.

In forming a partial converse or inverse, the basic figure, such as a triangle, is kept and not interchanged or denied.

In (b), the partial converse is formed by interchanging (1) and (3). Stated in words, the partial converse is: "If the bisector of an angle of a triangle is an altitude, then the triangle is isosceles." Another partial converse may be formed by interchanging (2) and (3).

In (c), the partial inverse is formed by replacing (1) and (3) by their negatives, (1') and (3'). Stated in words, the partial inverse is: "If two sides of a triangle are unequal, the line that bisects their included angle is not an altitude to the third side." Another partial inverse may be formed by negating (2) and (3).

4.1 FORMING PARTIAL CONVERSES and PARTIAL INVERSES
of a THEOREM

Form partial converses and partial inverses of the statement "equal supplementary angles are right angles."

Solution:

Partial Converses: (1) Equal right angles are supplementary.

                               (2) Supplementary right angles are equal.

Partial Inverses: (1) Equal angles that are not supplementary are not right angles.

                               (2) Supplementary angles that are not equal are not right angles.

5. *Necessary and Sufficient Conditions*

In logic and in geometry, it is often important to determine whether the conditions in the hypothesis of a statement are necessary or sufficient to justify its conclusion. This is done by ascertaining the truth or falsity of the statement and its converse, according to the following principles.

Pr. 1:   *If a statement and its converse are both true,* then the conditions in the hypothesis of the statement are necessary and sufficient for its conclusion.

        For example, the statement "If angles are right angles, then they are equal and supplementary" is true and its converse, "If angles are equal and supplementary, then they are right angles", is also true. Hence being right angles is necessary and sufficient for the angles to be equal and supplementary.

Pr. 2:   *If a statement is true and its converse is false,* then the conditions in the hypothesis of the statement are sufficient but not necessary for its conclusion.

        The statement "If angles are right angles, then they are equal" is true and its converse, "If angles are equal, then they are right angles," is false. Hence being right angles is sufficient for the angles to be equal. However, the angles need not be right angles in order to be equal.

Pr. 3:   *If a statement is false and its converse is true,* then the conditions in the hypothesis are necessary but not sufficient for its conclusion.

        The statement "If angles are supplementary, then they are right angles" is false and its converse, "If angles are right angles, then they are supplementary", is true. Hence angles need to be supplementary to be right angles, but being supplementary is not sufficient for angles to be right angles.

Pr. 4:   *If a statement and its converse are both false,* then the conditions in the hypothesis are neither necessary nor sufficient for its conclusion.

        Thus the statement "If angles are supplementary, then they are equal" is false and its converse, "If angles are equal, then they are supplementary", is false. Hence being supplementary is neither necessary nor sufficient for the angles to be equal.

Answer table for question: Are the conditions in the hypothesis of a statement
necessary or sufficient to justify its conclusion?

	STATEMENT	CONVERSE	SUFFICIENT?	NECESSARY?
Pr. 1	True	True	Yes	Yes
Pr. 2	True	False	Yes	No
Pr. 3	False	True	No	Yes
Pr. 4	False	False	No	No

5.1 DETERMINING NECESSARY and SUFFICIENT CONDITIONS

In each, state whether the conditions in the hypothesis are necessary or sufficient to
justify the conclusion.

(a) A regular polygon is equilateral and equiangular.

(b) An equiangular polygon is regular.

(c) A regular polygon is equilateral.

(d) An equilateral polygon is equiangular.

Solution:

(a) Since the statement and its converse are both true, the conditions are necessary and sufficient.

(b) Since the statement is false and its converse is true, the conditions are necessary but not sufficient.

(c) Since the statement is true and its converse is false, the conditions are sufficient but not necessary.

(d) Since both the statement and its converse are false, the conditions are neither necessary nor sufficient.

6. Other Forms of Reasoning

A. Reasoning by Analogy

Reasoning by analogy is a form of reasoning that leads to the conclusion that objects
or situations that are alike in some ways are also alike in other ways.

Thus an analogy is drawn in each of the following.

(1) Food is to the body as fuel is to an engine.

(2) Since the plural of tomato is tomatoes, the plural of piano is pianoes.

(3) Since fasting a full day cured my stomach ailment, it should cure your stomach ailment as well.

(4) Mars, like the Earth, is inhabited.

(5) Since the diagonals of a regular pentagon are equal, the diagonals of a regular hexagon are also equal.

Note that reasoning by analogy does not always produce correct results. Statement (2) is false,
since the plural of piano is pianos. Also (5) is false, and (3) and (4) are uncertain. Thus, objects that
are alike in some situations need not be alike in others.

B. Reasoning by Generalization

Reasoning by generalization is a form of reasoning that leads to the conclusion that
a truth applicable to a limited number of cases is true generally. This conclusion is not
necessarily true, as in each of the following generalizations.

(1) All birds fly, since the hawk, eagle, crow and pigeon fly.

(2) This drug has cured 150 cases in a row; it should be successful in all cases.

Statement (1) is false since the ostrich and penguin don't fly.

The process of coming to a general conclusion based on a number of particular cases is called inductive reasoning. Conclusions reached by inductive reasoning are never certain but are probable, and with more cases and a better choice of cases, the general conclusion becomes more probable.

C. Circular Reasoning

In correct reasoning, theorems proved are used to prove new theorems, and the process of deduction moves forward. However, if a theorem is used to prove itself or a previous theorem, the process of deduction stands still or moves backward, and logical sequence is violated.

Circular reasoning is incorrect reasoning that leads, not forward, but backward to a conclusion which was previously reached in the reasoning sequence.

Thus the following sequence of area theorems illustrates circular reasoning. Here, A can be used to prove B, B to prove C, and C to prove D. But it is incorrect to use theorem D to prove theorem A.

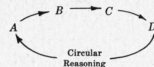

A: Area of a rectangle $= bh$
B: Area of parallelogram $= bh$
C: Area of a triangle $= \frac{1}{2}bh$
D: Area of a right triangle $= \frac{1}{2}bh$

A theorem may not be used to prove itself. The following proof is an example of this type of incorrect reasoning.

Given: $\triangle ABC$
$\qquad AB = AC$
To Prove: $\angle B = \angle C$

PROOF: Statements	Reasons
1. $AB = AC$	1. Given
2. $\angle B = \angle C$	2. Base angles of an isosceles triangle are equal.

The actual problem is to prove the reason given in the second step. By quoting it in the proof, it is being used to prove itself.

6.1 CORRECTING FAULTY REASONING

Criticize the reasoning in each of the following.

(a) Since my father and grandfather have been successful musicians, I will make it my career.

(b) Since an equilateral triangle is equiangular, an equilateral quadrilateral is also equiangular.

(c) I have never had less than an A in all my mathematics courses in high school, therefore I am certain to get A's in all my mathematics courses in college as well.

Solution:

(a) Members of the same family need not have the same talent or the same success in pursuing a career.

(b) The fact that polygons are equilateral does not make them equiangular. The rhombus is an example of an equilateral polygon whose angles need not be equal.

(c) This is a generalization, and hence the conclusion is not certain, although it is more or less probable.

6.2 CIRCULAR REASONING

Criticize the reasoning in the following proof.

Given: Quadrilateral $ABCD$
$\qquad BC = AD$, $BC \parallel AD$
To Prove: $ABCD$ is a parallelogram.

PROOF: Statements	Reasons
1. $BC = AD$, $BC \parallel AD$ 2. $ABCD$ is a parallelogram.	1. Given 2. If two sides of a quadrilateral are equal and parallel, then the quadrilateral is a parallelogram.

Solution:

The reasoning is faulty because a principle is being used to prove itself. The principle in this case is the reason given in the second step.

Supplementary Problems

1. State the terms to be defined in order to answer each question properly.   *(1.1)*

   (*a*) Is a 12 year old entitled to half-price admission to the movies as a child?

   (*b*) Are opera glasses luxury items which are subject to a Federal tax?

   (*c*) Is a person wearing a sweater improperly dressed at a dance?

   (*d*) Is a person guilty of smoking if he is holding a lighted cigarette in a train?

   (*e*) Is a quadrilateral a trapezoid if two pairs of its opposite sides are parallel?

   (*f*) Is a triangle a regular polygon if its sides are equal?

2. State the order in which the terms in each of the following sets should be defined: (*a*) jewelry, wedding ring, ornament, ring; (*b*) automobile, vehicle, commercial automobile, truck; (*c*) quadrilateral, rhombus, polygon, parallelogram; (*d*) obtuse triangle, obtuse angle, angle, isosceles obtuse triangle.   *(1.2)*

3. Correct each of the following definitions: *(1.3)*

   (*a*) A regular polygon is an equilateral polygon.

   (*b*) An isosceles triangle is a triangle having at least two equal sides and angles.

   (*c*) A pentagon is a geometric figure having five sides.

   (*d*) A rectangle is a parallelogram whose angles are right angles.

   (*e*) An inscribed angle is an angle formed by two chords.

   (*f*) A parallelogram is a quadrilateral whose opposite sides are equal and parallel.

   (*g*) An obtuse angle is an angle larger than a right angle.

4. State the nature and logical sequence of the statements in each of the following sets.   *(2.1)*

   (*a*) (1) Two triangles are similar if two angles of one equal two angles of the other.

   (2) Similar triangles are triangles whose corresponding angles are equal and whose corresponding sides are in proportion.

   (3) In a right triangle, the square of the hypotenuse equals the sum of the squares of the other two sides.

   (4) In a right triangle, if an altitude is drawn to the hypotenuse, each leg is a mean proportional between the hypotenuse and the segment of the hypotenuse adjacent to the leg.

   (*b*) (1) The area of a rectangle equals the product of its base and altitude.

   (2) The area of a polygon is the number of square units contained in its surface.

   (3) The area of a trapezoid equals one-half the product of its altitude and the sum of its bases.

   (4) The area of a triangle equals one-half the product of its base and altitude.

   (5) The area of a parallelogram equals the product of its base and altitude.

5. For each of the following, state the definition and its converse and indicate whether or not the converse is true: (*a*) An obtuse angle, (*b*) a parallelogram, (*c*) parallel lines, (*d*) complementary angles, (*e*) a chord of a circle, (*f*) an inscribed angle of a circle, (*g*) an integer, (*h*) a positive number.   *(3.1)*

6. State the negative of each: (*a*) $x + 2 = 4$. (*b*) $3y \neq 15$. (*c*) She loves you. (*d*) His mark was more than 65. (*e*) Joe is heavier than Dick. (*f*) $a + b \neq c$. *(3.2)*

7. State the inverse of each statement and indicate whether or not it is true. *(3.3)*

(*a*) A square has equal diagonals.

(*c*) A bachelor is an unmarried person.

(*b*) An equiangular triangle is equilateral.

(*d*) Zero is not a positive number.

8. State the converse, inverse and contrapositive of each of the following statements. Indicate the truth or falsity of each form, and check the logical equivalence of the statement and its contrapositive and also of the converse and inverse. *(3.4)*

(*a*) If two sides of a triangle are equal, the angles opposite these sides are equal.

(*b*) Congruent triangles are similar triangles.

(*c*) If two lines intersect, then they are not parallel.

(*d*) A Senator of the United States is a member of its Congress.

9. Form partial converses and partial inverses of the following theorem: A diameter perpendicular to a chord bisects its minor arc. *(4.1)*

10. Form partial converses and partial inverses of each of the following: *(4.1)*

(*a*)

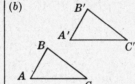

Given: Quad. *ABCD*
(1) $BC = AD$
(2) $BC \parallel AD$
To Prove: (3) $AB = CD$

(*b*)

Given: $\triangle ABC$, $\triangle A'B'C'$
(1) $AB = A'B'$
(2) $\angle B = \angle B'$
(3) $BC = B'C'$
To Prove: (4) $AC = A'C'$

11. In each, state whether the conditions in the hypothesis are necessary or sufficient to justify the conclusion. *(5.1)*

(*a*) Senators of the United States are elected members of Congress, two from each state.

(*b*) Elected members of Congress are senators of the United States.

(*c*) Elected persons are government officials.

(*d*) If a man lives in New York City, then he lives in New York State.

(*e*) A bachelor is an unmarried man.

(*f*) A bachelor is an unmarried person.

(*g*) A quadrilateral having two pairs of equal sides is a parallelogram.

12. Criticize each of the following: *(6.1)*

(*a*) The President elected in 1960 will die in office, since those elected in 1860, 1880, 1900, 1920 and 1940 died in office.

(*b*) Since the plural of wife, leaf and calf end in ves, it must be true that if a noun ends in f or fe, its plural must end in ves.

(*c*) I expect to pass in mathematics because my mathematics teacher hasn't failed anyone during the past year.

(*d*) All my neighbors have trouble with their color television sets. Anyone who buys a color television set must expect to have trouble also.

(*e*) Our rivals have beaten us in football ever since we started to play them. We will surely lose again this year.

(*f*) Those reducing pills have helped everyone I know who tried to reduce. They must work with everybody.

13. Criticize the reasoning in each of the following: *(6.2)*

(*a*) I know why his blood circulation is so poor; his hands and feet are always cold.

(*b*) On the test, if I am asked to prove that the base angles of an isosceles triangle are equal, I will simply use the theorem "If two sides of a triangle are equal, then the angles opposite these sides are equal."

Chapter 15

Constructions

A. Introduction

Geometric figures are constructed by means of straightedge and compasses. Since deductive reasoning is the basis for constructions, measuring instruments such as the ruler or protractor are not permitted. However, the ruler may be used as a straightedge by disregarding its markings.

In constructions, it is advisable to plan ahead by making a sketch of the situation. Such a sketch will serve to reveal the needed construction steps. Construction lines should be made light to distinguish them from the required figure.

B. List of Constructions

The following is a list of the constructions contained in this chapter. An asterisk before the construction indicates that it is required for examination purposes by the syllabus of the State of New York.

1. To construct a line segment equal to a given line segment.
2. To construct an angle equal to a given angle.
*3. To bisect a given angle.
*4. To construct a line perpendicular to a given line through a given point on the line.
*5. To bisect a given line segment.
*6. To construct a line perpendicular to a given line through a given point outside the line.
7. To construct a triangle given its three sides.
8. To construct an angle of 60°.
9. To construct a triangle given two sides and the included angle.
10. To construct a triangle given two angles and the included side.
11. To construct a triangle given two angles and a side not included.
12. To construct a right triangle given its hypotenuse and a leg.
*13. To construct a line parallel to a given line through a given point.
*14. To construct a tangent to a given circle through a given point on the circle.
*15. To construct a tangent to a given circle through a given point outside the circle.
*16. To circumscribe a circle about a triangle.
17. To locate the center of a given circle.
*18. To inscribe a circle in a given triangle.
*19. To divide a line segment into any number of equal parts.
20. To divide a given line segment into parts having a given ratio.
21. To construct the fourth proportional to three given line segments.
22. To construct the mean proportional to two given line segments.
*23. To inscribe a square in a given circle.
24. To inscribe a regular octagon in a given circle.
*25. To inscribe a regular hexagon in a given circle.
*26. To inscribe an equilateral triangle in a given circle.
*27. To construct a triangle similar to a given triangle on a given line segment as base.

1. *Duplicating Lines and Angles*

Construction 1: To construct a line segment equal to a given line segment.

Given: Line segment AB

To Construct: A line segment equal to AB.

Construction: On working line w, with any point C as a center and a radius equal to AB, construct an arc intersecting w at D. Then CD is the required line segment.

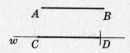

Construction 2: To construct an angle equal to a given angle.

Given: $\angle A$

To Construct: An angle equal to $\angle A$.

Construction:

1. With A as center and a convenient radius, construct an arc (1) intersecting the sides of $\angle A$ at B and C.

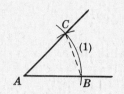

2. With A', a point on working line w, as center and the same radius, construct arc (2) intersecting w at B'.

3. With B' as center and a radius equal to BC, construct arc (3) intersecting arc (2) at C'.

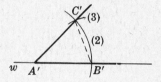

4. Draw $A'C'$. Then $\angle A'$ is the required angle. ($\triangle ABC \cong \triangle A'B'C'$ by s.s.s. = s.s.s.; hence $\angle A = \angle A'$.)

1.1 COMBINING LINE SEGMENTS

Given line segments a and b, construct line segments equal to
(a) $a + 2b$, (b) $2(a + b)$, (c) $b - a$.

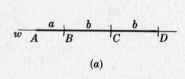

Construction: Use construction 1.

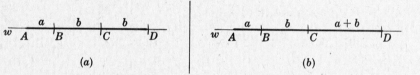

(a)   (b)   (c)

(a) On a working line w, construct a line segment AB equal to line segment a. From B, construct a line segment equal to b, to point C; and from C construct a line segment equal to b, to point D. Then AD is the required line segment.

(b) and (c) are constructed similarly. In (b), $AD = 2(a + b)$. In (c), $AC = b - a$.

1.2 COMBINING ANGLES

Given $\triangle ABC$, construct angles equal to (a) $2A$, (b) $A + B + C$, (c) $B - A$.

Construction: Use construction 2.

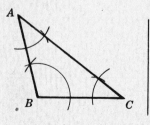 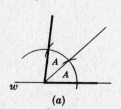 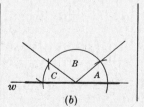 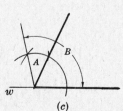

(a)   (b)   (c)

(b) Using working line w as one side, duplicate $\angle A$. Construct $\angle B$ adjacent to $\angle A$, as shown. Then construct $\angle C$ adjacent to $\angle B$. Then the exterior sides of the copied angles A and C form the required angle. Note that the angle is a straight angle.

(a) and (c) are constructed similarly.

2. Constructing Bisectors and Perpendiculars

Construction 3: To bisect a given angle.

Given: $\angle A$

To Construct: The bisector of $\angle A$.

Construction: With A as center and a convenient radius, construct an arc intersecting the sides of $\angle A$ at B and C. With B and C as centers and equal radii, construct arcs intersecting in D. Draw AD. Then AD is the required bisector. ($\triangle ABD \cong \triangle ADC$ by s.s.s. = s.s.s.; hence $\angle 1 = \angle 2$.)

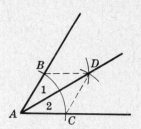

Construction 4: To construct a line perpendicular to a given line through a given point on the line.

Given: Line w and point P on w.

To Construct: A perpendicular to w at P.

Construction: Using construction 3, bisect the straight angle at P. Then DP is the required perpendicular.

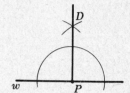

Construction 5: To bisect a given line segment. (To construct the perpendicular bisector of a given line segment.)

Given: Line segment AB.

To Construct: Perpendicular bisector of AB.

Construction: With A as center and a radius of more than half AB, construct arc (1). With B as center and the same radius, construct arc (2) intersecting arc (1) at C and D. Draw CD. CD is the required perpendicular bisector of AB. (Principle: Two points each equidistant from the ends of a line determine the perpendicular bisector of the line.)

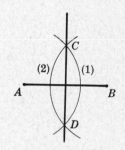

Construction 6: To construct a line perpendicular to a given line through a given external point.

Given: Line w and point P outside of w.

To Construct: A perpendicular to w through P.

Construction: 1. With P as center and a sufficiently long radius, construct an arc intersecting w at B and C.

               2. With B and C as centers and equal radii more than half of BC, construct arcs intersecting at A. Draw PA. Then PA is the required perpendicular. (Points P and A are each equidistant from B and C.)

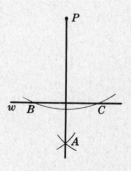

2.1 CONSTRUCTING SPECIAL LINES in a TRIANGLE

In scalene $\triangle ABC$, construct (a) a perpendicular bisector of AB and (b) a median to AB. In $\triangle DEF$, D is an obtuse angle. Construct (c) the altitude to DF and (d) the bisector of $\angle E$.

Construction:

(a) Use construction 5 to obtain PQ, the perpendicular bisector of AB.

(b) Point M is the midpoint of AB. Then CM is the median to AB.

(c) Use construction 6 to obtain EG, the altitude to DF (extended).

(d) Use construction 3 to bisect $\angle E$. EH is the required line.

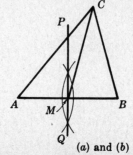

(a) and (b)

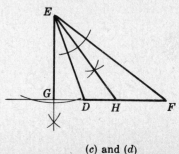

(c) and (d)

2.2 CONSTRUCTING BISECTORS and PERPENDICULARS to OBTAIN REQUIRED ANGLES

(a) Construct angles of 90°, 45°, and 135°.

(b) Given $\angle A$, construct an angle equal to $90° + A$.

Construction:

(a)

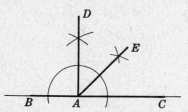

(b)

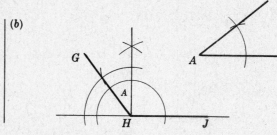

(a) $\angle DAB = 90°$, $\angle CAE = 45°$, $\angle BAE = 135°$ (b) $\angle GHJ = 90° + A$

3. *Constructing a Triangle*

A. *Determining a Triangle*

A triangle is determined by a set of given data which fixes its size and shape. Since the parts needed for congruent triangles fix the size and shape of a triangle, a triangle is determined when the given data consists of (1) three sides, (2) two sides and the angle included by these sides, (3) two angles and a side included by these angles, (4) two angles and a side not included by these angles, (5) the hypotenuse and either leg of a right triangle.

B. *Sketching Triangles to be Constructed*

Before making the actual construction, it is very helpful to make a preliminary sketch of the required triangle. In this sketch:

(1) Show the position of each of the given parts of the triangle.

(2) Make the given parts heavy, the remaining parts light.

(3) Approximate the size of the given parts.

(4) Use small letters for sides to agree with the large letters for the angles opposite them.

Thus, make a sketch such as the following before constructing a triangle given two angles and an included side:

Given Parts   Preliminary Sketch

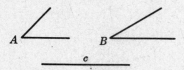

Construction 7: To construct a triangle given its three sides.

Given: Sides a, b and c of $\triangle ABC$.

To Construct: $\triangle ABC$

Construction: 1. On a working line w, construct $AC = b$.

2. With A as center and c as radius, construct arc (1).

3. With C as center and a as radius, construct arc (2) intersecting (1) at B.

4. Draw BC and AB. $\triangle ABC$ is the required triangle.

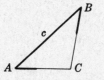

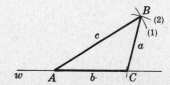

Construction 8: To construct an angle of 60°.

Given: Line w.

To Construct: An angle of 60°.

Construction: Using a convenient length as a side, construct an equilateral triangle using construction 7. Then any angle of the equilateral triangle is the required angle.

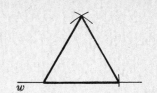

Construction 9: To construct a triangle given two sides and the included angle.

Given: $\angle A$, b and c.

To Construct: $\triangle ABC$

Construction: On a working line w, construct $AC = b$. At A, construct $\angle A$ with one side AC. On the other side of $\angle A$, construct $AB = c$. Draw BC. Then the required triangle is $\triangle ABC$.

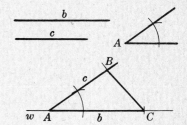

Construction 10: To construct a triangle given two angles and the included side.

Given: $\angle A$, $\angle C$ and b.

To Construct: $\triangle ABC$

Construction: On a working line w, construct $AC = b$. At A construct $\angle A$ with one side on AC, and at C construct $\angle C$ with one side on AC. Extend the new sides of the angles until they meet, at B.

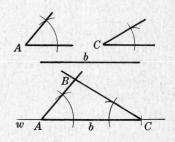

Construction 11: To construct a triangle given two angles and a side not included.

Given: $\angle A$, $\angle B$ and b.

To Construct: $\triangle ABC$

Construction: On working line w, construct $AC = b$. At C construct an angle equal to $\angle A + \angle B$ so that the extension of AC will be one side of the angle. The remainder of the straight angle at C will be $\angle C$. At A construct $\angle A$ with one side on AC. The intersection of the new sides of the angles is B.

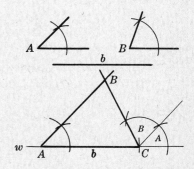

Construction 12: To construct a right triangle given its hypotenuse and a leg.

Given: Hypotenuse c and leg b of right triangle ABC.

To Construct: Right triangle ABC.

Construction: On a working line w, construct $AC = b$. At C construct a perpendicular to AC. With A as center and a radius of c, construct an arc intersecting the perpendicular at B.

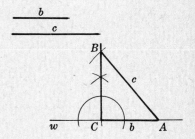

3.1 CONSTRUCTING a TRIANGLE

Construct an isosceles triangle; given the base and an arm.

Construction:

Use construction 7 since three sides of the triangle are known.

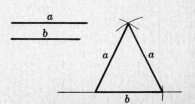

3.2 CONSTRUCTING ANGLES BASED on the CONSTRUCTION of the 60° ANGLE

Construct an angle of (a) 120°, (b) 30°, (c) 150°, (d) 105°, (e) 75°.

Construction:

(a)

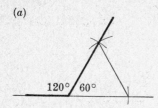

(b) and (c)

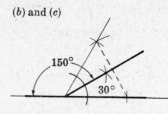

(d) and (e)

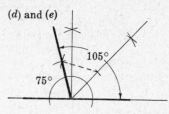

(d) Use constructions 3, 4 and 8 to construct 105° as 60° + $\frac{1}{2}$(90°). The others are constructed similarly.

4. Constructing Parallel Lines

Construction 13: To construct a line parallel to a given line through a given point.

Given: AB and point P.

To Construct: A line through P parallel to AB.

Construction: Draw a line RS through P intersecting AB in Q. Construct $\angle SPD = \angle PQB$. Then CD is the required parallel. (If two corresponding angles are equal, the lines cut by the transversal are parallel.)

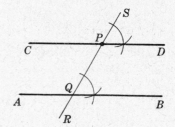

4.1 CONSTRUCTING a PARALLELOGRAM

Construct a parallelogram given two adjacent sides and a diagonal.

Construction:

Three vertices of the parallelogram are obtained by constructing $\triangle ABD$ by construction 7. The fourth vertex, C, is obtained by constructing $\triangle BCD$ upon diagonal BD by construction 7.

Also, C may be obtained by constructing $BC \parallel AD$ and $DC \parallel AB$.

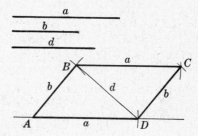

5. Circle Constructions

Construction 14: To construct a tangent to a given circle through a given point on the circle.

Given: Circle O and point P on the circle.

To Construct: A tangent to circle O at P.

Construction: Draw radius OP and construct $AB \perp OP$ at P. AB is the required tangent. (A line perpendicular to a radius at its outer extremity is a tangent to the circle.)

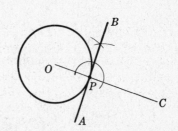

Construction 15: To construct a tangent to a given circle through a given point outside the circle.

Given: Circle O and point P outside the circle.

To Construct: A tangent to circle O from P.

Construction: Draw OP and make OP the diameter of a new circle Q. Connect P to the intersections of circles O and Q. Then PA and PB are tangents. ($\angle OAP$ and $\angle OBP$ are right angles, since angles inscribed in semicircles are right angles.)

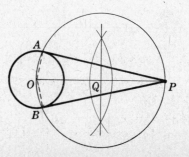

Construction 16: To circumscribe a circle about a triangle.

Given: △ABC

To Construct: The circumscribed circle of △ABC.

Construction: Construct the perpendicular bisectors of two sides of the triangle. Their intersection is the center of the required circle and the distance to any vertex is the radius. (Any point on the perpendicular bisector of a line is equidistant from the ends of the line.)

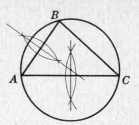

Construction 17: To locate the center of a given circle.

Given: A circle.

To Construct: The center of the given circle.

Construction: Select any three points A, B and C on the circle. Construct the perpendicular bisectors of line segments AB and AC. The intersection of these perpendicular bisectors is the center of the circle.

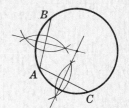

Construction 18: To inscribe a circle in a given triangle.

Given: △ABC

To Construct: The circle inscribed in △ABC.

Construction: Construct the bisectors of two of the angles of △ABC. Their intersection is the center of the required circle and the distance (perpendicular) to any side is the radius. (Any point on the bisector of an angle is equidistant from the sides of the angle.)

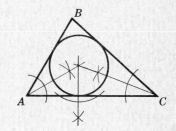

5.1 CONSTRUCTING TANGENTS

A secant from a point P outside circle O meets the circle in B and A. Construct a triangle circumscribed about the circle so that two of its sides meet in P and the third side is tangent to the circle at A.

Construction:

Use constructions 14 and 15.

At A construct a tangent to circle O. From P construct tangents to circle O intersecting the first tangent in C and D. The required triangle is △PCD.

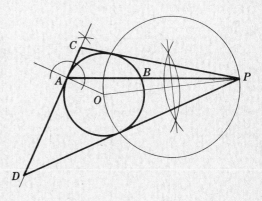

5.2 CONSTRUCTING CIRCLES

Construct the circumscribed and inscribed circles of an isosceles triangle.

Construction:

Use constructions 16 and 18.

The bisector of ∠E is also the perpendicular bisector of DF. Then the center of each circle is on EG. I, the center of the inscribed circle, is found by constructing the bisector of ∠D or ∠F. C, the center of the circumscribed circle, is found by constructing the perpendicular bisector of DE or EF.

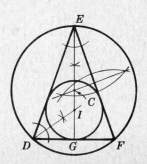

6. Dividing a Line Segment and Constructing Proportionals

Construction 19: To divide a line segment into any number of equal parts.

Given: Line segment AB.

To Construct: Divide AB into any number of equal parts.

Construction: On a line AP, cut off the required number of equal segments. Then connect B to the endpoint of the last segment and construct parallels to BC. The points of intersection of these parallels and AB divide AB into the required number of segments. (If three or more parallel lines cut off equal segments on one transversal, they cut off equal segments on any other transversal.)

Construction 20: To divide a given line segment into parts having a given ratio.

Given: AB, m and n.

To Construct: Divide AB into two segments whose ratio is $m:n$.

Construction: On a line AG, construct $AC = m$ and $CD = n$. Draw BD and construct $CE \parallel BD$. Then AE and EB are the required lines. (A line parallel to one side of a triangle divides the other two sides into proportional segments.)

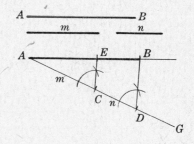

Construction 21: To construct the fourth proportional to three given line segments.

Given: Segments a, b and c.

To Construct: The fourth proportional to a, b, c.

Construction: On a working line AP, construct $AD = a$ and $DB = b$. On another line AQ, construct $AE = c$. Draw DE and construct $BC \parallel DE$. Then EC is the required line. (Same principle as for Construction 20.)

Construction 22: To construct the mean proportional to two given line segments.

Given: Segments a and b.

To Construct: The mean proportional to a and b.

Construction: On a working line, construct $AB = a$ and $BC = b$. Construct a semicircle using AC as diameter. At B erect a perpendicular to AC, meeting the semicircle at D. Then DB is the required line. (In a right triangle, the altitude on the hypotenuse is the mean proportional between the segments of the hypotenuse.)

6.1 DIVIDING a LINE SEGMENT into PROPORTIONAL PARTS

Find two-fifths of line segment AB.

Construction:

Use construction 19.

On another line, AP, construct five equal segments. Draw BH. Through the endpoint E of the second segment of AH, construct a line parallel to BH, meeting AB in C. Then AC is two-fifths of AB.

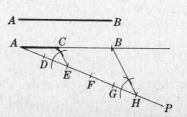

6.2   CONSTRUCTING a FOURTH PROPORTIONAL and a MEAN PROPORTIONAL

(a) Given segments a, b and c. Construct a segment x such that $x = \dfrac{ab}{c}$.

(b) Given segments p and q. Construct a segment x such that $x = \sqrt{pq}$.

Construction:

(a) 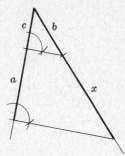   (b)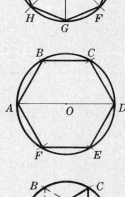

(a)   Use construction 21.

     Construct x as a fourth proportional to c, a and b. Since $\dfrac{c}{a} = \dfrac{b}{x}$, $x = \dfrac{ab}{c}$.

(b)   Use construction 22.

     Construct x as a mean proportional to p and q. Since $\dfrac{p}{x} = \dfrac{x}{q}$, $x^2 = pq$ and $x = \sqrt{pq}$.

7.   *Inscribing and Circumscribing Regular Polygons*

Construction 23: To inscribe a square in a given circle.

Given: Circle O

To Construct: A square inscribed in circle O.

Construction: Draw a diameter and construct another diameter perpendicular to it. Join the endpoints of the diameters ,to form the required square.

Construction 24: To inscribe a regular octagon in a given circle.

Given: Circle O

To Construct: A regular octagon inscribed in circle O.

Construction: As in construction 23, construct perpendicular diameters. Then bisect the angles formed by these diameters, dividing the circle into eight equal arcs. The chords of these arcs are the sides of the required regular octagon.

Construction 25: To inscribe a regular hexagon in a given circle.

Given: Circle O

To Construct: A regular hexagon inscribed in circle O.

Construction: Draw diameter AD and, using A and D as centers, construct arcs having the same radius as circle O. Construct the required regular hexagon by joining consecutive points in which these arcs intersect the circle.

Construction 26: To inscribe an equilateral triangle in a given circle.

Given: Circle O

To Construct: An equilateral triangle inscribed in circle O.

Construction: Inscribed equilateral triangles are obtained by joining alternately the six points of division obtained in construction 25.

7.1 CIRCUMSCRIBING a REGULAR POLYGON

Given a circle, circumscribe a square about the circle.

Construction:

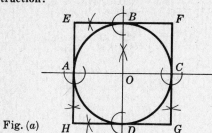 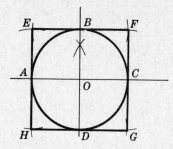

Fig. (a)   Fig. (b)

Using construction 23, obtain four points that divide the circle into four equal parts. Then at each of these points construct tangents to the circle. The tangents are the sides of the required square. See Fig. (a).

Alternate Method. Using each of the four points dividing the circle into four equal arcs as centers and the radius of the given circle as the radius, construct arcs whose intersections with each other are the vertices of the square. See Fig. (b).

8. *Constructing Similar Triangles*

Construction 27: To construct a triangle similar to a given triangle on a given line segment as base.

Given: $\triangle ABC$ and line segment $A'C'$.

To Construct: $\triangle A'B'C' \sim \triangle ABC$ on $A'C'$ as a base.

Construction: On $A'C'$, construct $\angle A' = \angle A$ and $\angle C' = \angle C$ using construction 2. Extend the other sides until they meet, at B. (If two angles of one triangle equal two angles of another triangle, the triangles are similar.)

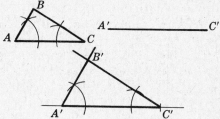

8.1 CONSTRUCTING SIMILAR TRIANGLES

Construct a triangle similar to a given triangle with a base twice as long as the base of the given triangle.

Construction:

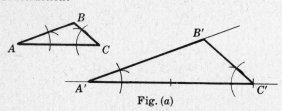

Fig. (a)   Fig. (b)

See Fig. (a). Construct $A'C'$ twice as long as AC and then use construction 27.

Alternate Method. See Fig. (b). Extend two sides of $\triangle ABC$ to twice their lengths and join the endpoints.

9. *Transforming a Polygon into an Equivalent Polygon*

To transform a polygon is to change it into a polygon having the same area. Thus a rectangle may be transformed into an equivalent square, and a pentagon transformed into an equivalent triangle.

The following principle is very useful in transforming polygons: Two triangles are equal if they have a common base and their vertices lie on a line parallel to the base.

Thus since $PQ \parallel AB$, scalene $\triangle ACD$ = isosceles $\triangle ADB$ = right $\triangle AEB$. (Triangles having the same base and equal altitudes are equal in area.)

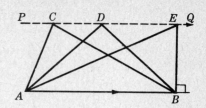

9.1 TRANSFORMING POLYGONS

(a) Transform quadrilateral $ABCD$ into a triangle.

(b) Construct a square equal to rectangle $ABCD$.

Construction:

(a)

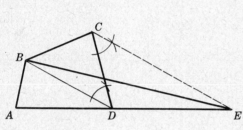

(b)

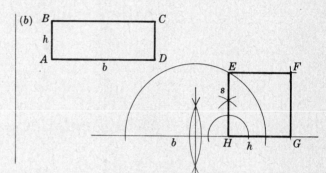

(a) Draw diagonal BD and construct a line through C parallel to BD, intersecting AD (extended) at E. Draw BE. Then $\triangle ABE$ is the required triangle. Since $\triangle BCD$ and $\triangle BDE$ have common base BD and their vertices lie on CE, which is parallel to BD, they are equal. By adding each of these triangles to $\triangle ABD$, quadrilateral $ABCD$ and triangle ABE are proved equal.

(b) If b and h are the base and altitude of rectangle $ABCD$, construct the mean proportional s to b and h. Then s is the side of the required square. Since $b:s = s:h$, $s^2 = bh$ and the area of square $EFGH$ equals the area of rectangle $ABCD$.

Supplementary Problems

1. Given line segments a and b, construct a line segment whose length equals (a) $a+b$, (b) $a-b$, (c) $2a+b$, (d) $a+3b$, (e) $2(a+b)$, (f) $2(3b-a)$. (1.1)

2. Given line segments a, b and c, construct a line segment whose length equals (a) $a+b+c$, (b) $a+c-b$, (c) $a+2(b+c)$, (d) $b+2(a-c)$, (e) $3(b+c-a)$. (1.1)

3. Given angles A and B, construct an angle equal to
(a) $A+B$, (b) $A-B$, (c) $2B-A$, (d) $2A-B$, (e) $2(A-B)$.
(1.2)

4. Given angles A, B and C, construct an angle equal to
(a) $A+C$, (b) $B+C-A$, (c) $2C$, (d) $B-C$, (e) $2(A-B)$.
(1.2)

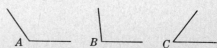

5. Given triangle ABC, construct an angle equal to   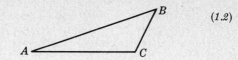   (1.2)

   (a) $A + B$ (c) $A + B + C$ (e) $2B$

   (b) $C - A$ (d) $2(C - B)$ (f) $3A$

6. In a right triangle, construct (a) the bisector of the right angle, (b) the perpendicular bisector of the hypotenuse, (c) the median to the hypotenuse.   (2.1)

7. For each kind of triangle (acute, right and obtuse), show that the following sets of lines are concurrent, that is, they intersect in one point: (a) the angle bisectors, (b) the medians, (c) the altitudes, (d) the perpendicular bisectors.   (2.1)

8. Given $\triangle ABC$, construct   (2.2)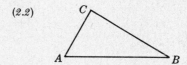

   (a) the supplement of $\angle A$,

   (b) the complement of $\angle B$,

   (c) the complement of $\frac{1}{2}\angle C$.

9. Construct an angle equal to (a) $22\frac{1}{2}°$, (b) $67\frac{1}{2}°$, (c) $112\frac{1}{2}°$.   (2.2)

10. Given an acute angle, construct its (a) supplement, (b) complement, (c) half of its supplement, (d) half of its complement.   (2.2)

11. By actual construction, illustrate that the difference between the supplement and the complement of a given acute angle equals a right angle.   (2.2)

12. Construct a right triangle given its (a) legs, (b) hypotenuse and a leg, (c) leg and an acute angle adjacent to the leg, (d) leg and an acute angle opposite the leg, (e) hypotenuse and an acute angle.   (3.1)

13. Construct an isosceles triangle, given (a) an arm and a vertex angle, (b) an arm and a base angle, (c) an arm and the altitude to the base, (d) the base and the altitude to the base.   (3.1)

14. Construct an isosceles right triangle, given (a) a leg, (b) the hypotenuse, (c) the altitude to the hypotenuse.   (3.1)

15. Construct a triangle, given (a) two sides and the median to one of them, (b) two sides and the altitude to one of them, (c) an angle, the angle bisector of the given angle and a side adjacent to the given angle.   (3.1)

16. Construct angles of $15°$ and $165°$.   (3.2)

17. Given $\angle A$, construct angles equal to (a) $A + 60°$, (b) $A + 30°$, (c) $A + 120°$.   (3.2)

18. Construct a parallelogram, given (a) two adjacent sides and an angle, (b) the diagonals and the acute angle at their intersection, (c) the diagonals and a side, (d) two adjacent sides and the altitude to one of them, (e) a side, an angle and the altitude to the given side.   (4.1)

19. Circumscribe a triangle about a given circle, if the points of tangency are given.   (5.1)

20. Secant AB passes through the center of circle O shown in the adjacent figure. Circumscribe a quadrilateral about the circle so that A and B are opposite vertices.   (5.1)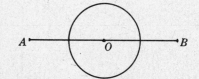

21. Circumscribe and inscribe circles about (a) an acute triangle, (b) an obtuse triangle.   (5.2)

22. Circumscribe a circle about (a) a right triangle, (b) a rectangle, (c) a square.   (5.2)

23. Construct the inscribed and circumscribed circles of an equilateral triangle.   (5.2)

24. Locate the center of a circle drawn around the outside of a half-dollar piece.

25. Divide a line 3 inches long into (*a*) two equal parts, (*b*) three equal parts, (*c*) four equal parts.

26. Divide a line into two parts such that (*a*) one part is three times the other, (*b*) the ratio of the two parts is $1:2$, (*c*) one part is three-fifths of the given line.   (*6.1*)

27. In triangle ABC, AB is divided into two segments p and q. Divide AC into two segments having the ratio $p:q$.   (*6.1*)

28. Given two line segments p and q, construct x so that (*a*) $x = p^2/q$, (*b*) $x = 2q^2/p$, (*c*) $x^2 = 2p^2$ (*d*) $x = \sqrt{3pq}$.   (*6.2*)

29. Given three line segments p, q and r, construct x so that (*a*) $px = qr$, (*b*) $p:x = q:r$, (*c*) $p:q = x:r$, (*d*) $x = 2pr/q$, (*e*) $x = r^2/2q$.   (*6.2*)

30. In a given circle, inscribe (*a*) a square, (*b*) a regular octagon, (*c*) a regular 16-gon, (*d*) a regular hexagon, (*e*) an equilateral triangle, (*f*) a regular dodecagon.

31. Circumscribe each of the following about a given circle:
(*a*) a regular octagon, (*b*) a regular hexagon, (*c*) an equilateral triangle.   (*7.1*)

32. Construct a triangle similar to a given triangle with a base (*a*) three times as long, (*b*) half as long, (*c*) one and a half times as long.   (*8.1*)

33. Transform a given triangle into an equivalent triangle having an angle of (*a*) 90°, (*b*) 45°, (*c*) 60°, (*d*) 30°, (*e*) 120°, (*f*) 135°.   (*9.1*)

34. Transform a pentagon into an equivalent (*a*) quadrilateral, (*b*) triangle.   (*9.1*)

35. Construct a square equal to (*a*) a given rectangle, (*b*) a given parallelogram, (*c*) twice a given triangle.   (*9.1*)

36. Construct a square equal to (*a*) twice a given square, (*b*) half a given square, (*c*) the sum of two given squares.   (*9.1*)

Chapter 16

Proofs of Required Theorems

The syllabus of New York State lists the proofs of fifteen theorems which are required for examination purposes. These theorems are considered the most important in the logical sequence in geometry. The proof of each of the following theorems appears on the following pages.

List of Required Theorems

1. If two sides of a triangle are equal, the angles opposite these sides are equal. (Base angles of an isosceles triangle are equal.)
2. The sum of the angles in a triangle equals a straight angle.
3. If two angles of a triangle are equal, the sides opposite these angles are equal.
4. Two right triangles are congruent if the hypotenuse and a leg of one are equal to the corresponding parts of the other.
5. A diameter perpendicular to a chord bisects the chord and its arcs.
6. An angle inscribed in a circle is measured by one-half its intercepted arc.
7. An angle formed by two chords intersecting inside a circle is measured by one-half the sum of the intercepted arcs.
8a. An angle formed by two secants intersecting outside a circle is measured by one-half the difference of the intercepted arcs.
8b. An angle formed by a tangent and a secant intersecting outside a circle is measured by one-half the difference of its intercepted arcs.
8c. An angle formed by two tangents intersecting outside a circle is measured by one-half the difference of its intercepted arcs.
9. If three angles of one triangle are equal to three angles of another triangle, the triangles are similar.
10. If the altitude is drawn to the hypotenuse of a right triangle, (a) the two triangles thus formed are similar to the given triangle and to each other, (b) each leg of the given triangle is the mean proportional between the hypotenuse and the projection of that leg upon the hypotenuse.
11. The square of the hypotenuse of a right triangle equals the sum of the squares of the other two sides.
12. The area of a parallelogram equals the product of one side and the altitude to that side.
13. The area of a triangle is equal to one-half the product of one side and the altitude to that side.
14. The area of a trapezoid is equal to one-half the product of the altitude and the sum of the bases.
15. The area of a regular polygon is equal to one-half the product of its perimeter and apothem.

1. If two sides of a triangle are equal, the angles opposite these
   sides are equal. (Base angles of an isosceles triangle are equal.)

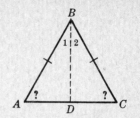

   Given: $\triangle ABC$, $AB = BC$

   To Prove: $\angle A = \angle C$

   Plan: By drawing the bisector of the vertex angle, the angles to be proved
   equal become corresponding angles of congruent triangles.

PROOF:　　　Statements	Reasons
1. Draw BD bisecting $\angle B$.	1. An angle may be bisected.
2. $\angle 1 = \angle 2$	2. To bisect is to divide into two equal parts.
3. $AB = BC$	3. Given
4. $BD = BD$	4. Identity
5. $\triangle ADB \cong \triangle BDC$	5. s.a.s. = s.a.s.
6. $\angle A = \angle C$	6. Corresponding parts of congruent triangles are equal.

2. The sum of the angles in a triangle equals a straight angle.

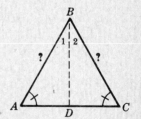

   Given: $\triangle ABC$

   To Prove: $\angle A + \angle B + \angle C = 1$ straight angle

   Plan: By drawing a line through one vertex parallel to the opposite side,
   a straight angle is formed whose parts can be proved equal to the
   angles of the triangle.

PROOF:　　　Statements	Reasons
1. Through B, draw $DE \parallel AC$.	1. Through an external point, a line can be drawn parallel to a given line.
2. $\angle DBE$ is a straight angle.	2. A straight angle is an angle whose sides lie on the same straight line, but extend in opposite directions from the vertex.
3. $\angle DBA + \angle ABC + \angle CBE = 1$ straight angle	3. The whole equals the sum of all its parts.
4. $\angle A = \angle DBA$, $\angle C = \angle CBE$	4. Alternate interior angles of parallel lines are equal.
5. $\angle A + \angle B + \angle C = 1$ straight angle	5. Substitution

3. If two angles of a triangle are equal, the sides opposite these
   angles are equal.

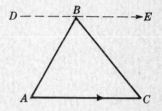

   Given: $\triangle ABC$, $\angle A = \angle C$

   To Prove: $AB = BC$

   Plan: By drawing the bisector of $\angle B$, the sides to be proved equal become
   corresponding sides of congruent triangles.

PROOF:　　　Statements	Reasons
1. Draw BD bisecting $\angle B$.	1. An angle may be bisected.
2. $\angle 1 = \angle 2$	2. To bisect is to divide into two equal parts.
3. $\angle A = \angle C$	3. Given
4. $BD = BD$	4. Identity
5. $\triangle BDA \cong \triangle BDC$	5. s.a.a. = s.a.a.
6. $AB = BC$	6. Corresponding parts of congruent triangles are equal.

4. Two right triangles are congruent if the hypotenuse and a leg of one are equal to the corresponding parts of the other.

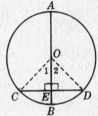

Given: Right $\triangle ABC$ with right angle at C.
Right $\triangle DEF$ with right angle at F.
$AB = DE$, $BC = EF$

To Prove: $\triangle ABC = \triangle DEF$

Plan: Move the two given triangles together so that BC coincides with EF, forming an isosceles triangle. The triangles are proved congruent by using required theorem 1 and s.a.a. = s.a.a.

PROOF: Statements	Reasons
1. $BC = EF$	1. Given
2. Move triangles ABC and DEF together so that BC coincides with EF, and A and D are on opposite sides of BC.	2. A geometric figure may be moved without changing size or shape. Equal lines may be made to coincide.
3. $\angle C$ and $\angle F$ are right angles.	3. Given
4. $\angle ACD$ is a straight angle.	4. The whole equals the sum of its parts.
5. AD is a straight line.	5. The sides of a straight angle lie in a straight line.
6. $AB = DE$	6. Given
7. $\angle A = \angle D$	7. If two sides of a triangle are equal, the angles opposite these sides are equal.
8. $\triangle ABC = \triangle DEF$	8. s.a.a. = s.a.a.

5. A diameter perpendicular to a chord bisects the chord and its arcs.

Given: Circle O, diameter $AB \perp CD$

To Prove: $CE = ED$, $\overarc{BC} = \overarc{BD}$, $\overarc{AC} = \overarc{AD}$

Plan: Congruent triangles are formed when radii are drawn to C and D, proving $CE = ED$. Equal central angles are used to prove $\overarc{BC} = \overarc{BD}$, and then the subtraction axiom is used to prove $\overarc{AC} = \overarc{AD}$.

PROOF: Statements	Reasons
1. Draw OC and OD.	1. A straight line may be drawn between two points.
2. $OC = OD$	2. Radii of a circle are equal.
3. $AB \perp CD$	3. Given
4. $\angle OEC$ and $\angle OED$ are right angles.	4. Perpendiculars form right angles.
5. $OE = OE$	5. Identity
6. $\triangle OEC \cong \triangle OED$	6. hy. leg = hy. leg
7. $CE = ED$, $\angle 1 = \angle 2$	7. Corresponding parts of congruent triangles are equal.
8. $\overarc{CB} = \overarc{BD}$	8. In a circle, equal central angles have equal arcs.
9. $\overarc{ACB} = \overarc{ADB}$	9. A diameter bisects a circle.
10. $\overarc{AC} = \overarc{AD}$	10. If equals are subtracted from equals, the differences are equal.

6. An angle inscribed in a circle is measured by one-half its intercepted arc.

Case I: The center of the circle is on one side of the angle.

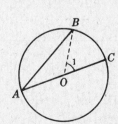

Given: $\angle A$ is inscribed in circle O.
O is on side AC.

To Prove: $\angle A \cong \frac{1}{2} \overarc{BC}$

Plan: By drawing radius OB, isosceles $\triangle AOB$ is formed. Thus $\angle A$ is proved to be equal to one-half central $\angle 1$ which is measured by $\overarc{BC}$.

PROOF: Statements	Reasons
1. Draw OB.	1. A straight line can be drawn between two points.
2. $AO = OB$	2. Radii of a circle are equal.
3. $\angle A = \angle B$	3. If two sides of a triangle are equal, the angles opposite these sides are equal.
4. $\angle A + \angle B = \angle 1$	4. In a triangle, an exterior angle equals the sum of the two nonadjacent interior angles.
5. $\angle A + \angle A = 2\angle A = \angle 1$	5. Substitution
6. $\angle A = \frac{1}{2}\angle 1$	6. Halves of equals are equal.
7. $\angle 1 \stackrel{\circ}{=} \overset{\frown}{BC}$	7. A central angle is measured by its intercepted arc.
8. $\angle A \stackrel{\circ}{=} \frac{1}{2}\overset{\frown}{BC}$	8. Substitution

Case II: The center is inside the angle.

Given: $\angle BAC$ inscribed in circle O.
O is inside $\angle BAC$.

To Prove: $\angle BAC \stackrel{\circ}{=} \frac{1}{2}\overset{\frown}{BC}$

Plan: By drawing a diameter, $\angle BAC$ is divided into two angles which can be measured by applying Case I.

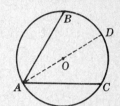

PROOF: Statements	Reasons
1. Draw diameter AD.	1. A straight line may be drawn between two points.
2. $\angle BAD \stackrel{\circ}{=} \frac{1}{2}\overset{\frown}{BD}$, $\quad \angle DAC \stackrel{\circ}{=} \frac{1}{2}\overset{\frown}{DC}$	2. An inscribed angle is measured by one-half its intercepted arc if the center of the circle is on one side.
3. $\angle BAC \stackrel{\circ}{=} \frac{1}{2}\overset{\frown}{BD} + \frac{1}{2}\overset{\frown}{DC}$ or $\angle BAC \stackrel{\circ}{=} \frac{1}{2}(\overset{\frown}{BD} + \overset{\frown}{DC})$	3. If equals are added to equals, the sums are equal.
4. $\angle BAC \stackrel{\circ}{=} \frac{1}{2}\overset{\frown}{BC}$	4. Substitution

Case III: The center is outside the angle.

Given: $\angle BAC$ inscribed in circle O.

To Prove: $\angle BAC \stackrel{\circ}{=} \frac{1}{2}\overset{\frown}{BC}$

Plan: By drawing a diameter, $\angle BAC$ becomes the difference of two angles which can be measured by applying Case I.

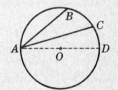

PROOF: Statements	Reasons
1. Draw diameter AD.	1. A straight line may be drawn between two points.
2. $\angle BAD \stackrel{\circ}{=} \frac{1}{2}\overset{\frown}{BD}$, $\quad \angle CAD \stackrel{\circ}{=} \frac{1}{2}\overset{\frown}{CD}$	2. An inscribed angle is measured by one-half its intercepted arc if the center of the circle is on one side.
3. $\angle BAC \stackrel{\circ}{=} \frac{1}{2}\overset{\frown}{BD} - \frac{1}{2}\overset{\frown}{CD}$ or $\angle BAC \stackrel{\circ}{=} \frac{1}{2}(\overset{\frown}{BD} - \overset{\frown}{CD})$	3. If equals are subtracted from equals, the differences are equal.
4. $\angle BAC \stackrel{\circ}{=} \frac{1}{2}\overset{\frown}{BC}$	4. Substitution

7. **An angle formed by two chords intersecting inside a circle is measured by one-half the sum of the intercepted arcs.**

Given: $\angle 1$ formed by chords AB and CD intersecting at point E inside circle O.

To Prove: $\angle 1 \stackrel{\circ}{=} \frac{1}{2}(\overset{\frown}{AC} + \overset{\frown}{BD})$

Plan: By drawing chord AD, $\angle 1$ becomes an exterior angle of a triangle whose nonadjacent interior angles are inscribed angles measured by $\frac{1}{2}\overset{\frown}{AC}$ and $\frac{1}{2}\overset{\frown}{BD}$.

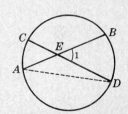

PROOF: Statements	Reasons
1. Draw AD.	1. A straight line may be drawn between two points.
2. $\angle 1 = \angle A + \angle D$	2. An exterior angle of a triangle equals the sum of the nonadjacent interior angles.
3. $\angle A \doteq \frac{1}{2}\overset{\frown}{BD}$, $\angle D \doteq \frac{1}{2}\overset{\frown}{AC}$	3. An angle inscribed in a circle is measured by one-half its intercepted arc.
4. $\angle 1 \doteq \frac{1}{2}\overset{\frown}{BD} + \frac{1}{2}\overset{\frown}{AC} \doteq \frac{1}{2}(\overset{\frown}{BD} + \overset{\frown}{AC})$	4. Substitution

8a. An angle formed by two secants intersecting outside a circle is measured by one-half the difference of the intercepted arcs.

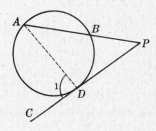

   Given: $\angle P$ formed by secants PBA and PDC intersecting in P, a point outside circle O.

   To Prove: $\angle P \doteq \frac{1}{2}(\overset{\frown}{AC} - \overset{\frown}{BD})$

   Plan: By drawing AD, $\angle 1$ becomes an exterior angle of $\triangle ADP$, of which $\angle P$ is a nonadjacent interior angle.

PROOF: Statements	Reasons
1. Draw AD.	1. A straight line may be drawn between two points.
2. $\angle P + \angle A = \angle 1$	2. An exterior angle of a triangle equals the sum of the nonadjacent interior angles.
3. $\angle P = \angle 1 - \angle A$	3. If equals are subtracted from equals, the differences are equal.
4. $\angle 1 \doteq \frac{1}{2}\overset{\frown}{AC}$, $\angle A \doteq \frac{1}{2}\overset{\frown}{BD}$	4. An angle inscribed in a circle is measured by one-half its intercepted arc.
5. $\angle P \doteq \frac{1}{2}\overset{\frown}{AC} - \frac{1}{2}\overset{\frown}{BD}$ or $\angle P \doteq \frac{1}{2}(\overset{\frown}{AC} - \overset{\frown}{BD})$	5. Substitution

8b. An angle formed by a secant and a tangent intersecting outside a circle is measured by one-half the difference of its intercepted arcs.

   Given: $\angle P$ formed by secant PBA and tangent PDC intersecting in P, a point outside circle O.

   To Prove: $\angle P \doteq \frac{1}{2}(\overset{\frown}{AD} - \overset{\frown}{BD})$

   Plan: By drawing chord AD, $\angle 1$ becomes an exterior angle of $\triangle ADP$, of which $\angle P$ and $\angle A$ are nonadjacent interior angles.

PROOF: Statements	Reasons
1. Draw AD.	1. A straight line may be drawn between two points.
2. $\angle P + \angle A = \angle 1$	2. An exterior angle of a triangle equals the sum of the nonadjacent interior angles.
3. $\angle P = \angle 1 - \angle A$	3. If equals are subtracted from equals, the differences are equal.
4. $\angle 1 \doteq \frac{1}{2}\overset{\frown}{AD}$	4. An angle formed by a tangent and a chord is measured by one-half its intercepted arc.
5. $\angle A \doteq \frac{1}{2}\overset{\frown}{BD}$	5. An inscribed angle is measured by one-half its intercepted arc.
6. $\angle P \doteq \frac{1}{2}\overset{\frown}{AD} - \frac{1}{2}\overset{\frown}{BD}$ or $\angle P \doteq \frac{1}{2}(\overset{\frown}{AD} - \overset{\frown}{BD})$	6. Substitution

8c. An angle formed by two tangents intersecting outside a circle is measured by one-half the difference of its intercepted arcs.

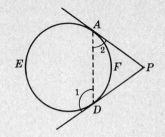

Given: $\angle P$ formed by tangents PA and PD intersecting in P, a point outside circle O.

To Prove: $\angle P \stackrel{\circ}{=} \frac{1}{2}(\overset{\frown}{AED} - \overset{\frown}{AFD})$

Plan: By drawing chord AD, $\angle 1$ becomes an exterior angle of $\triangle ADP$, of which $\angle P$ and $\angle 2$ are nonadjacent interior angles.

PROOF: Statements	Reasons
1. Draw AD.	1. A straight line may be drawn between two points.
2. $\angle P + \angle 2 = \angle 1$	2. An exterior angle of a triangle equals the sum of the nonadjacent interior angles.
3. $\angle P = \angle 1 - \angle 2$	3. If equals are subtracted from equals, the differences are equal.
4. $\angle 1 \stackrel{\circ}{=} \frac{1}{2}\overset{\frown}{AED}, \quad \angle 2 \stackrel{\circ}{=} \frac{1}{2}\overset{\frown}{AFD}$	4. An angle formed by a tangent and a chord is measured by one-half its intercepted arc.
5. $\angle P \stackrel{\circ}{=} \frac{1}{2}\overset{\frown}{AED} - \frac{1}{2}\overset{\frown}{AFD}$ or $\angle P \stackrel{\circ}{=} \frac{1}{2}(\overset{\frown}{AED} - \overset{\frown}{AFD})$	5. Substitution

9. If three angles of one triangle are equal to three angles of another triangle, the triangles are similar.

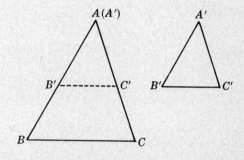

Given: $\triangle ABC$ and $\triangle A'B'C'$.
  $\angle A = \angle A'$, $\angle B = \angle B'$, $\angle C = \angle C'$.

To Prove: $\triangle ABC \sim \triangle A'B'C'$

Plan: To prove the triangles similar, it must be shown that corresponding sides are in proportion. This is done by placing the triangles so that a pair of equal angles coincide, and then repeating this so that another pair of equal angles coincide.

PROOF: Statements	Reasons
1. $\angle A = \angle A'$	1. Given
2. Place $\triangle A'B'C'$ on $\triangle ABC$ so that $\angle A'$ coincides with $\angle A$.	2. A geometric figure may be moved without change of size or shape. Equal angles may be made to coincide.
3. $\angle B = \angle B'$	3. Given
4. $B'C' \parallel BC$	4. Two lines are parallel if their corresponding angles are equal.
5. $\dfrac{A'B'}{AB} = \dfrac{A'C'}{AC}$	5. A line parallel to one side of a triangle divides the other two sides proportionately.
6. In like manner, by placing $\triangle A'B'C'$ on $\triangle ABC$ so that $\angle B'$ coincides with $\angle B$, $$\dfrac{A'B'}{AB} = \dfrac{B'C'}{BC}$$	6. Reasons 1 to 5.
7. $\dfrac{A'B'}{AB} = \dfrac{A'C'}{AC} = \dfrac{B'C'}{BC}$	7. Things (ratios) equal to the same thing are equal to each other.
8. $\triangle A'B'C' \sim \triangle ABC$	8. Two polygons are similar if their corresponding angles are equal and if their corresponding sides are in proportion.

10. If the altitude is drawn to the hypotenuse of a right triangle, (*a*) the two triangles thus formed are similar to the given triangle and to each other, (*b*) each leg of the given triangle is the mean proportional between the hypotenuse and the projection of that leg upon the hypotenuse.

> **Given:** $\triangle ABC$ with a right angle at C, altitude CD to hypotenuse AB.
>
> **To Prove:** (*a*) $\triangle ADC \sim \triangle BDC \sim \triangle ABC$
> (*b*) $c:a = a:p, \quad c:b = b:q$
>
> **Plan:** The triangles are similar since they have a right angle and a pair of equal acute angles. The proportions follow from the similar triangles.

PROOF: Statements	Reasons
1. $\angle C$ is a right angle.	1. Given
2. CD is the altitude to AB.	2. Given
3. $CD \perp AB$	3. An altitude to a side of a triangle is perpendicular to that side.
4. $\angle CDB$ and $\angle CDA$ are right angles.	4. Perpendiculars form right angles.
5. $\angle A = \angle A, \ \angle B = \angle B$	5. Identity
6. $\triangle ADC \sim \triangle ABC, \ \triangle BDC \sim \triangle ABC$	6. Right triangles are similar if an acute angle of one equals an acute angle of the other.
7. $\triangle ADC \sim \triangle BDC$	7. Triangles similar to the same triangle are similar to each other.
8. $c:a = a:p, \ c:b = b:q$	8. Corresponding sides of similar triangles are in proportion.

11. The square of the hypotenuse of a right triangle equals the sum of the squares of the other two sides.

> **Given:** Right $\triangle ABC$, with a right angle at C.
> Legs are a and b, and hypotenuse is c.
>
> **To Prove:** $c^2 = a^2 + b^2$
>
> **Plan:** Draw $CD \perp AB$ and apply theorem 10.

PROOF: Statements	Reasons
1. Draw $CD \perp AB$.	1. Through an external point, a line may be drawn perpendicular to a given line.
2. $\dfrac{c}{a} = \dfrac{a}{p}, \ \dfrac{c}{b} = \dfrac{b}{q}$	2. If the altitude is drawn to the hypotenuse of a right triangle, either leg is the mean proportional between the hypotenuse and the projection of that leg upon the hypotenuse.
3. $a^2 = cp, \ b^2 = cq$	3. In a proportion, the product of the means equals the product of the extremes.
4. $a^2 + b^2 = cp + cq = c(p + q)$	4. If equals are added to equals, the sums are equal.
5. $c = p + q$	5. The whole equals the sum of its parts.
6. $a^2 + b^2 = c(c) = c^2$	6. Substitution

12. The area of a parallelogram equals the product of one side and the altitude to that side.

> **Given:** $\square ABCD$, base $AD = b$, altitude $BE = h$.
>
> **To Prove:** Area of $ABCD = bh$
>
> **Plan:** By dropping a perpendicular to the base, extended, a rectangle is formed having the same base and altitude as the parallelogram. By adding congruent triangles to a common area, the rectangle and parallelogram are proved equal.

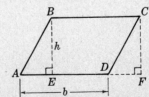

PROOF: Statements	Reasons
1. Draw $CF \perp AD$ (extended).	1. Through an external point, a line may be drawn perpendicular to a given line.
2. $CF \parallel BE$	2. Lines perpendicular to the same line are parallel.
3. $BC \parallel AD$	3. Opposite sides of a parallelogram are parallel.
4. $\angle CFD$ and $\angle BEA$ are right angles.	4. Perpendiculars form right angles.
5. $BCFE$ is a rectangle.	5. A parallelogram having a right angle is a rectangle.
6. $AB = CD$, $CF = BE$	6. Opposite sides of a parallelogram are equal.
7. $\triangle ABE \cong \triangle CFD$	7. hy. leg = hy. leg
8. quad. $BCDE$ = quad. $BCDE$	8. Identity
9. $\triangle ABE + $ quad. $BCDE$ $= \triangle CFD + $ quad. $BCDE$ or rectangle $BCFE = \square ABCD$	9. If equals are added to equals, the sums are equal.
10. Area of rectangle $BCFE = bh$	10. The area of a rectangle equals the product of its base and altitude.
11. Area of $\square ABCD = bh$	11. Substitution

13. The area of a triangle is equal to one-half the product of one side and the altitude to that side.

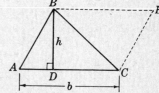

    Given: $\triangle ABC$, base $AC = b$, altitude $BD = h$.

    To Prove: Area of $\triangle ABC = \frac{1}{2}bh$

    Plan: By drawing $BE \parallel AC$ and $EC \parallel AB$, a parallelogram is formed having the same base and altitude as the triangle. Then the triangle is half the parallelogram.

PROOF: Statements	Reasons
1. Draw $BE \parallel AC$, $CE \parallel AB$.	1. Through an external point, a line may be drawn parallel to a given line.
2. $ABEC$ is a parallelogram with base b and altitude h.	2. A quadrilateral is a parallelogram if its opposite sides are parallel.
3. $\triangle ABC = \frac{1}{2} \square ABEC$	3. A diagonal divides a parallelogram into two congruent triangles.
4. Area of $\square ABEC = bh$	4. The area of a parallelogram equals the product of its base and altitude.
5. Area of $\triangle ABC = \frac{1}{2}bh$	5. Substitution

14. The area of a trapezoid is equal to one-half the product of the altitude and the sum of its bases.

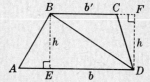

    Given: Trapezoid $ABCD$, altitude $BE = h$, base $AD = b$, base $BC = b'$.

    To Prove: Area of $ABCD = \frac{1}{2}h(b + b')$

    Plan: By drawing a diagonal, the trapezoid is divided into two triangles having common altitude h and bases b and b'.

PROOF: Statements	Reasons
1. Draw BD.	1. A straight line may be drawn between two points.
2. Draw $DF \perp BC$ (extended).	2. Through an external point, a line may be drawn perpendicular to a given line.
3. $DF = BE = h$	3. Parallel lines are everywhere equidistant.
4. Area of $\triangle ABD = \frac{1}{2}bh$ Area of $\triangle BCD = \frac{1}{2}b'h$	4. The area of a triangle equals one-half the product of its base and altitude.
5. Area of $ABCD = \frac{1}{2}bh + \frac{1}{2}b'h$ $= \frac{1}{2}h(b + b')$	5. If equals are added to equals, the sums are equal.

15. The area of a regular polygon is equal to one-half the product of its perimeter and apothem.

    Given: Regular polygon $ABCDE\ldots$ having center O, apothem r, perimeter p.

    To Prove: Area of $ABCDE\ldots = \frac{1}{2}rp$

    Plan: By joining each vertex to the center, congruent triangles are obtained, the sum of whose areas equals the area of the regular polygon.

PROOF: Statements	Reasons
1. Draw $OA, OB, OC, OD, OE, \ldots$	1. A straight line may be drawn between two points.
2. r is the altitude of each triangle formed.	2. Apothems of a regular polygon are equal.

3. Area of $\triangle AOB = \frac{1}{2}ar$
        $\triangle BOC = \frac{1}{2}br$
        $\triangle COD = \frac{1}{2}cr$
     $\ldots\ldots\ldots\ldots$
	3. The area of a triangle equals one-half the product of its base and altitude.

4. Area of regular polygon $ABCDE\ldots$
    $= \frac{1}{2}ar + \frac{1}{2}br + \frac{1}{2}cr + \cdots$
    $= \frac{1}{2}r(a + b + c + \cdots)$
	4. If equals are added to equals, the sums are equal.

5. $p = a + b + c + \cdots$	5. The whole equals the sum of its parts.
6. Area of $ABCDE\ldots = \frac{1}{2}rp$	6. Substitution

Chapter 17

Extending Plane Geometry into Solid Geometry

1. Understanding Solid Geometry

A. Solids

A *solid* is an enclosed portion of space bounded by plane and curved surfaces.

Thus, the *pyramid* ◭ , the *cube* ⬓ , the *cone* ◮ , the *cylinder* ⬛ and the *sphere* ◓ are solids.

A solid has three dimensions: Length, width and thickness.

A solid may be represented or illustrated by a box, a brick, a block, a crystal or a ball. However, these are representations of solids but are not the ideal solids which are the concern of Solid Geometry. In Solid Geometry the geometric properties of solids are studied, such as their shape, their size, their relationships of parts and the relationships among solids. In Solid Geometry, physical properties of solids such as their color, weight or smoothness are disregarded.

Kinds of Solids:

1. A *polyhedron* is a solid bounded by plane (flat) surfaces only. Thus the pyramid and cube are polyhedrons. The cone, cylinder and sphere are not polyhedrons since each has a curved surface.

   The bounding surfaces of the polyhedron are its faces, the lines of intersection of the faces are its edges, and the points of intersection of its edges are its vertices. A diagonal of the polyhedron joins two vertices not in the same face.

   Thus the polyhedron shown in the above figure has six faces. Two of them are triangles and the other four are quadrilaterals. Note *AF*, a diagonal of the polyhedron. The shaded polygon *HJKL* is a *section of the polyhedron* formed by the intersection of the solid and the plane passing through it.

   The angle between any two intersecting faces is a *dihedral angle*. The angle between the covers of a book when opened is a dihedral angle. As the book is opened wider, the dihedral angle grows from one that is acute to one that is right, obtuse or straight. The dihedral angle can be measured by measuring the *plane angle* between two lines, one in each face, and in a plane perpendicular to the intersection between the faces.

2. A *prism* is a polyhedron two of whose faces are parallel polygons and whose remaining faces are parallelograms. The bases of a prism are its parallel polygons. These may have any number of sides. The lateral (side) faces are par-

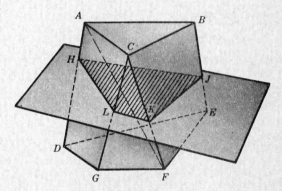

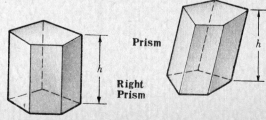

allelograms. The distance between the two bases is h. This line is at right angles to each base.

A *right prism* is a prism whose lateral faces are rectangles. The distance, h, is the height of any of the lateral faces.

3. A *rectangular solid* (box) is a prism bounded by six rectangles. The rectangular solid can be formed from the pattern of six rectangles folded along the dotted lines. The length (l), the width (w) and the height (h) are its dimensions.

4. A *cube* is a rectangular solid bounded by six squares. The cube can be formed from the pattern of six squares folded along the dotted lines. Each equal dimension is represented by e in the diagram.

A *cubic unit* is a cube whose edge is 1 unit. Thus, a cubic inch is a cube whose edge is 1 inch.

5. A *parallelopiped* is a prism bounded by six parallelograms. Hence the rectangular solid and cube are special parallelopipeds.

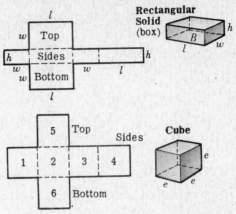

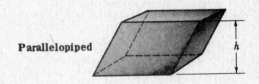

Parallelopiped

Relationships among polygons in Plane Geometry and the corresponding relationships among the previously mentioned polyhedrons in Solid Geometry are shown in the following table:

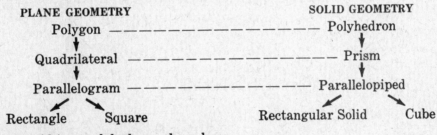

PLANE GEOMETRY   SOLID GEOMETRY

Polygon ————————————————— Polyhedron

Quadrilateral ———————————————— Prism

Parallelogram ———————————————— Parallelopiped

Rectangle   Square   Rectangular Solid   Cube

6. A *pyramid* is a polyhedron whose base is a polygon and whose other faces meet at a point, its vertex. The base (B) may have any number of sides. However, the other faces must be triangles. The distance from the vertex to the base is equal to the altitude or height (h), a line from the vertex at right angles to the base.

A *regular pyramid* is a pyramid whose base is a regular polygon and whose altitude joins the vertex and the center of the base.

A *frustum of a pyramid* is the part of a pyramid which remains if the top of the pyramid is cut off by a plane parallel to the base. Note that its lateral faces are trapezoids.

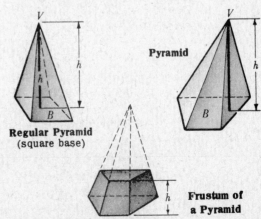

Pyramid

Regular Pyramid
(square base)

**Frustum of
a Pyramid**

7. A *circular cone* is a solid whose base is a circle and whose lateral surface comes to a point. (*A circular cone will be referred to as a cone.*)

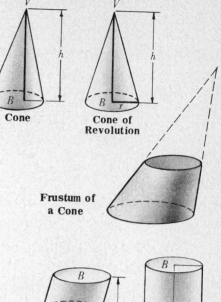

Cone   Cone of Revolution

A *cone of revolution* is formed by revolving a right triangle about one of its legs. This leg becomes the altitude or height (*h*) of the cone and the other becomes the radius (*r*) of the base.

A *frustum of a cone* is the part of a cone which remains if the top of the cone is cut off by a plane parallel to the base.

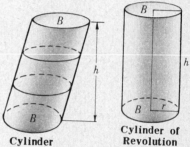

Frustum of a Cone

8. A *circular cylinder* is a solid whose bases are parallel circles and whose cross-sections parallel to the bases are also circles. (*A circular cylinder will be referred to as a cylinder.*)

A *cylinder of revolution* is formed by revolving a rectangle about one of its two dimensions as an axis. This dimension becomes the height (*h*) of the cylinder and the other becomes the radius (*r*) of the base.

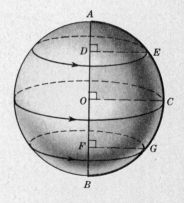

Cylinder   Cylinder of Revolution

9. A *sphere* is a solid such that every point on its surface is at an equal distance from the same point, its center.

A sphere is formed by revolving a semicircle about its diameter as an *axis*. The outer end of the radius perpendicular to the axis generates a *great circle* while the outer ends of the other chords perpendicular to the diameter generate *small circles*.

Thus, sphere *O* is formed by the rotation of semicircle *ACB* about *AB* as an axis. In the process, point *C* generates a great circle while points *E* and *G* generate small circles.

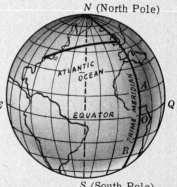

Locating points on the earth's surface is better understood if you think of the earth as a sphere formed by rotating semicircle *NOS*, running through Greenwich, England (near London), about *NS* as an axis. Point *O*, halfway between *N* and *S*, generates the *equator*, *EQ*. Points *A* and *B* generate *parallels of latitude* which are small circles on the earth's surface parallel to the equator. Each position of the rotating semicircle is a semi-meridian. (A meridian is a great circle passing through the North and South Poles.) The meridian through Greenwich is called the Prime Meridian.

Using the intersection of the equator and the Prime Meridian as the origin, the location of New York City is 40°48½′ North Latitude and 73°57½′ West Longitude. The line shown on the map is an arc of a great circle through New York City and London. Such an arc is the shortest distance between two points on the earth's surface. Find this line by stretching a rubber band tightly between New York City and London on a globe of the Earth.

10. *Regular Polyhedrons*

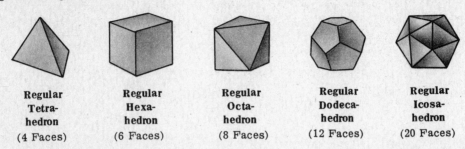

| Regular
Tetra-
hedron
(4 Faces) | Regular
Hexa-
hedron
(6 Faces) | Regular
Octa-
hedron
(8 Faces) | Regular
Dodeca-
hedron
(12 Faces) | Regular
Icosa-
hedron
(20 Faces) |

Regular polyhedrons are solids having faces which are regular polygons and having the same number of faces meeting at each vertex. There are only five such solids, as shown. Note that their faces are equilateral triangles, squares or regular pentagons.

Regular hexagons cannot be faces of a regular polyhedron. If three regular hexagons have a common vertex, the sum of the three interior angles at that vertex would equal 3(120°) or 360°. The result would be that the three regular hexagons would lie in the same plane.

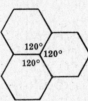

B. Extension of Plane Geometry Principles to Space Geometry Principles

In space geometry, "sphere" corresponds to "circle" in plane geometry. Similarly, "plane" corresponds to "straight line". By interchanging circle and sphere, or straight line and plane, note how each of the following *dual statements* can be interchanged. In so doing many plane geometry principles with which you are acquainted become space geometry principles.

RELATED DUAL STATEMENTS

1. Every point on a *circle* is at a radius distance from its center.

2. Two intersecting straight *lines* determine a *plane*.

1. Every point on a *sphere* is at a radius distance from its center.

2. Two intersecting *planes* determine a straight *line*.

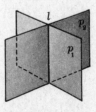

3. Two *lines* perpendicular to the same *plane* are parallel.

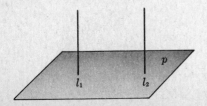

3. Two *planes* perpendicular to the same *line* are parallel.

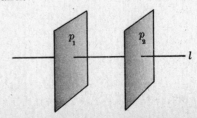

Be sure in obtaining dual statements that there is a complete interchange. If the interchange is incomplete as in the following pair of statements, there is no duality.

4. The locus of a point at a given distance from a given line is a pair of lines parallel to the given line and at the given distance from it.

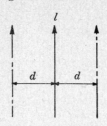

4. The locus of a point at a given distance from a given line is a circular cylindrical surface having the given line as axis and the given distance as radius.

Unlike a cylinder, a cylindrical surface is not limited in extent nor does it have bases. Similarly, a conical surface is unlimited in extent and has no base.

I. *Extension of Distance Principles*

DISTANCE IN A PLANE

1. The distance from a point to a *line* is the length of the perpendicular from the point to the *line*.

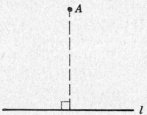

2. The distance between two parallel *lines* is the length of a perpendicular between them.

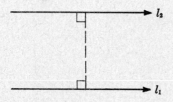

3. The distance from a point to a *circle* is the external segment of the secant from the point through the center of the *circle*.

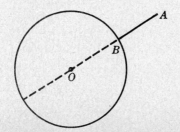

DISTANCE IN SPACE

1. The distance from a point to a *plane* is the length of the perpendicular from the point to the *plane*.

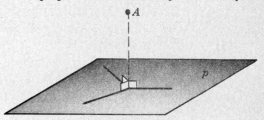

2. The distance between two parallel *planes* is the length of a perpendicular between them.

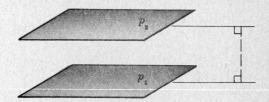

3. The distance from a point to a *sphere* is the external segment of the secant from the point through the center of the *sphere*.

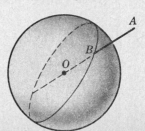

4. Parallel *lines* are everywhere equidistant.

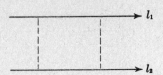

4. Parallel *planes* are everywhere equidistant.

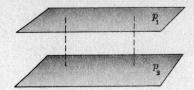

5. Any point on the *line* which is the perpendicular bisector of a line segment is equidistant from the ends of the segment.

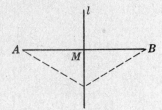

5. Any point on the *plane* which is the perpendicular bisector of a line segment is equidistant from the ends of the segment.

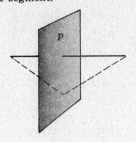

6. Any point on the line which is the bisector of the angle between two *lines* is equidistant from the sides of the angle.

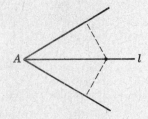

6. Any point on the plane which is the bisector of the dihedral angle between two *planes* is equidistant from the sides of the angle.

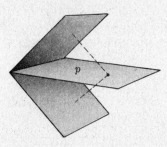

II. *Extension of Locus Principles*

LOCUS IN A PLANE

1. The locus of a point at a given distance from a given point is a *circle* having the given point as center and the given distance as radius.

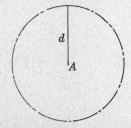

LOCUS IN SPACE

1. The locus of a point at a given distance from a given point is a *sphere* having the given point as center and the given distance as radius.

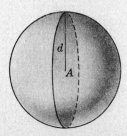

2. The locus of a point at a given distance from a given *line* is a *pair of lines* parallel to the given *line* and at the given distance from it.

2. The locus of a point at a given distance from a given *plane* is a pair of *planes* parallel to the given *plane* and at the given distance from it.

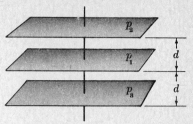

3. The locus of a point equidistant from two points is the *line* which is the perpendicular bisector of the line segment joining the two points.

3. The locus of a point equidistant from two points is the *plane* which is the perpendicular bisector of the line segment joining the two points.

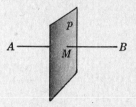

4. The locus of a point equidistant from the *lines* which are the sides of an angle is the *line* which is the bisector of the angle between them.

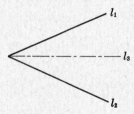

4. The locus of a point equidistant from the *planes* which are the sides of a dihedral angle is the *plane* which is the bisector of the angle between them.

5. The locus of a point equidistant from two parallel lines is the *line* parallel to them and midway between them.

5. The locus of a point equidistant from two parallel planes is the *plane* parallel to them and midway between them.

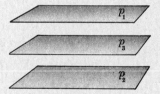

6. The locus of a point equidistant from two concentric *circles* is the *circle* midway between the given *circles* and concentric to them.

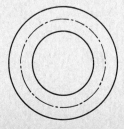

6. The locus of a point equidistant from two concentric *spheres* is the *sphere* midway between the given *spheres* and concentric to them.

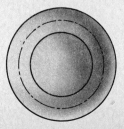

7. The locus of the vertex of a right triangle having a given hypotenuse is the *circle* having the hypotenuse as its diameter.

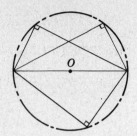

7. The locus of the vertex of a right triangle having a given hypotenuse is the *sphere* having the hypotenuse as its diameter.

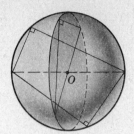

C. *Extension of Coordinate Geometry Using Three Perpendicular Axes*

(Arrows indicate positive direction; dotted lines indicate negative direction)

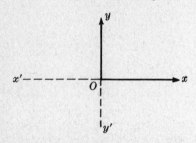

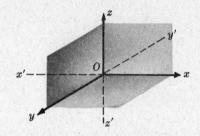

COORDINATE GEOMETRY IN A PLANE

1. Coordinates: $P_1(x_1, y_1)$, $P_2(x_2, y_2)$

2. Midpoint of P_1P_2:

$$x_M = \frac{x_1 + x_2}{2}, \qquad y_M = \frac{y_1 + y_2}{2}$$

3. Distance P_1P_2:

$$d = \sqrt{(x_2 - x_1)^2 + (y_2 - y_1)^2}$$

4. Equation of a circle having the origin as a center and a radius r.

$$x^2 + y^2 = r^2$$

COORDINATE GEOMETRY IN SPACE

1. Coordinates: $P_1(x_1, y_1, z_1)$, $P_2(x_2, y_2, z_2)$

2. Midpoint of P_1P_2:

$$x_M = \frac{x_1 + x_2}{2}, \qquad y_M = \frac{y_1 + y_2}{2}, \qquad z_M = \frac{z_1 + z_2}{2}$$

3. Distance P_1P_2:

$$d = \sqrt{(x_2 - x_1)^2 + (y_2 - y_1)^2 + (z_2 - z_1)^2}$$

4. Equation of a sphere having the origin as a center and a radius r.

$$x^2 + y^2 + z^2 = r^2$$

2. *Areas of Solids: Square Measure*

Area Formulas for Solids, using T for the total area of the solid:

1. Total area of the six squares of a *cube*:
   $T = 6e^2$

   $A = e^2$ for each face.

2. Total area of the six rectangles of a *rectangular solid*:
   $T = 2lw + 2lh + 2wh$

   $A = lw$ for top or bottom faces
   $A = lh$ for front or back faces
   $A = wh$ for left or right faces

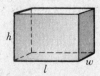

3. Total area of a *sphere*:
   $A = 4\pi r^2$

4. Total area of a *cylinder of revolution*:
   $T = 2\pi rh + 2\pi r^2$
   $T = 2\pi r(r + h)$

2.1 FINDING TOTAL AREAS of SOLIDS

Find, to the nearest integer, the total area of

(*a*) **a cube with an edge of 5 in.**

(*b*) **a rectangular solid with dimensions of 10 ft., 7 ft. and 4½ ft.**

(*c*) **a sphere with a radius of 1.1 yd.**

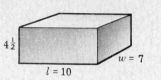 

Solutions:

(*a*) $T = 6e^2$
       $T = 6(5^2)$
          $= 150$
   Ans. 150 sq. in.

(*b*) $T = 2lw + 2lh + 2wh$
       $T = 2(10)(7) + 2(10)(4\frac{1}{2}) + 2(7)(4\frac{1}{2})$
          $= 293$
   Ans. 293 sq. ft.

(*c*) $T = 4\pi r^2$
       $T = 4(3.14)(1.1^2)$
          $= 15.1976$
   Ans. 15 sq. yd.

3. *Volumes of Solids: Cubic Measure*

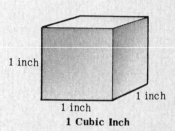

1 Cubic Inch

A *cubic unit* is a cube whose edge is 1 unit. Thus, a cubic inch is a cube whose side is 1 inch.

The *volume of a solid* is the number of cubic units that it contains. Thus, a box 5 units long, 3 units wide and 4 units high has a volume of 60 cubic units; that is, it has a capacity or space large enough to contain 60 cubes, 1 unit on a side.

In volume formulas, the volume is in *cubic units*, the unit being the same as that used for the dimensions. Thus, if the edge of a cube is 3 yards, its volume is 27 cubic yards.

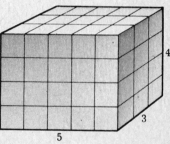

<u>**Volume Formulas**</u>, using V for the volume of the solid, B for the area of a base and h for the distance between the bases or between the vertex and a base:

1. Rectangular Solid: $V = lwh$

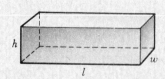

3. Cylinder: $V = Bh$ or $V = \pi r^2 h$

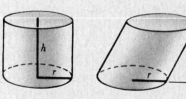

2. Prism: $V = Bh$

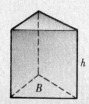 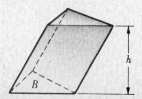

4. Cube: $V = e^3$

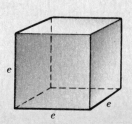

5. Pyramid: $V = \frac{1}{3}Bh$ **6. Cone:** $V = \frac{1}{3}Bh$ or $V = \frac{1}{3}\pi r^2 h$ **7. Sphere:** $V = \frac{4}{3}\pi r^3$

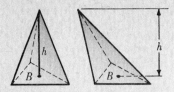  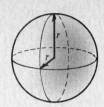

3.1 RELATIONS AMONG CUBIC UNITS

Find the volume V of (*Hint:* A cubic unit is a cube whose edge is 1 unit.)

(a) a cubic foot in cubic inches

(b) a cubic yard in cubic feet

(c) a liter (cubic decimeter) in cubic centimeters.

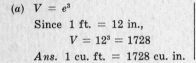

1 Cubic Unit

Solutions:

(a) $V = e^3$
  Since 1 ft. = 12 in.,
  $V = 12^3 = 1728$
  Ans. 1 cu. ft. = 1728 cu. in.

(b) $V = e^3$
  Since 1 yd. = 3 ft.,
  $V = 3^3 = 27$
  Ans. 1 cu. yd. = 27 cu. ft.

(c) $V = e^3$
  Since 1 cu. dm. = 10 cu. cm.,
  $V = 10^3 = 1000$
  Ans. 1 liter = 1000 cu. cm.

3.2 FINDING VOLUMES of CUBES

Find the volume of a cube (V) in cu. ft. whose edge is (a) 4 in., (b) 4 ft., (c) 4 yd., (d) 4 rd.

Solutions: (To find volume in cu. ft., express side in ft.)

(a) $V = e^3$
  Since 4 in. = $\frac{1}{3}$ ft.,
  $V = (\frac{1}{3})^3 = \frac{1}{27}$
  Ans. $\frac{1}{27}$ cu. ft.

(b) $V = e^3$
  $V = 4^3 = 64$
  Ans. 64 cu. ft.

(c) $V = e^3$
  Since 4 yd. = 12 ft.,
  $V = 12^3 = 1728$
  Ans. 1728 cu. ft.

(d) $V = e^3$
  Since 4 rd. = $4(16\frac{1}{2})$ or 66 ft.
  $V = 66^3 = 287,496$
  Ans. 287,496 cu. ft.

3.3 FINDING VOLUMES of RECTANGULAR SOLID, PRISM and PYRAMID

Find the volume of

(a) a rectangular solid having a length of 6 in., a width of 4 in. and a height of 1 ft.

(b) a prism having a height of 15 yd. and a triangular base of 120 sq. ft.

(c) a pyramid having a height of 8 yd. and a square base whose side is $4\frac{1}{2}$ yd.

Solutions:

(a) $V = lwh$
  $V = 6(4)(12) = 288$
  Ans. 288 cu. in.

(b) $V = Bh$
  $V = 120(45) = 5400$
  Ans. 5400 cu. ft. or 200 cu. yd.

(c) $V = \frac{1}{3}Bh$
  $V = \frac{1}{3}(\frac{9}{2})^2(8) = 54$
  Ans. 54 cu. yd.

3.4 FINDING VOLUMES of SPHERE, CYLINDER and CONE

Find the volume, to the nearest integer, of

(a) a sphere with a radius of 10 in.

(b) a cylinder with a height of 4 yd. and a base whose radius is 2 ft.

(c) a cone with a height of 2 ft. and a base whose radius is 2 yd.

Solutions: (Let $\pi = 3.14$)

(a) $V = \frac{4}{3}\pi r^3$
  $= \frac{4}{3}(3.14)10^3$
  $= 4186\frac{2}{3}$
  Ans. 4187 cu. in.

(b) $V = \pi r^2 h$
  $= (3.14)(2^2)12$
  $= 150.72$
  Ans. 151 cu. ft.

(c) $V = \frac{1}{3}\pi r^2 h$
  $= \frac{1}{3}(3.14)(6^2)(2)$
  $= 75.36$
  Ans. 75 cu. ft.

3.5 DERIVING FORMULAS from $V = Bh$

From $V = Bh$, the volume formula for a prism or cylinder, derive the volume formulas for each of the following:

(a) 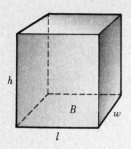   (b)   (c) 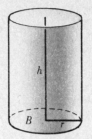   (d)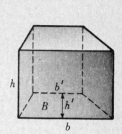

Rectangular Solid   **Cube**   **Cylinder of Revolution**   **Right Prism with a Trapezoid for a Base**

Solutions:

(a) $V = Bh$
Since $B = lw$,
$V = (lw)h$
Ans. $V = lwh$

(b) $V = Bh$
Since $B = e^2$,
and $h = e$,
$V = (e^2)e$
Ans. $V = e^3$

(c) $V = Bh$
Since $B = \pi r^2$,
$V = (\pi r^2)h$
Ans. $V = \pi r^2 h$

(d) $V = Bh$
Since $B = \dfrac{h'}{2}(b + b')$,
$V = \dfrac{h'}{2}(b + b')h$
Ans. $V = \dfrac{hh'}{2}(b + b')$

3.6 FORMULAS for COMBINED VOLUMES

State the formula for the volume of each solid:

(a) 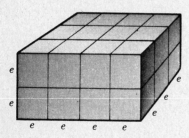   (b) 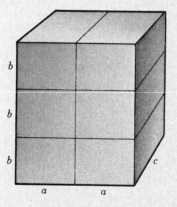   (c)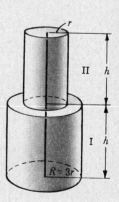

Solutions:

(a) $V = lwh$
Now, $l = 4e,\ w = 3e,\ h = 2e$
Hence, $V = (4e)(3e)(2e)$
Ans. $V = 24e^3$

(b) $V = lwh$
Now, $l = 2a,\ w = c,\ h = 3b$
Hence, $V = (2a)(c)(3b)$
Ans. $V = 6abc$

(c) $V = $ cyl. I $+$ cyl. II
$V = \pi R^2 h + \pi r^2 h$
Now, $R = 3r$
Hence, $V = \pi(3r)^2 h + \pi r^2 h$
Ans. $V = 10\pi r^2 h$

Supplementary Problems

1. Find, to the nearest integer, using $\pi = 3.14$, the total area of   *(2.1)*

    (a) a cube with an edge of 7 yd.

    (b) a rectangular solid with dimensions of 8 ft., $6\frac{1}{2}$ ft. and 14 ft.

    (c) a sphere with a radius of 30 in.

    (d) a cylinder of revolution with a radius of 10 rd. and a height of $4\frac{1}{2}$ rd. [*Hint.* Use $T = 2\pi r(r+h)$]

2. Find the volume of   *(3.1)*

    (a) a cubic yard in cubic inches

    (b) a cubic rod in cubic yards

    (c) a cubic meter in cubic centimeters (1 meter = 100 cm.).

3. Find, to the nearest cubic inch, the volume of a cube whose edge is   *(3.2)*

    (a) 3 in.   (b) $4\frac{1}{2}$ in.   (c) 7.5 in.   (d) .3 ft.   (e) 1 ft. 2 in.

4. Find, to the nearest integer, the volume of   *(3.3)*

    (a) a rectangular solid whose length is 3 in., width $8\frac{1}{2}$ in. and height 8 in.

    (b) a prism having a height of 2 ft. and a square base whose side is 3 yd.

    (c) a pyramid having a height of 2 yd. and a base whose area is 6.4 sq. ft.

5. Find, to the nearest integer, the volume of   *(3.4)*

    (a) a sphere with a radius of 6 in.

    (b) a cylinder having a height of 10 ft. and a base whose radius is 2 yd.

    (c) a cone having a height of 3 yd. and a base whose radius is 1.4 ft.

6. From $V = \frac{1}{3}Bh$, the volume formula for a pyramid or cone, derive volume formulas for each of the following: *(3.5)*

    (a) 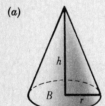 (b) 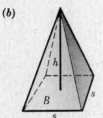 (c) 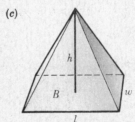 (d)

    Cone   Pyramid with a   Pyramid with a   Cone where
              Square Base   Rectangular Base $h = 2r$

7. State a formula for the volume of each solid: *(3.6)*

    (a) 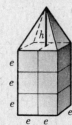 (b) 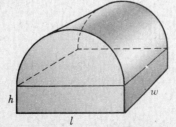 (c)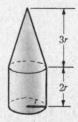

Formulas for Reference

ANGLE FORMULAS

1. Complement of $a°$	1. $c = 90 - a$	
2. Supplement of $a°$	2. $s = 180 - a$	
3. Sum of angles of a triangle	3. $S = 180$	
4. Sum of angles of a quadrilateral	4. $S = 360$	
5. Sum of exterior angles of an n-gon	5. $S = 360$	
6. Sum of interior angles of an n-gon	6. $S = 180(n - 2)$	
7. Each interior angle of an equiangular or regular n-gon	7. $S = \dfrac{180(n - 2)}{n}$	
8. Each exterior angle of an equiangular or regular n-gon	8. $S = \dfrac{360}{n}$	
9. Central $\angle O$ intercepting an arc of $a°$	9. $\angle O = a°$	
10. Inscribed $\angle A$ intercepting an arc of $a°$	10. $\angle A = \frac{1}{2}a°$	
11. $\angle A$ formed by a tangent and a chord and intercepting an arc of $a°$	11. $\angle A = \frac{1}{2}a°$	
12. $\angle A$ formed by two intersecting chords and intercepting arcs of $a°$ and $b°$	12. $\angle A = \frac{1}{2}(a° + b°)$	
13. $\angle A$ formed by two intersecting tangents, two intersecting secants or by an intersecting tangent and secant and intercepting arcs of $a°$ and $b°$	13. $\angle A = \frac{1}{2}(a° \neg b°)$	
14. $\angle A$ inscribed in a semicircle	14. $\angle A = 90°$	
15. Opposite $\angle A$ and B of an inscribed quadrilateral	15. $\angle A = 180° - \angle B$	

AREA FORMULAS

1. Area of a rectangle	1. $K = bh$	
2. Area of a square	2. $K = s^2$,	$K = \frac{1}{2}d^2$
3. Area of a parallelogram	3. $K = bh$,	$K = ab \sin C$
4. Area of a triangle	4. $K = \frac{1}{2}bh$,	$K = \frac{1}{2}ab \sin C$
5. Area of a trapezoid	5. $K = \frac{1}{2}h(b + b')$,	$K = hm$
6. Area of an equilateral triangle	6. $K = \frac{1}{4}s^2\sqrt{3}$,	$K = \frac{1}{3}h^2\sqrt{3}$
7. Area of a rhombus	7. $K = \frac{1}{2}dd'$	
8. Area of a regular polygon	8. $K = \frac{1}{2}pr$	
9. Area of a circle	9. $K = \pi r^2$,	$K = \frac{1}{4}\pi d^2$
10. Area of a sector	10. $K = \dfrac{n}{360}(\pi r^2)$,	$K = \dfrac{\pi n r^2}{360}$
11. Area of a minor segment	11. $K = $ area of sector $-$ area of triangle	

CIRCLE INTERSECTION FORMULAS

1.	2.	3.
	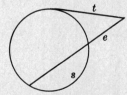	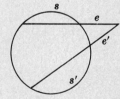
Intersecting Chords $ab = cd$	Intersecting Tangent and Secant $\dfrac{s}{t} = \dfrac{t}{e}, \quad t^2 = se$	Intersecting Secants $se = s'e'$

RIGHT TRIANGLE FORMULAS

1.	Law of Pythagoras	1. $c^2 = a^2 + b^2$
2.	Leg opposite 30° Leg opposite 45° Leg opposite 60°	2. $b = \frac{1}{2}c$ $b = \frac{1}{2}c\sqrt{2}, \quad b = a$ $b = \frac{1}{2}c\sqrt{3}, \quad b = a\sqrt{3}$
3.	Altitude of equilateral triangle Side of equilateral triangle	3. $h = \frac{1}{2}s\sqrt{3}$ $s = \frac{2}{3}h\sqrt{3}$
4.	Side of square Diagonal of square	4. $s = \frac{1}{2}d\sqrt{2}$ $d = s\sqrt{2}$
5.	Altitude upon hypotenuse Leg of right triangle	5. $\dfrac{p}{h} = \dfrac{h}{q}, \quad h^2 = pq, \quad h = \sqrt{pq}$ $\dfrac{c}{a} = \dfrac{a}{p}, \quad a^2 = pc, \quad a = \sqrt{pc}$ $\dfrac{c}{b} = \dfrac{b}{q}, \quad b^2 = qc, \quad b = \sqrt{qc}$

COORDINATE GEOMETRY FORMULAS

1.	Midpoint M Distance P_1P_2 Slope of P_1P_2	1. $x_M = \dfrac{x_1 + x_2}{2}, \quad y_M = \dfrac{y_1 + y_2}{2}$ $d = \sqrt{(x_2 - x_1)^2 + (y_2 - y_1)^2}$ $m = \dfrac{y_2 - y_1}{x_2 - x_1}, \quad m = \dfrac{\Delta y}{\Delta x}, \quad m = \tan i$
2.	Slopes of parallels, L_1 and L_2 Slopes of perpendiculars, L_1 and L'	2. Same slope, m $mm' = -1$ $m' = -\dfrac{1}{m}, \quad m = -\dfrac{1}{m'}$
3.	Equation of L_1, parallel to x-axis Equation of L_2, parallel to y-axis	3. $y = k'$ $x = k$
4.	Equation of L_1 with slope m and y-intercept b Equation of L_2 with slope m passing through the origin Equation of L_1 with x-intercept a and y-intercept b Equation of L_3 with slope m and passing through (x_1, y_1)	4. $y = mx + b$ $y = mx$ $\dfrac{x}{a} + \dfrac{y}{b} = 1$ $y - y_1 = m(x - x_1)$
5.	Equation of circle with center at origin and radius r	5. $x^2 + y^2 = r^2$

TABLE OF TRIGONOMETRIC FUNCTIONS

Angle	Sine	Cosine	Tangent	Angle	Sine	Cosine	Tangent
1°	.0175	.9998	.0175	46°	.7193	.6947	1.0355
2°	.0349	.9994	.0349	47°	.7314	.6820	1.0724
3°	.0523	.9986	.0524	48°	.7431	.6691	1.1106
4°	.0698	.9976	.0699	49°	.7547	.6561	1.1504
5°	.0872	.9962	.0875	50°	.7660	.6428	1.1918
6°	.1045	.9945	.1051	51°	.7771	.6293	1.2349
7°	.1219	.9925	.1228	52°	.7880	.6157	1.2799
8°	.1392	.9903	.1405	53°	.7986	.6018	1.3270
9°	.1564	.9877	.1584	54°	.8090	.5878	1.3764
10°	.1736	.9848	.1763	55°	.8192	.5736	1.4281
11°	.1908	.9816	.1944	56°	.8290	.5592	1.4826
12°	.2097	.9781	.2126	57°	.8387	.5446	1.5399
13°	.2250	.9744	.2309	58°	.8480	.5299	1.6003
14°	.2419	.9703	.2493	59°	.8572	.5150	1.6643
15°	.2588	.9659	.2679	60°	.8660	.5000	1.7321
16°	.2756	.9613	.2867	61°	.8746	.4848	1.8040
17°	.2924	.9563	.3057	62°	.8829	.4695	1.8807
18°	.3090	.9511	.3249	63°	.8910	.4540	1.9626
19°	.3256	.9455	.3443	64°	.8988	.4384	2.0503
20°	.3420	.9397	.3640	65°	.9063	.4226	2.1445
21°	.3584	.9336	.3839	66°	.9135	.4067	2.2460
22°	.3746	.9272	.4040	67°	.9205	.3907	2.3559
23°	.3907	.9205	.4245	68°	.9272	.3746	2.4751
24°	.4067	.9135	.4452	69°	.9336	.3584	2.6051
25°	.4226	.9063	.4663	70°	.9397	.3420	2.7475
26°	.4384	.8988	.4877	71°	.9455	.3256	2.9042
27°	.4540	.8910	.5095	72°	.9511	.3090	3.0777
28°	.4695	.8829	.5317	73°	.9563	.2924	3.2709
29°	.4848	.8746	.5543	74°	.9613	.2756	3.4874
30°	.5000	.8660	.5774	75°	.9659	.2588	3.7321
31°	.5150	.8572	.6009	76°	.9703	.2419	4.0108
32°	.5299	.8480	.6249	77°	.9744	.2250	4.3315
38°	.5446	.8387	.6494	78°	.9781	.2079	4.7046
34°	.5592	.8290	.6745	79°	.9816	.1908	5.1446
35°	.5736	.8192	.7002	80°	.9848	.1736	5.6713
36°	.5878	.8090	.7265	81°	.9877	.1564	6.3138
37°	.6018	.7986	.7536	82°	.9903	.1392	7.1154
38°	.6157	.7880	.7813	83°	.9925	.1219	8.1443
39°	.6293	.7771	.8098	84°	.9945	.1045	9.5144
40°	.6428	.7660	.8391	85°	.9962	.0872	11.4301
41°	.6561	.7547	.8693	86°	.9976	.0698	14.3007
42°	.6691	.7431	.9004	87°	.9986	.0523	19.0811
43°	.6820	.7314	.9325	88°	.9994	.0349	28.6363
44°	.6947	.7193	.9657	89°	.9998	.0175	57.2900
45°	.7071	.7071	1.0000	90°	1.0000	.0000	

TABLE OF SQUARES AND SQUARE ROOTS

N	N²	√N	N	N²	√N	N	N²	√N
1	1	1.000	51	2601	7.141	101	10201	10.050
2	4	1.414	52	2704	7.211	102	10404	10.100
3	9	1.732	53	2809	7.280	103	10609	10.149
4	16	2.000	54	2916	7.348	104	10816	10.198
5	25	2.236	55	3025	7.416	105	11025	10.247
6	36	2.449	56	3136	7.483	106	11236	10.296
7	49	2.646	57	3249	7.550	107	11449	10.344
8	64	2.828	58	3364	7.616	108	11664	10.392
9	81	3.000	59	3481	7.681	109	11881	10.440
10	100	3.162	60	3600	7.746	110	12100	10.488
11	121	3.317	61	3721	7.810	111	12321	10.536
12	144	3.464	62	3844	7.874	112	12544	10.583
13	169	3.606	63	3969	7.937	113	12769	10.630
14	196	3.742	64	4096	8.000	114	12996	10.677
15	225	3.873	65	4225	8.062	115	13225	10.724
16	256	4.000	66	4356	8.124	116	13456	10.770
17	289	4.123	67	4489	8.185	117	13689	10.817
18	324	4.243	68	4624	8.246	118	13924	10.863
19	361	4.359	69	4761	8.307	119	14161	10.909
20	400	4.472	70	4900	8.367	120	14400	10.954
21	441	4.583	71	5041	8.426	121	14641	11.000
22	484	4.690	72	5184	8.485	122	14884	11.045
23	529	4.796	73	5329	8.544	123	15129	11.091
24	576	4.899	74	5476	8.602	124	15376	11.136
25	625	5.000	75	5625	8.660	125	15625	11.180
26	676	5.099	76	5776	8.718	126	15876	11.225
27	729	5.196	77	5929	8.775	127	16129	11.269
28	784	5.292	78	6084	8.832	128	16384	11.314
29	841	5.385	79	6241	8.888	129	16641	11.358
30	900	5.477	80	6400	8.944	130	16900	11.402
31	961	5.568	81	6561	9.000	131	17161	11.446
32	1024	5.657	82	6724	9.055	132	17424	11.489
33	1089	5.745	83	6889	9.110	133	17689	11.533
34	1156	5.831	84	7056	9.165	134	17956	11.576
35	1225	5.916	85	7225	9.220	135	18225	11.619
36	1296	6.000	86	7396	9.274	136	18496	11.662
37	1369	6.083	87	7569	9.327	137	18769	11.705
38	1444	6.164	88	7744	9.381	138	19044	11.747
39	1521	6.245	89	7921	9.434	139	19321	11.790
40	1600	6.325	90	8100	9.487	140	19600	11.832
41	1681	6.403	91	8281	9.539	141	19881	11.874
42	1764	6.481	92	8464	9.592	142	20164	11.916
43	1849	6.557	93	8649	9.644	143	20449	11.958
44	1936	6.633	94	8836	9.695	144	20736	12.000
45	2025	6.708	95	9025	9.747	145	21025	12.042
46	2116	6.782	96	9216	9.798	146	21316	12.083
47	2209	6.856	97	9409	9.849	147	21609	12.124
48	2304	6.928	98	9604	9.899	148	21904	12.166
49	2401	7.000	99	9801	9.950	149	22201	12.207
50	2500	7.071	100	10000	10.000	150	22500	12.247

Answers
to Supplementary Problems

Chapter 1

1. (a) point, (b) line, (c) surface, (d) surface, (e) line, (f) point
2. (a) AE, DE (c) AD, BE, CE, EF
   (b) ED, CD, BD, FD (d) F
3. (a) $AB = 16$, (b) $AE = 10\frac{1}{2}$, (c) AF bisects BG, AG bisects FC
4. (a) 18, (b) 90°, (c) 50°, (d) 130°, (e) 230°
5. (a) $\angle CBE$, (b) $\angle AEB$, (c) $\angle ABE$, (d) $\angle ABC$, $\angle BCD$, $\angle BED$, (e) $\angle AED$
6. (a) 130°, (b) 120°, (c) 75°, (d) 132°
7. (a) 75°, (b) 40°, (c) $10\frac{1}{3}°$ or 10°20′, (d) 9°11′
8. (a) 90°, (b) 120°, (c) 135°, (d) 270°, (e) 180°
9. (a) 90°, (b) 60°, (c) 15°, (d) 165°
10. (a) $AB \perp BC$ and $AC \perp CD$, (b) 129°, (c) 102°, (d) 51°, (e) 129°
11. (a) $\triangle ABC$, hypotenuse AB, legs AC and BC
    $\triangle ACD$, hypotenuse AC, legs AD and CD
    $\triangle BCD$, hypotenuse BC, legs BD and CD
    (b) $\triangle DAB$ and $\triangle ABC$
    (c) $\triangle AEB$, legs AE and BE, base AB, vertex angle $\angle AEB$
    $\triangle CED$, legs DE and CE, base CD, vertex angle $\angle CED$
12. (a) $AR = BR$ and $\angle PRA = \angle PRB$, (b) $\angle ABF = \angle CBF$, (c) $\angle CGA = \angle CGD$, (d) $AM = MD$
13. (a) vert. ∠   (c) adj. ∠   (e) comp. ∠
    (b) comp. adj. ∠ (d) supp. adj. ∠ (f) vert. ∠
14. (a) 25°, 65°   (d) 61°, 119°   (f) 56°, 84°
    (b) 18°, 72°   (e) 50°, 130°   (g) 90°, 90°
    (c) 60°, 120°
15. (a) 48°, 27° (b) 65°, 25° (c) 148°, 32°

Chapter 2

1. (a) A is H, (b) P is D, (c) R is S, (d) E is K, (e) A is G, (f) triangles are geometric figures, (g) a rectangle is a quadrilateral
2. (a) Numbers ending in an even number are divisible by 2. (b) Gold and silver are metals. (c) Those who pass the examination will pass the course. (d) $\angle a$ and $\angle d$ are straight angles. (e) Polygons have as many angles as sides.
3. (a) $a = c = f$, (b) $g = 15$, (c) $f = a$, (d) $a = h$, (e) $b = e$, (f) $\angle 1 = \angle 3 = \angle 4$, (g) $\angle 4 = 55°$, (h) $\angle 1 = \angle 4$, (i) $\angle 2 = \angle 4$, (j) $\angle 1 = \angle 2 = \angle 3$
4. (a) 130, (b) 4, (c) Yes, (d) $x = 8\frac{1}{2}$, (e) $y = 15$, (f) $x = 6$, (g) $x = \pm 6$, (h) $\angle b = 63°$, (i) $\angle y = 50°$
5. (a) $AC = 12$, $AE = 11$, $AF = 15$, $DF = 9$
   (b) $\angle ADC = 92°$, $\angle BAE = 68°$, $\angle FAD = 86°$, $\angle BAD = 128°$

(c) $\overparen{AC} = 150°$, $\overparen{BD} = 135°$, $\overparen{CE} = 150°$
$\overparen{ABD} = 215°$, $\overparen{BCE} = 220°$, $\overparen{ABE} = 300°$

6. (a) $AB = DF$ (c) $\overparen{AC} = \overparen{BF}$ (e) $\angle ECA = \angle DCB$
   (b) $AB = AC$ (d) $\overparen{AC} = \overparen{DF}$ (f) $\angle BAD = \angle BCD$
7. (a) If equals are divided by equals, the quotients are equal.
   (b) Doubles of equals are equal.
   (c) If equals are multiplied by equals, the products are equal.
   (d) Halves of equals are equal.
8. (a) If equals are divided by equals, the quotients are equal.
   (b) If equals are multiplied by equals, the products are equal.
   (c) Doubles of equals are equal.
   (d) Halves of equals are equal.
9. (a) Their new rates of pay per hour will be the same.
   (b) Those stocks have the same value now.
   (c) The classes have the same registers now.
   (d) $100°\text{C} = 212°\text{F}$.
   (e) Their parts will be the same length.
   (f) He has a total of $10,000 in Banks A, B and C.
   (g) Their values are the same.
10. (a) Vertical angles are equal.
    (b) All straight angles are equal.
    (c) Supplements of equal angles are equal.
    (d) Perpendiculars form right angles and all right angles are equal.
    (e) Complements of equal angles are equal.
11. In each, (H) indicates the hypothesis and (C) indicates the conclusion.
    (a) (H) Stars, (C) twinkle.
    (b) (H) Jet planes, (C) are the speediest.
    (c) (H) Water, (C) boils at 212° Fahrenheit.
    (d) (H) If it is the American flag, (C) its colors are red, white and blue.
    (e) (H) If you fail to do homework in the subject, (C) you cannot learn geometry.
    (f) (H) If the umpire calls a fourth ball, (C) a batter goes to first base.
    (g) (H) If A is B's brother and C is B's daughter, (C) then A is C's uncle.
    (h) (H) An angle bisector, (C) divides the angle into two equal parts.
    (i) (H) If it is divided into three equal parts, (C), a line is trisected.
    (j) (H) A pentagon, (C) has five sides and five angles.
    (k) (H) Some rectangles, (C) are squares.
    (l) (H) If their sides are made longer, (C) angles do not become larger.

(m) (H) If they are equal and supplementary, (C) angles are right angles.

(n) (H) If one of its sides is not a straight line, (C), the figure cannot be a polygon.

12. (a) An acute angle is half a right angle. Not necessarily true.

(b) A triangle having one obtuse angle is an obtuse triangle. True.

(c) If the batter is out, then the umpire called a third strike. Not necessarily true.

(d) If you are shorter than I, then I am taller than you. True.

(e) If our weights are unequal, then I am heavier than you. Not necessarily true.

Chapter 3

1. (a) $\triangle I \cong \triangle II \cong \triangle III$, s.a.s. = s.a.s.
   (b) $\triangle I \cong \triangle III$, a.s.a. = a.s.a.
   (c) $\triangle I \cong \triangle II \cong \triangle III$, s.s.s. = s.s.s.

2. (a) a.s.a. = a.s.a.　　(e) a.s.a. = a.s.a.
   (b) s.a.s. = s.a.s.　　(f) s.a.s. = s.a.s.
   (c) s.s.s. = s.s.s.　　(g) s.a.s. = s.a.s.
   (d) s.a.s. = s.a.s.　　(h) a.s.a. = a.s.a.

3. (a) $AD = DC$, (b) $\angle ABD = \angle DBC$, (c) $\angle 1 = \angle 4$,
   (d) $BE = ED$, (e) $BD = AC$, (f) $\angle BAD = \angle CDA$

4. (a) $\angle 1 = \angle 3$, $\angle 2 = \angle 4$, $BD = BE$.
   (b) $AB = AC$, $BD = DC$, $\angle B = \angle C$.
   (c) $\angle E = \angle C$, $\angle A = \angle F$, $\angle EDF = \angle ABC$.

5. (a) $x = 19$, $y = 8$
   (b) $x = 4$, $y = 12$　　(c) $x = 48$, $y = 12$

8. (a) $\angle b = \angle d$, $\angle E = \angle G$
   (b) $\angle A = \angle 1 = \angle 4$, $\angle 2 = \angle C$
   (c) $\angle 1 = \angle 5$, $\angle 4 = \angle 6$, $\angle EAD = \angle EDA$

9. (a) $BE = EC$　　(b) $AB = BD = AD$, $BC = CD$
   (c) $BD = DE$, $EF = FC$, $AB = AC$

Chapter 4

1. (a) $x = 105$, $y = 75$　　(g) $x = 60$, $y = 120$
   (b) $x = 60$, $y = 40$　　(h) $x = 90$, $y = 35$
   (c) $x = 85$, $y = 95$　　(i) $x = 30$, $y = 40$
   (d) $x = 50$, $y = 50$　　(j) $x = 80$, $y = 10$
   (e) $x = 65$, $y = 65$　　(k) $x = 30$, $y = 150$
   (f) $x = 40$, $y = 30$　　(l) $x = 85$, $y = 95$

2. (a) $x = 22$, $y = 102$　　(b) $x = 40$, $y = 100$
   (c) $x = 80$, $y = 40$

3. (a) Each angle is 105°. (b) Each angle is 70°.
   (c) Angles are 72° and 108°.

7. (a) 25, (b) 9, (c) 20, (d) 8

8. (a) 8, (b) 10, (c) 2, (d) 14

10. (a) P is equidistant from B and C. P is on $\perp$ bisector of BC.
    Q is equidistant from A and B. Q is on $\perp$ bisector of AB.
    R is equidistant from A, C and D. R is on $\perp$ bisectors of AD and CD.

    (b) P is equidistant from AB and AD. P is on bisector of $\angle A$.
    Q is equidistant from AB and BC. Q is on bisector of $\angle B$.
    R is equidistant from BC, CD and AD. R is on the bisectors of $\angle C$ and $\angle D$.

11. (a) P is equidistant from AD, AB and BC. Q is equidistant from AD and AB and also equidistant from A and D. R is equidistant from AB and BC and also equidistant from A and D.

(b) P is equidistant from AD and CD and also equidistant from B and C. Q is equidistant from A, B and C. R is equidistant from AD and CD and also equidistant from A and B.

12. (a) $x = 50$, $y = 110$　　(d) $x = 51$, $y = 112$
    (b) $x = 65$, $y = 65$　　(e) $x = 52$, $y = 40$
    (c) $x = 30$, $y = 100$　　(f) $x = 120$, $y = 90$

13. (a) $x = 55$, $y = 125$　　(d) $x = 100$, $y = 30$
    (b) $x = 80$, $y = 90$　　(e) $x = 30$, $y = 120$
    (c) $x = 56$, $y = 68$　　(f) $x = 90$, $y = 30$

14. (a) 18°, 54°, 108°　　(d) 36°, 72°, 108°, 144°
    (b) 40°, 50°, 90°　　(e) 50°, 75°
    (c) 36°, 36°, 108°　　(f) 100°, 60° and 20°

16. (a) Since $x = 45$, each angle = 60°.
    (b) Since $x = 25$, $x + 15 = 40$ and $3x - 35 = 40$; that is, two angles are each 40°.
    (c) If $2x$, $3x$ and $5x$ represent the angles, $x = 18$ and $5x = 90$; that is, one of the angles = 90°.
    (d) If x and $5x - 10$ represent the unknown angles, $x = 21$ and $5x - 10 = 95$; that is, one of the angles is 95°.

17. (a) 7 st. $\angle$, 30 st. $\angle$　　(c) 30, 12, 27, 202
    (b) 1620°, 5400°, 180,000°　　(d) No, no, yes

18. (a) 20°, 18°, 9°　(b) 160°, 162°, 171°
    (c) 3, 9, 20, 180　(d) 3, 12, 36, 72, 360　(e) 8

19. (a) 65°, 90°, 95°, 110°　　(b) 140°, 100°, 60°, 60°
    (c) 50°, 100°, 110°, 130°, 150°

20. (a) $\triangle I \cong \triangle III$ by hy. leg = hy. leg
    (b) $\triangle I \cong \triangle III$ by s.a.a. = s.a.a.

Chapter 5

1. (a) $x = 15$, $y = 25$　　(b) $x = 20$, $y = 130$
   (c) $x = 20$, $y = 140$

4. (a) $\square EFGH$　　(c) $\square GHKJ, HILK, GILJ$
   (b) $\square ABCD$ and $EBFD$　　(d) $\square ACHB, CEFH$

5. (a) Two sides are = and ∥. (b) Opposite sides are equal. (c) Opposite angles are equal. (d) AD and BC are equal and parallel. ($AD = EF = BC$.)

6. (a) $x = 6$, $y = 12$　　(c) $x = 120$, $y = 30$
   (b) $x = 5$, $y = 9$　　(d) $x = 15$, $y = 45$

7. (a) $x = 14$, $y = 6$　　(c) $x = 8$, $y = 5$
   (b) $x = 18$, $y = 4\frac{1}{2}$　　(d) $x = 3$, $y = 9$

10. (a) $x = 5$, $y = 7$　　(d) $x = 8$, $y = 4$
    (b) $x = 10$, $y = 35$　　(e) $x = 25$, $y = 25$
    (c) $x = 2\frac{1}{2}$, $y = 17\frac{1}{2}$　　(f) $x = 11$, $y = 118$

13. (a) $x = 6$, $y = 40$　　(b) $x = 3$, $y = 5\frac{1}{2}$
    (c) $x = 8\frac{1}{3}$, $y = 22$

14. (a) $x = 28$, $y = 25\frac{1}{2}$. (b) $x = 12$. Since y does not join midpoints, Pr. 3 does not apply. (c) $x = 19$, $y = 23\frac{1}{2}$.

15. (a) $m = 19$, (b) $b' = 36$, (c) $b = 73$

16. (a) $x = 11$, $y = 33$　　(b) $x = 32$, $y = 26$
    (c) $x = 12$, $y = 36$

17. (a) $22\frac{1}{2}$　　(b) 70

18. (a) 21, (b) 30, (c) 14, (d) 26

Chapter 6

5. (a) square　　　　(c) trapezoid
   (b) isosceles triangle　　(d) right triangle

6. (a) 140°, (b) 60°, (c) 90°, (d) $(180 - x)°$, (e) $x°$,
   (f) $(90 + x)°$

7. (a) 100°　　(c) 54°, 27°　　(e) 35°
   (b) 50°, 80°　　(d) 45°　　(f) 45°

8. (a) $x = 22$, (b) $y = 6$, (c) $AB + CD = 22$, (d) Perimeter $= 44$, (e) $x = 21$, (f) $r = 14$

9. (a) 0, (b) 40, (c) 33, (d) 7

10. (a) tangent externally, (b) tangent internally, (c) the circles are 5 units apart, (d) overlapping

11. (a) concentric, (b) tangent internally, (c) tangent externally, (d) outside of each other, (e) the smaller entirely inside the larger, (f) overlapping

13. (a) 40, (b) 90, (c) 170, (d) 180, (e) $2x$, (f) $180 - x$, (g) $2x - 2y$

14. (a) 20, (b) 45, (c) 85, (d) 90, (e) 130, (f) 174, (g) x, (h) $90 - \frac{1}{2}x$, (i) $x - y$

15. (a) 85, (b) 170, (c) c, (d) $2i$, (e) 60, (f) 30

16. (a) 60, 120, 180　　(c) 100, 120, 140
    (b) 80, 120, 160　　(d) 36, 144, 180

17. (a) $\angle x = 136°$, (b) $\widehat{y} = 111°$, (c) $\angle x = 130°$, (d) $\angle y = 126°$, (e) $\angle x = 110°$, (f) $\widehat{y} = 77°$

18. (a) $135°$, (b) $90°$, (c) $(180 - x)°$, (d) $(90 + x)°$, (e) $100°$, (f) $80°$, (g) $55°$, (h) $72°$

19. (a) $85°$, (b) $y°$, (c) $110°$, (d) $95°$, (e) $72°$, (f) $50°$, (g) $145°$, (h) $87°$

20. (a) 50, (b) 60

21. $\widehat{x} = 65°$, $\widehat{y} = 65°$　(b) $\angle x = 90°$, $\angle y = 55°$ (c) $\angle x = 37°$, $\angle y = 50°$

22. (a) 19, (b) 45, (c) 69, (d) 90, (e) 125, (f) 167, (g) $\frac{1}{2}x$, (h) $180 - \frac{1}{2}x$, (i) $x + y$

23. (a) 110, (b) 135, (c) 180, (d) 270, (e) $180 - 2x$, (f) $360 - 2x$, (g) $2x - 2y$, (h) $7x$

24. (a) 45, (b) 60, (c) 30, (d) 18

25. (a) $\widehat{x} = 120°$, $\angle y = 60°$　(b) $\angle x = 62°$, $\angle y = 28°$ (c) $\angle x = 46°$, $\angle y = 58°$

26. (a) $75°$, (b) $75°$, (c) $115°$, (d) $100°$, (e) $140°$, (f) $230°$, (g) $80°$, (h) $48°$

27. (a) $85°$, (b) $103°$, (c) $80°$, (d) $72°$, (e) $90°$, (f) $110°$, (g) $130°$, (h) $110°$

28. (a) $\widehat{x} = 68°$, $\angle y = 95°$　(b) $\angle x = 90°$, $\angle y = 120°$ (c) $\widehat{x} = 34°$, $\widehat{y} = 68°$

29. (a) $30°$, (b) $37°$, (c) $20°$, (d) $36°$, (e) $120°$, (f) $130°$, (g) $94°$, (h) $25°$

30. (a) $45°$, (b) $75°$, (c) $50°$, (d) $36\frac{1}{2}°$, (e) $90°$, (f) $140°$, (g) $115°$, (h) $45°$, (i) $80°$

31. (a) $20°$, (b) $85°$, (c) $(180 - x)°$, (d) $(90 + x)°$, (e) $90°$, (f) $25°$, (g) $42°$, (h) $120°$, (i) $72°$, (j) $110°$, (k) $145°$, (l) $(180 - y)°$, (m) $240°$, (n) $(180 + x)°$, (o) $270°$

32. (a) $\widehat{x} = 43°$, $\angle y = 43°$　(b) $\widehat{x} = 190°$, $\angle y = 55°$ (c) $\widehat{x} = 140°$, $\angle y = 40°$

33. (a) $120°$, (b) $150°$, (c) $180°$, (d) $50°$, (e) $22\frac{1}{2}°$, (f) $45°$

34. (a) $\widehat{x} = 150°$, $\widehat{y} = 40°$　(c) $\widehat{x} = 252°$, $\widehat{y} = 108°$ (b) $\widehat{x} = 190°$, $\widehat{y} = 70°$

35. (a) $25°$, (b) $39°$, (c) $50°$, (d) $30°$, (e) $40°$, (f) $76°$, (g) $45°$, (h) $95°$, (i) $75°$, (j) $120°$

36. (a) $74°$, (b) $90°$, (c) $55°$, (d) $60°$, (e) $40°$, (f) $37°$, (g) $84°$, (h) $110°$, (i) $66°$, (j) $98°$, (k) $75°$, (l) $79°$

37. (a) $\angle x = 120°$, $\angle y = 60°$
    (b) $\angle x = 45°$, $\angle y = 22\frac{1}{2}°$
    (c) $\angle x = 36°$, $\angle y = 72°$

38. (a) $\widehat{x} = 40°$, $\angle y = 80°$　(b) $\widehat{x} = 45°$, $\angle y = 67\frac{1}{2}°$ (c) $\angle x = 78°$, $\angle y = 103°$

Chapter 7

1. (a) 4, (b) $\frac{1}{3}$, (c) $\frac{6}{5}$, (d) $\frac{10}{7}$, (e) $\frac{9}{7}$, (f) 2, (g) $\frac{1}{5}$, (h) $\frac{3}{7}$, (i) $\frac{2}{3}$, (j) $\frac{7}{8}$, (k) 2, (l) $\frac{5}{7}$, (m) 20, (n) $\frac{1}{3}$, (o) 3

2. (a) 6, (b) $\frac{14}{5}$, (c) $\frac{1}{7}$, (d) $\frac{3}{2}$, (e) 3, (f) $\frac{7}{2}$, (g) 2, (h) 4, (i) $\frac{1}{20}$, (j) 8, (k) $\frac{5}{3}$, (l) $\frac{9}{2}$

3. (a) $2 : 3 : 10$　　　(d) $1 : 4 : 7$　　　(g) $50 : 5 : 1$
   (b) $12 : 6 : 1$　　　(e) $4 : 3 : 1$　　　(h) $6 : 2 : 1$
   (c) $5 : 2 : 1$　　　(f) $8 : 2 : 1$　　　(i) $8 : 2 : 1$

4. (a) $\frac{6}{7}$　　(e) 6　　(i) $\frac{2}{7}$　　(m) 3
   (b) 12　　(f) $\frac{16}{9}$　　(j) 11　　(n) $\frac{3}{20}$
   (c) $\frac{13}{3}$　　(g) $\frac{1}{3}$　　(k) $\frac{4}{5}$　　(o) $\frac{1}{2}$
   (d) $\frac{1}{4}$　　(h) $\frac{3}{2}$　　(l) 60　　(p) 14

5. (a) $\frac{1}{8}$, (b) $3c$, (c) $\frac{d}{2}$, (d) $\frac{2r}{D}$, (e) $\frac{b}{a}$, (f) $\frac{4}{S}$, (g) $\frac{S}{6}$, (h) $\frac{3r}{2t}$, (i) $1 : 4 : 10$, (j) $3 : 2 : 1$, (k) $x^2 : x : 1$, (l) $6 : 5 : 4 : 1$

6. (a) $5x$ and $4x$; sum $= 9x$
   (b) $9x$ and x; sum $= 10x$
   (c) $2x$, $5x$ and $11x$; sum $= 18x$
   (d) x, $2x$, $2x$, $3x$ and $7x$; sum $= 15x$

7. (a) $5x + 4x = 45$, $x = 5$, $25°$ and $20°$.
   (b) $5x + 4x = 90$, $x = 10$, $50°$ and $40°$.
   (c) $5x + 4x = 180$, $x = 20$, $100°$ and $80°$.
   (d) $5x + 4x + x = 180$, $x = 18$, $90°$ and $72°$.

8. (a) $7x + 6x = 91$, $x = 7$, $49°$, $42°$ and $35°$.
   (b) $7x + 5x = 180$, $x = 15$, $105°$, $90°$ and $75°$.
   (c) $7x + 3x = 90$, $x = 9$, $63°$, $54°$ and $45°$.
   (d) $7x + 6x + 5x = 180$, $x = 10$, $70°$, $60°$ and $50°$.

9. (a) 16, (b) 16, (c) ±6, (d) $\pm2\sqrt{5}$, (e) ±5, (f) 2, (g) bc/a, (h) $\pm6y$

10. (a) 21, (b) $4\frac{2}{3}$, (c) ±6, (d) $\pm5\sqrt{3}$, (e) 8, (f) ±4, (g) 3, (h) $\pm\sqrt{ab}$

11. (a) 15, (b) 3, (c) 6, (d) $2\frac{2}{3}$, (e) $3\frac{1}{3}$, (f) 30, (g) 32, (h) $6a$

12. (a) 6, (b) 6, (c) 3, (d) $4b$, (e) $\sqrt{10}$, (f) $\sqrt{27}$ or $3\sqrt{3}$, (g) $\sqrt{pq}$

13. (a) $\frac{c}{b} = \frac{d}{x}$, (b) $\frac{a}{p} = \frac{q}{x}$, (c) $\frac{h}{a} = \frac{a}{x}$, (d) $\frac{3}{7} = \frac{1}{x}$, (e) $\frac{c}{a} = \frac{b}{x}$

14. (a) $\frac{x}{y} = \frac{1}{2}$, (b) $\frac{x}{y} = \frac{3}{4}$, (c) $\frac{x}{y} = \frac{1}{2}$, (d) $\frac{x}{y} = \frac{h}{a}$, (e) $\frac{x}{y} = b$

15. Only (b) is not a proportion since $3(12) \neq 5(7)$; that is, $36 \neq 35$.

16. (a) $\frac{x}{2} = \frac{9}{3}$, $x = 6$, (b) $\frac{x}{1} = \frac{4}{5}$, $x = \frac{4}{5}$, (c) $\frac{x}{a} = \frac{b}{2}$, $x = \frac{ab}{2}$, (d) $\frac{x}{5} = \frac{1}{10}$, $x = \frac{1}{2}$, (e) $\frac{x}{20} = \frac{5}{4}$, $x = 25$

17. (a) d, (b) 35, (c) 5, (d) 4

18. (a) 21, (b) 3/2, (c) 5

19. (a) 16, (b) $6\frac{2}{3}$, (c) 10

20. (a) Yes, since $\frac{15}{10} = \frac{18}{12}$.
    (b) No, since $\frac{10}{13} \neq \frac{7}{9}$.　　(c) Yes, since $\frac{3x}{5x} = \frac{36}{60}$.

21. (a) 12, (b) 8, (c) 60

22. (a) 15, (b) 15, (c) $6\frac{1}{2}$

24. (a) 35° (b) 53°

25. (a) $a = 16$ (b) $b = 15$ (c) $c = 126$

27. (a) $\angle ABE = \angle EDC,\ \angle BAE = \angle DCE$;
 (also vert. $\angle$ at E)
 (b) $\angle BAF = \angle FEC,\ \angle B = \angle D$;
 (also $\angle EAD = \angle BFA$)
 (c) $\angle A = \angle EDF,\ \angle F = \angle BCA$
 (d) $\angle A = \angle A,\ \angle B = \angle C$
 (e) $\angle C = \angle D,\ \angle CAB = \angle CAD$
 (f) $\angle A = \angle A,\ \angle C = \angle DBA$

28. (a) $\angle D = \angle B,\ \angle AED = \angle FGB$
 (b) $\angle ADB = \angle ABC,\ \angle A = \angle A$
 (c) $\angle ABC = \angle AED,\ \angle BAE = \angle EDA$

29. (a) $\angle C = \angle F,\ \frac{14}{20} = \frac{21}{30}$
 (b) $\angle A = \angle A,\ \frac{10}{25} = \frac{6}{15}$ (c) $\angle B = \angle B,\ \frac{16}{28} = \frac{20}{35}$

30. (a) $\frac{6}{18} = \frac{8}{24} = \frac{10}{30}$ (b) $\frac{24}{36} = \frac{28}{42} = \frac{30}{45}$ (c) $\frac{12}{18} = \frac{16}{24} = \frac{18}{27}$

32. (a) $q = 20$ (b) $p = 8$ (c) $b = 7$ (d) $a = 12$
 (e) $AB = 35$ (f) $d = 2\frac{1}{4}$

33. (a) 8, (b) 6, (c) $26\frac{2}{3}$

34. (a) 42 ft, (b) 66 ft

37. (a) 8 : 5, (b) 3 : 5, (c) halved (in each case)

38. (a) 15 (b) 60 (c) 25, 35, 40 (d) 4 (e) 6, 3

39. (a) 3 : 7 (b) 7 : 2 (c) quadrupled (d) 7

43. (a) 5 (b) 14 (c) 6 (d) 5 (e) 12 (f) 13
 (g) 48 (h) 2

44. 30, 18

45. (a) 8 (b) 6 (c) 12 (d) 5 (e) 7 (f) 12 (g) 30
 (h) $7\frac{1}{2}$ (i) 5 (j) 8

46. (a) 8 (b) 13 (c) 21 (d) 6 (e) 9 (f) 14
 (g) 3 (h) 8

47. (a) $a = 4,\ h = \sqrt{12}$ or $2\sqrt{3}$, (b) $c = 9,\ h = \sqrt{20}$ or
 $2\sqrt{5}$, (c) $q = 4$ and $b = \sqrt{80}$ or $4\sqrt{5}$, (d) $p = 18$,
 $h = \sqrt{108} = 6\sqrt{3}$

48. (a) 25, (b) 39, (c) $\sqrt{41}$, (d) 10, (e) $7\sqrt{2}$

49. (a) $b = 16$, (b) $a = 2\sqrt{7}$, (c) $a = 8$, (d) $b = 2\sqrt{3}$,
 (e) $b = 5\sqrt{2}$, (f) $b = \sqrt{3}$

50. (a) 9, 12 (b) 10, 24 (c) 80, 150 (d) $2\sqrt{5},\ 4\sqrt{5}$

51. (a) 41, (b) $5\sqrt{5}$

52. (a) 12, (b) $10\sqrt{2}$ (c) $5\sqrt{5}$

53. All except (h)

54. (a) Yes (b) No, since $(2x)^2 + (3x)^2 \neq (4x)^2$

55. (a) 8 (b) 6 (c) $\sqrt{19}$ (d) $5\sqrt{3}$

56. (a) 15 (b) $2\sqrt{5}$ (c) 6

57. (a) 16 (b) 30 (c) $4\sqrt{3}$ (d) 10

58. (a) 10 (b) 12 (c) 28 (d) 15

59. (a) 5 (b) 20 (c) 15 (d) 25

60. (a) 12 (b) 24

61. 12

62. 30

63. (a) 10 and $10\sqrt{3}$ (b) $7\sqrt{3}$ and 14 (c) 5 and 10

64. (a) $11\sqrt{3}$; $a\sqrt{3}$ (b) 48; $16\sqrt{3}$

65. (a) 25 and $25\sqrt{3}$ (b) 35 and $35\sqrt{3}$

66. (a) 28, $8\sqrt{3}$ (b) 17, $14\sqrt{3}$

67. (a) $17\sqrt{2}$; $a\sqrt{2}$ (b) $34\sqrt{2}$; 30

68. (a) $20\sqrt{2}$ (b) $40\sqrt{2}$

69. (a) 45, $13\sqrt{2}$ (b) 11, $27\sqrt{2}$ (c) $15\sqrt{2}$, 55

70. $4\sqrt{2}$

71. $6\sqrt{2}$, $5\sqrt{2}$

Chapter 8

1. (a) .4226, .7431, .8572, .9998 (e) cosine
 (b) .9659, .6157, .2756, .0349 (f) tangent
 (c) .0699, .6745, 1.4281, 19.0811
 (d) sine and tangent

2. (a) $x = 20°$ (d) $A' = 21°$ (g) $W = 19°$
 (b) $A = 29°$ (e) $y = 45°$ (h) $B' = 67°$
 (c) $B = 71°$ (f) $Q = 69°$

3. (a) 26°, (b) 47°, (c) 69°, (d) 8°, (e) 40°, (f) 74°,
 (g) 7°, (h) 27°, (i) 80°, (j) 13° since $\sin x = .2200$,
 (k) 45° since $\sin x = .707$
 (l) 59° since $\cos x = .5200$
 (m) 68° since $\cos x = .3750$
 (n) 30° since $\cos x = .866$
 (o) 16° since $\tan x = .2857$
 (p) 10° since $\tan x = .1732$

4. (a) $\sin A = \frac{4}{5},\ \cos A = \frac{3}{5},\ \tan A = \frac{4}{3}$
 (b) $\sin A = \frac{3}{5},\ \cos A = \frac{4}{5},\ \tan A = \frac{3}{4}$
 (c) $\sin A = \frac{\sqrt{7}}{4},\ \cos A = \frac{3}{4},\ \tan A = \frac{\sqrt{7}}{3}$

5. (a) $A = 27°$ since $\cos A = .8900$
 (b) $A = 58°$ since $\sin A = .8500$
 (c) $A = 52°$ since $\tan A = 1.2800$

6. (a) $B = 42°$ since $\sin B = .6700$
 (b) $B = 74°$ since $\cos B = .2800$
 (c) $B = 68°$ since $\tan B = 2.500$
 (d) $B = 30°$ since $\tan B = .577$

8. (a) 23°, 67° (c) 16°, 74°
 (b) 28°, 62° (d) 10°, 80°

9. (a) $x = 188,\ y = 313$
 (b) $x = 174,\ y = 250$ (c) $x = 123,\ y = 182$

10. (a) 82 ft (b) 88 ft

11. 156 ft

12. (a) 2530 ft (b) 2560 ft

13. (a) 21 in. (b) 79 in.

14. 14

15. 16 and 18 in.

16. 31 ft

17. 15 yd

18. (a) 1050 ft (b) 9950 ft

19. 7°

20. 282 ft

21. (a) 81° (b) 45°

22. (a) 22 ft (b) 104 ft

23. 754 ft

24. 404 ft

25. (a) 295 ft (b) 245 ft (c) 960 ft

26. (a) 234 ft (b) 343 ft

27. (a) 96 ft (b) 166 ft

28. (a) 9.1 (b) 22

Chapter 9

1. (a) 99 sq. in., (b) 3 sq. ft. or 432 sq. in., (c) 500,
 (d) 120, (e) $36\sqrt{3}$, (f) $100\sqrt{3}$, (g) 300, (h) 150

2. (a) 48, (b) 432, (c) $25\sqrt{3}$, (d) 240

3. (a) 7 and 4, (b) 12 and 6 (c) 9 and 6, (d) 6 and 2,
 (e) 10 and 7, (f) 20 and 8

4. (*a*) 1296 sq. in., (*b*) $30\frac{1}{4}$ sq. yd., (*c*) 100 square decimeters (100 sq. dm.)

5. (*a*) 225, (*b*) $12\frac{1}{4}$, (*c*) 3.24, (*d*) $64a^2$, (*e*) 121, (*f*) $6\frac{1}{4}$, (*g*) $9b^2$, (*h*) 32, (*i*) $40\frac{1}{2}$, (*j*) 64

6. (*a*) 128, (*b*) 72, (*c*) 100, (*d*) 49, (*e*) 400

7. (*a*) 1600, (*b*) 400, (*c*) 100, (*d*) 14,400

8. (*a*) 9, (*b*) 36, (*c*) $9\sqrt{2}$, (*d*) $4\frac{1}{2}$, (*e*) $\frac{9}{2}\sqrt{2}$

9. (*a*) $2\frac{1}{2}$, (*b*) 52, (*c*) 10, (*d*) $5\sqrt{2}$, (*e*) 6, (*f*) 4

10. (*a*) 16 sq. ft., (*b*) 6 sq. ft. or 864 sq. in., (*c*) 70, (*d*) 1.62

11. (*a*) $3x^2$, (*b*) x^2+3x, (*c*) x^2-25, (*d*) $12x^2+11x+2$

12. (*a*) 150, (*b*) $54\sqrt{2}$, (*c*) $56\sqrt{3}$, (*d*) 60, (*e*) 11, (*f*) 55

13. (*a*) 36, (*b*) 15, (*c*) 16

14. (*a*) $2\frac{2}{3}$, (*b*) 20, (*c*) 9, (*d*) 3, (*e*) 15, (*f*) 12, (*g*) 8, (*h*) 7

15. (*a*) 11 sq. in., (*b*) 3 sq. ft. or $\frac{2}{3}$ sq. yd., (*c*) $4x-28$, (*d*) $10x^2$, (*e*) $2x^2+18x$, (*f*) $\frac{1}{2}(x^2-16)$, (*g*) x^2-9

16. (*a*) 10, (*b*) $5\sqrt{2}$, (*c*) $24\sqrt{3}$, (*d*) $62\frac{1}{2}$, (*e*) 29, (*f*) 46

17. (*a*) 15, (*b*) 64, (*c*) $18\sqrt{3}$, (*d*) $8\sqrt{3}$, (*e*) 24, (*f*) 20

18. (*a*) 84, (*b*) 48, (*c*) 30, (*d*) 120, (*e*) 148, (*f*) 423, (*g*) $8\sqrt{3}$, (*h*) 9

19. (*a*) 24, (*b*) 2, (*c*) 4

20. (*a*) 8, (*b*) 10, (*c*) 8, (*d*) 18, (*e*) $9\frac{3}{5}$, (*f*) $12\frac{1}{2}$, (*g*) 12, (*h*) 18

21. (*a*) $25\sqrt{3}$, (*b*) $36\sqrt{3}$, (*c*) $12\sqrt{3}$, (*d*) $25\sqrt{3}$, (*e*) $b^2\sqrt{3}$, (*f*) $4x^2\sqrt{3}$, (*g*) $3r^2\sqrt{3}$

22. (*a*) $2\sqrt{3}$, (*b*) $\frac{49}{2}\sqrt{3}$, (*c*) $24\sqrt{3}$, (*d*) $18\sqrt{3}$

23. (*a*) $24\sqrt{3}$, (*b*) $54\sqrt{3}$, (*c*) $150\sqrt{3}$

24. (*a*) 15, (*b*) 8, (*c*) 12, (*d*) 5

25. (*a*) 140, (*b*) 69, (*c*) 225, (*d*) $60\sqrt{2}$, (*e*) 94

26. (*a*) 150, (*b*) 204, (*c*) 39, (*d*) $64\sqrt{3}$, (*e*) 160

27. (*a*) 4, (*b*) 7, (*c*) 18 and 9, (*d*) 9 and 6, (*e*) 10 and 5

28. (*a*) 17 and 9, (*b*) 23 and 13, (*c*) 17 and 11, (*d*) 5, (*e*) 13

29. (*a*) 36, (*b*) $38\frac{1}{2}$, (*c*) $12\sqrt{3}$, (*d*) $12x^2$, (*e*) 120, (*f*) 96, (*g*) 18, (*h*) $\frac{49}{2}\sqrt{2}$, (*i*) $32\sqrt{3}$, (*j*) $98\sqrt{3}$

30. (*a*) 737, (*b*) 14, (*c*) 77

31. (*a*) 10, (*b*) 12 and 9, (*c*) 20 and 10, (*d*) 5, (*e*) $\sqrt{10}$

32. 12

37. (*a*) 1 : 49, (*b*) 49 : 4, (*c*) 1 : 3, (*d*) 1 : 25, (*e*) $81 : x^2$, (*f*) 9 : *x*, (*g*) 1 : 2

38. (*a*) 49 : 100, (*b*) 4 : 9, (*c*) 25 : 36, (*d*) 1 : 9, (*e*) 9 : 4, (*f*) 1 : 2

39. (*a*) 10 : 1, (*b*) 1 : 7, (*c*) 20 : 9, (*d*) 5 : 11, (*e*) 2 : *y*, (*f*) 3*x* : 1, (*g*) $\sqrt{3}:2$, (*h*) $1:\sqrt{2}$, (*i*) $x:\sqrt{5}$, (*j*) $\sqrt{x}:4$

40. (*a*) 6 : 5, (*b*) 3 : 7, (*c*) $\sqrt{3}:1$, (*d*) $\sqrt{5}:2$, (*e*) $\sqrt{3}:3$ or $1:\sqrt{3}$

41. (*a*) 100, (*b*) $12\frac{1}{2}$, (*c*) 12, (*d*) 100, (*e*) 105, (*f*) 18, (*g*) $20\sqrt{3}$

42. (*a*) 12, (*b*) 63, (*c*) 48, (*d*) $2\frac{1}{2}$, (*e*) 45

Chapter 10

1. (*a*) 200, (*b*) 24.5, (*c*) 112, (*d*) 13, (*e*) 9, (*f*) $3\frac{1}{3}$, (*g*) 4.5

2. (*a*) $12\frac{1}{2}$, (*b*) 23.47, (*c*) $7\sqrt{3}$, (*d*) 18.5, (*e*) $3\sqrt{2}$

3. (*a*) 24°, (*b*) 24°, (*c*) 156°

4. (*a*) 40°, (*b*) 9, (*c*) 140°

5. (*a*) 15°, (*b*) 15°, (*c*) 24

6. (*a*) 5°, (*b*) 72, (*c*) 175°

7. (*a*) Regular octagon, (*b*) Regular hexagon, (*c*) Equilateral triangle, (*d*) Regular decagon, (*e*) Square, (*f*) Regular dodecagon (12 sides).

9. (*a*) 9, (*b*) 30, (*c*) $6\sqrt{3}$, (*d*) 6, (*e*) $13\sqrt{3}$, (*f*) 6, (*g*) $20\sqrt{3}$, (*h*) 60

10. (*a*) $18\sqrt{2}$, (*b*) $7\sqrt{2}$, (*c*) 40, (*d*) $8\sqrt{2}$, (*e*) 3.4, (*f*) 28, (*g*) $5\sqrt{2}$, (*h*) $2\sqrt{2}$

11. (*a*) $30\sqrt{3}$, (*b*) 14, (*c*) 27, (*d*) 18, (*e*) $8\sqrt{3}$, (*f*) $4\sqrt{3}$, (*g*) $48\sqrt{3}$, (*h*) 42, (*i*) 6, (*j*) 10, (*k*) $\frac{5}{2}\sqrt{3}$, (*l*) $3\sqrt{3}$

12. (*a*) 817, (*b*) 3078

13. (*a*) $54\sqrt{3}$, (*b*) $96\sqrt{3}$, (*c*) $600\sqrt{3}$

14. (*a*) 576, (*b*) 324, (*c*) 100

15. (*a*) $36\sqrt{3}$, (*b*) $27\sqrt{3}$, (*c*) $\frac{16}{3}\sqrt{3}$, (*d*) $144\sqrt{3}$, (*e*) $3\sqrt{3}$, (*f*) $48\sqrt{3}$

16. (*a*) 10, (*b*) 10, (*c*) $5\sqrt{3}$

17. (*a*) 18, (*b*) $9\sqrt{3}$, (*c*) $6\sqrt{3}$, (*d*) $3\sqrt{3}$

18. (*a*) 1 : 8, (*b*) 4 : 9, (*c*) 9 : 10, (*d*) 8 : 11, (*e*) 3 : 1, (*f*) 2 : 5, (*g*) $4\sqrt{2}:3$, (*h*) 5 : 2

19. (*a*) 5 : 2, (*b*) 1 : 5, (*c*) 1 : 3, (*d*) 3 : 4, (*e*) 5 : 1

20. (*a*) 5 : 1, (*b*) 4 : 7, (*c*) *x* : 2, (*d*) $\sqrt{2}:1$, (*e*) $\sqrt{3}:y$, (*f*) $\sqrt{x}:3\sqrt{2}$ or $\sqrt{2x}:6$

21. (*a*) 1 : 4, (*b*) 1 : 25, (*c*) 36 : 1, (*d*) 9 : 100, (*e*) 49 : 25

22. (*a*) 12π, (*b*) 14π, (*c*) 10π, (*d*) $2\pi\sqrt{3}$

23. (*a*) 9π, (*b*) 25π, (*c*) 64π, (*d*) $\frac{1}{4}\pi$, (*e*) 18π

24. (*a*) $C=10\pi$, $A=25\pi$ (*b*) $r=8$, $A=64\pi$ (*c*) $r=4$, $C=8\pi$

25. (*a*) 12π, (*b*) 4π, (*c*) 7π, (*d*) 26π, (*e*) $8\pi\sqrt{3}$, (*f*) 3π

26. (*a*) 98π, (*b*) 18π, (*c*) 32π, (*d*) 25π, (*e*) 72π, (*f*) 100π

27. (*a*) (1) $C=8\pi$, $A=16\pi$; (2) $C=4\sqrt{3}\,\pi$, $A=12\pi$ (*b*) (1) $C=16\pi$, $A=64\pi$; (2) $C=8\sqrt{3}\,\pi$, $A=48\pi$ (*c*) (1) $C=12\pi$, $A=36\pi$; (2) $C=6\pi$, $A=9\pi$ (*d*) (1) $C=16\pi$, $A=64\pi$; (2) $C=8\pi$, $A=16\pi$ (*e*) (1) $C=20\sqrt{2}\,\pi$, $A=200\pi$; (2) $C=20\pi$, $A=100\pi$ (*f*) (1) $C=6\sqrt{2}\,\pi$, $A=18\pi$; (2) $C=6\pi$, $A=9\pi$

28. (*a*) 10 ft., (*b*) 17 ft., (*c*) $3\sqrt{5}$ ft. or 6.7 ft.

29. (*a*) 2π, (*b*) 10π, (*c*) 8, (*d*) 11π, (*e*) 6π, (*f*) 10π

30. (*a*) 3π, (*b*) $12\frac{1}{2}$, (*c*) 5π, (*d*) 2π, (*e*) π, (*f*) 4π

31. (*a*) 6π, (*b*) $\pi/6$, (*c*) $25\pi/6$, (*d*) 25π, (*e*) $4\frac{1}{2}$, (*f*) 13, (*g*) 24π, (*h*) $8\pi/3$

32. (*a*) 6π, (*b*) 20, (*c*) 3π, (*d*) 16π

33. (*a*) 120°, (*b*) 240°, (*c*) 36°, (*d*) 180°, (*e*) 135°, (*f*) $(180/\pi)°$ or 57.3° to nearest tenth

34. (*a*) 72°, (*b*) 270°, (*c*) 40°, (*d*) 150°, (*e*) 320°

35. (*a*) 90°, (*b*) 270°, (*c*) 45°, (*d*) 36°

36. (*a*) 12, (*b*) 9, (*c*) 10, (*d*) 6, (*e*) 5, (*f*) $3\sqrt{2}$

37. (*a*) 4, (*b*) 10, (*c*) 10 in. (*d*) 9

38. (*a*) $6\pi-9\sqrt{3}$, (*b*) $24\pi-36\sqrt{3}$, (*c*) $\frac{3}{2}\pi-\frac{9}{4}\sqrt{3}$, (*d*) $\dfrac{\pi r^2}{6}-\dfrac{r^2\sqrt{3}}{4}$, (*e*) $\dfrac{2\pi r^2}{3}-r^2\sqrt{3}$

39. (a) $4\pi - 8$, (b) $150\pi - 225\sqrt{3}$, (c) $24\pi - 36\sqrt{3}$,
(d) $16\pi - 32$, (e) $50\pi - 100$

40. (a) $\dfrac{64\pi}{3} - 16\sqrt{3}$, (b) $24\pi - 16\sqrt{2}$, (c) $\dfrac{80\pi}{3} - 16$

41. (a) $\dfrac{16\pi}{3} - 4\sqrt{3}$, (b) $\dfrac{8\pi}{3} - 4\sqrt{3}$, (c) $4\pi - 8$

42. (a) $12\pi - 9\sqrt{3}$, (b) $\dfrac{3}{2}\pi - \dfrac{9}{4}\sqrt{3}$, (c) $9\pi - 18$

43. (a) $200 - 25\pi/2$, (b) $48 + 26\pi$, (c) $25\sqrt{3} - 25\pi/2$,
(d) $100\pi - 96$, (e) $128 - 32\pi$, (f) $300\pi + 400$,
(g) 39π, (h) 100

44. (a) 36π, (b) $36\sqrt{3} + 18\pi$, (c) 14π

Chapter 11

1. The description of each locus is left for the student.

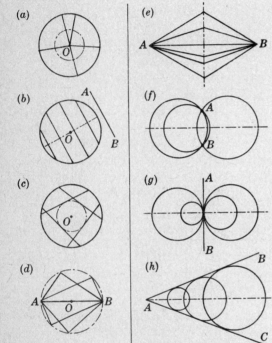

2. The diagrams are left for the student.
(a) The line parallel to the banks and midway between them.
(b) The perpendicular bisector of the line joining the two floats.
(c) The bisector of the angle between the roads.
(d) The pair of bisectors of the angles between the roads.

3. The diagrams are left for the student.
(a) A circle having the sun as its center and the fixed distance as its radius.
(b) A circle concentric to the coast, outside it and at the fixed distance from it.
(c) A pair of parallel lines on either side of the row and 20 ft. from it.
(d) A circle having the center of the clock as its center and the length of the clock hand as its radius.

4. (a) EF, (b) GH, (c) EF, (d) GH, (e) EF,
(f) GH, (g) AB, (h) a $90°$ arc from A to G with B as center

5. (a) AC, (b) BD, (c) BD, (d) AC, (e) E
6. In each case, the letter refers to the circumference of the circle. (a) A, (b) C, (c) B, (d) A, (e) C, (f) A and C, (g) B
7. The description of each locus is left for the student.

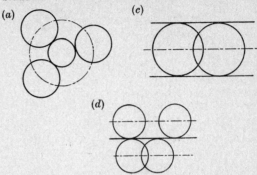

8. (a) EF, (b) GH, (c) line parallel to AD and EF midway between them, (d) EF, (e) BC, (f) GH
9. The explanation is left for the student.

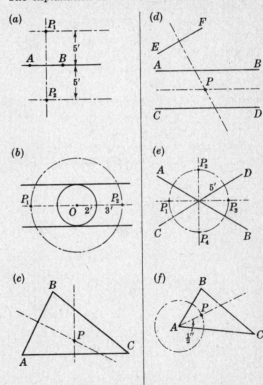

10. (a) The intersection of two of the angle bisectors.
(b) The intersection of two of the $\perp$ bisectors of the sides.
(c) The intersection of the $\perp$ bisector of AB and the bisector of $\angle B$.
(d) The intersection of the bisector of $\angle C$ and a circle with C as center and 5 as radius.
(e) The intersections of two circles, one with B as center and 5 as radius and the other with A as center and 10 as radius.

11.   (a) 1,   (b) 1,   (c) 4,   (d) 2,   (e) 2,   (f) 1

Chapter 12

1. $A(3, 0)$, $B(4, 3)$, $C(3, 4)$, $D(0, 2)$, $E(-2, 4)$, $F(-4, 2)$, $G(-1, 0)$, $H(-3\frac{1}{2}, -2)$, $I(-2, -3)$, $J(0, -4)$, $K(1\frac{1}{2}, -2\frac{1}{2})$, $L(4, -2\frac{1}{2})$

3.   Perimeter of square formed is 20 units, its area is 25 square units.

4.   Area of parallelogram $= 30$ sq. units.
    Area of $\triangle BCD = 15$ sq. units.

5.   (a) $(4,3)$,   (b) $(2\frac{1}{2}, 3\frac{1}{2})$,   (c) $(-4,6)$,   (d) $(7,-5)$,   (e) $(-10, -2\frac{1}{2})$,   (f) $(0,10)$,   (g) $(4,-1)$,   (h) $(-5, -2\frac{1}{2})$,   (i) $(5,5)$,   (j) $(-3,-10)$,   (k) $(5,6)$,   (l) $(0,-3)$

6.   (a) $(4,0)$, $(0,3)$, $(4,3)$
    (b) $(-3,0)$, $(0,5)$, $(-3,5)$
    (c) $(6,-2)$, $(0,-2)$, $(6,0)$
    (d) $(4,6)$, $(4,9)$, $(3,8)$
    (e) $(2,-3)$, $(-2,2)$, $(0,5)$
    (f) $(-\frac{1}{2},0)$, $(\frac{1}{2},\frac{1}{2})$, $(0,-1\frac{1}{2})$

7.   (a) $(0,2)$, $(1,7)$, $(4,5)$, $(3,0)$
    (b) $(-2,7)$, $(3,6)$, $(6,1)$, $(1,2)$
    (c) $(-1,2)$, $(3,3)$, $(3,-4)$, $(-1,-5)$
    (d) $(-2,-1)$, $(4,2\frac{1}{2})$, $(7,-4)$, $(1,-7\frac{1}{2})$

8.   (a) $(2,6)$, $(4,3)$,   (b) $(1,0)$, $(0,-2\frac{1}{2})$,   (c) common midpoint, $(2,2)$

9.   (a) $(-2,3)$,   (b) $(-3,-6)$,   (c) $(-1\frac{1}{2},-2)$,   (d) (a,b),   (e) $(2a,3b)$,   (f) $(a,b+c)$

10.   (a) $M(4,8)$,   (b) $A(-1,0)$,   (c) $B(6,-3)$

11.   (a) $B(2,3\frac{1}{2})$,   (b) $D(3,3)$,   (c) $A(-2,9)$

12.   (a) Prove that $ABCD$ is a parallelogram (since opposite sides are equal) and has a rt. $\angle$.
    (b) The point $(3,2\frac{1}{2})$ is the midpoint of each diagonal.
    (c) Yes, since the midpoint of each diagonal is their common point.

13.   (a) $D(3,2)$, $(1\frac{1}{2},1)$.   (b) $E(0,2)$, $(3,1)$.   (c) No, since the midpoint of each median is not a common point.

14.   (a) 5,   (b) 6,   (c) 10,   (d) 12,   (e) 5.4,   (f) 7.5,   (g) 9,   (h) a

15.   (a) $3,3,6$,   (b) $4,14,18$,   (c) $1,3,4$,   (d) $a,2a,3a$

16.   (a) 13,   (b) 5,   (c) 15,   (d) 5,   (e) 10,   (f) 15,   (g) $3\sqrt{2}$,   (h) $5\sqrt{2}$,   (i) $\sqrt{10}$,   (j) $2\sqrt{5}$,   (k) 4,   (l) $a\sqrt{2}$

18.   (a) $\triangle ABC$,   (b) $\triangle DEF$,   (c) $\triangle GHJ$,   (d) $\triangle KLM$ is *not* a rt. $\triangle$

19.   (a) $5\sqrt{2}$,   (b) $\sqrt{5}$,   (c) $\sqrt{65}$

21.   (a) 10,   (b) 5,   (c) $5\sqrt{2}$,   (d) 13,   (e) 4,   (f) 3

22.   (a) on,   (b) on,   (c) outside,   (d) on,   (e) inside,   (f) inside,   (g) on

23.   (a) 9/5,   (b) 5/9,   (c) 5/2,   (d) 3,   (e) 2,   (f) 1,   (g) 5,   (h) -2,   (i) -3,   (j) 3/2,   (k) -1,   (l) 1

24.   (a) 3,   (b) 4,   (c) $-\frac{1}{2}$,   (d) -7,   (e) 5,   (f) 0,   (g) 3,   (h) 5,   (i) -4,   (j) $-\frac{2}{3}$,   (k) -1,   (l) -2,   (m) 5,   (n) 6,   (o) -4,   (p) -8

25.   (a) $72°$,   (b) $18°$,   (c) $68°$,   (d) $22°$,   (e) $45°$,   (f) $0°$

26.   (a) .0875,   (b) .3057,   (c) .3640,   (d) .7002,   (e) 1,   (f) 3.2709,   (g) 11.430

27.   (a) $0°$,   (b) $25°$,   (c) $45°$,   (d) $55°$,   (e) $7°$,   (f) $27°$,   (g) $37°$,   (h) $53°$,   (i) $66°$

28.   (a) BC, BD, AD, AE (c) AF, CD
    (b) BF, CF, DE (d) AB, EF

29.   (a) 0,   (b) no slope,   (c) 5,   (d) -5,   (e) .5,   (f) $-.0005$

30.   (a) 0,   (b) no slope,   (c) no slope,   (d) 0,   (e) 5,   (f) -1,   (g) 2

31.   (a) 3/2,   (b) 7/3,   (c) -1,   (d) 6

32.   (a) -2,   (b) -1,   (c) $-\frac{1}{3}$,   (d) $-\frac{2}{5}$,   (e) -10,   (f) 1,   (g) $\frac{5}{4}$,   (h) $\frac{4}{13}$,   (i) no slope,   (j) 0

33.   (a) 0,   (b) -2,   (c) 3,   (d) -1

34.   (a) $-3/2$,   (b) 2/3,   (c) $-3/2$

35.   (a) $-\frac{1}{2}$,   (b) 1,   (c) 2,   (d) -1

36.   (a) $-3/2$,   (b) 1/4,   (c) 5/6

37.   (a) and (b)

38.   (a) 19,   (b) 9,   (c) 2

39.   (a) $x=-5$,   (b) $y=3\frac{1}{2}$,   (c) $y=3$ and $y=-3$,   (d) $y=-5$,   (e) $x=4$ and $x=-4$,   (f) $x=5$ and $x=-1$,   (g) $y=4$,   (h) $x=1$,   (i) $x=9$

40.   (a) $x=6$,   (b) $y=5$,   (c) $x=6$,   (d) $x=5$,   (e) $x=6$,   (f) $y=3$

41.   (a) $x=y$ (f) $x-y=2$ or
    (b) $y=x+5$ $y-x=2$
    (c) $x=y-4$ (g) $x=y$ and
    (d) $y-x=10$ $x=-y$
    (e) $x+y=12$ (h) $x+y=5$

42.   (a) Line having y-intercept 5, slope 2.
    (b) Line passing through $(2,3)$, slope 4.
    (c) Line passing through $(-2,-3)$, slope 5/4.
    (d) Line passing through origin, slope $\frac{1}{2}$.
    (e) Line having y-intercept 7, slope -1.
    (f) Line passing through origin, slope $\frac{1}{3}$.

43.   (a) $y=4x$,   (b) $y=-2x$,   (c) $y=\frac{3}{2}x$ or $2y=3x$,   (d) $y=-\frac{2}{5}x$ or $5y=-2x$,   (e) $y=0$

44.   (a) $y=4x+5$,   (b) $y=-3x+2$,   (c) $y=\frac{1}{3}x-1$ or $3y=x-3$,   (d) $y=3x+8$,   (e) $y=-4x-3$,   (f) $y=2x$ or $y-2x=0$

45.   (a) $\dfrac{y-4}{x-1}=2$ or $y=2x+2$
    (b) $\dfrac{y-3}{x+2}=2$ or $y=2x+7$
    (c) $\dfrac{y}{x+4}=2$ or $y=2x+8$
    (d) $\dfrac{y+7}{x}=2$ or $y=2x-7$

46.   (a) $y=4x$,   (b) $y=\frac{1}{2}x+3$,   (c) $\dfrac{y-2}{x-1}=3$,   (d) $\dfrac{y+2}{x+1}=\dfrac{1}{3}$,   (e) $y=2x$

47.   (a) Circle with center at origin and radius 7.
    (b) $x^2+y^2=16$,   (c) $x^2+y^2=64$ and $x^2+y^2=4$

48.   (a) $x^2+y^2=25$,   (b) $x^2+y^2=81$,   (c) $x^2+y^2=4$ or $x^2+y^2=144$

49.   (a) 3,   (b) 4/3,   (c) 2,   (d) $\sqrt{3}$

50.   (a) $x^2+y^2=16$,   (b) $x^2+y^2=121$,   (c) $x^2+y^2=4/9$ or $9x^2+9y^2=4$,   (d) $x^2+y^2=9/4$ or $4x^2+4y^2=9$,   (e) $x^2+y^2=5$,   (f) $x^2+y^2=3/4$ or $4x^2+4y^2=3$

51. (a) 10, (b) 10, (c) 20, (d) 20, (e) 7, (f) 25
52. (a) 16, (b) 12, (c) 20, (d) 24
53. (a) 10, (b) 12, (c) 22
54. (a) 5, (b) 13, (c) $7\frac{1}{2}$
55. (a) 6, (b) 10, (c) 1.2
56. (a) 15, (b) 49, (c) 53
57. (a) 30, (b) 49, (c) 88, (d) 24, (e) 16, (f) 18

Chapter 13

1. (a) <, (b) >, (c) >, (d) >, (e) >, (f) <
2. (a) > (b) > (c) < (d) >
3. (a) > (b) < (c) < (d) >
4. (a) more (b) less
5. (a) >, (b) >, (c) <, (d) >, (e) <, (f) <
6. (c), (d) and (e)
7. (a) 5 to 7 (c) 4 to 10 (e) 2 to 8
   (b) 6 to 10 (d) 3 to 9 (f) 1 to 13
8. (a) $\angle B, \angle A, \angle C$ (b) DF, EF, DE (c) $\angle 3, \angle 2, \angle 1$
9. (a) $\angle BAC > \angle ACD$ (b) $AB > BC$
10. (a) BC, AB, AC (c) $AD, AB = CD, BC$
    (b) $\angle BOC, \angle AOB, \angle AOC$ (d) OG, OH, OJ

Chapter 14

1. (a) child, (b) luxury items, (c) improperly dressed, (d) smoking, (e) trapezoid, (f) regular polygon
2. (a) ornament, jewelry, ring, wedding ring; (b) vehicle, automobile, commercial automobile, truck; (c) polygon, quadrilateral, parallelogram, rhombus; (d) angle, obtuse angle, obtuse triangle, isosceles obtuse triangle.
3. (a) A regular polygon is an equilateral and an equiangular polygon.
   (b) An isosceles triangle is a triangle having at least two equal sides.
   (c) A pentagon is a polygon having five sides.
   (d) A rectangle is a parallelogram having one right angle.
   (e) An inscribed angle is an angle formed by two chords and having its vertex on the circumference of the circle.
   (f) A parallelogram is a quadrilateral whose opposite sides are parallel.
   (g) An obtuse angle is an angle larger than a right angle and less than a straight angle.

4. (a) (2) is a definition; (1), (3) and (4) are theorems. Their logical sequence is (2), (1), (4), (3).
   (b) (2) is a definition; (1) is an assumed theorem or a postulate; (3), (4) and (5) are theorems. Their logical sequence is (2), (1), (5), (4), (3).
6. (a) $x + 2 \neq 4$. (b) $3y = 15$. (c) She does not love you. (d) His mark was not more than 65. (e) Joe is not heavier than Dick. (f) $a + b = c$.
7. (a) A non-square does not have equal diagonals. False (for example, when applied to a rectangle or a regular pentagon).
   (b) A non-equiangular triangle is not equilateral. True.
   (c) A person who is not a bachelor is a married person. This inverse is false when applied to an unmarried female.
   (d) A number that is not zero is a positive number. This inverse is false when applied to negative numbers.
11. (a) Necessary and sufficient.
    (b) Necessary but not sufficient.
    (c) Neither necessary nor sufficient.
    (d) Sufficient but not necessary.
    (e) Necessary and sufficient.
    (f) Sufficient but not necessary.
    (g) Necessary but not sufficient.
13. (a) This is an example of circular reasoning; the poor blood circulation is the cause of the cold hands and feet, not the result.
    (b) The two principles are the same; hence, a theorem is being used to prove itself.

Chapter 17

1. (a) $6(7^2)$ or 294 sq. yd.
   (b) $2(8)(6\frac{1}{2}) + 2(8)(14) + 2(6\frac{1}{2})(14)$ or 510 sq. ft.
   (c) $4(3.14)30^2$ or 11,304 sq. in.
   (d) $2(3.14)(10)(10 + 4\frac{1}{2})$ or 911 sq. rd.
2. (a) 36^3 or 46,656 cu. in.
   (b) $(5\frac{1}{2})^3$ or $166\frac{3}{8}$ cu. yd.
   (c) 100^3 or 1,000,000 cu. cm.
3. (a) 27 cu. in. (d) 47 cu. in.
   (b) 91 cu. in. (e) 2744 cu. in.
   (c) 422 cu. in.
4. (a) $3(8\frac{1}{2})(8)$ or 204 cu. in.
   (b) $2(9)(9)$ or 162 cu. ft.
   (c) $\frac{1}{3}(6)(6.4)$ or 13 cu. ft.
5. (a) 904 cu. in., (b) 1130 cu ft., (c) 18 cu. ft.
6. (a) $V = \frac{1}{3}\pi r^2 h$ (c) $V = \frac{1}{3}lwh$
   (b) $V = \frac{1}{3}s^2 h$ (d) $V = \frac{2}{3}\pi r^3$
7. (a) $6e^3 + \dfrac{2e^2 h}{3}$, (b) $lwh + \dfrac{\pi l^2 w}{8}$, (c) $3\pi r^3$

INDEX

Catalog

If you are interested in a list of SCHAUM'S
OUTLINE SERIES send your name
and address, requesting your free catalog, to:

SCHAUM'S OUTLINE SERIES, Dept. C
McGRAW-HILL BOOK COMPANY
1221 Avenue of Americas
New York, N.Y. 10020